NEUROMETHODS

Series Editor
Wolfgang Walz
University of Saskatchewan
Saskatoon, SK, Canada

Neuromethods publishes cutting-edge methods and protocols in all areas of neuroscience as well as translational neurological and mental research. Each volume in the series offers tested laboratory protocols, step-by-step methods for reproducible lab experiments and addresses methodological controversies and pitfalls in order to aid neuroscientists in experimentation. *Neuromethods* focuses on traditional and emerging topics with wide-ranging implications to brain function, such as electrophysiology, neuroimaging, behavioral analysis, genomics, neurodegeneration, translational research and clinical trials. *Neuromethods* provides investigators and trainees with highly useful compendiums of key strategies and approaches for successful research in animal and human brain function including translational "bench to bedside" approaches to mental and neurological diseases.

Translational Methods for Multiple Sclerosis Research

Edited by

Sergiu Groppa

Department of Neurology and Neuroimaging Center (NIC) of the Focus Program Translational Neuroscience (FTN), University Medical Center, Johannes Gutenberg University Mainz, Mainz, Germany

Sven G. Meuth

Department of Neurology, University of Düsseldorf, Düsseldorf, Germany

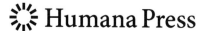 Humana Press

Editors
Sergiu Groppa
Department of Neurology
and Neuroimaging Center (NIC)
of the Focus Program Translational
Neuroscience (FTN)
University Medical Center, Johannes
Gutenberg University Mainz
Mainz, Germany

Sven G. Meuth
Department of Neurology
University of Düsseldorf
Düsseldorf, Germany

ISSN 0893-2336   ISSN 1940-6045   (electronic)
Neuromethods
ISBN 978-1-0716-1215-6   ISBN 978-1-0716-1213-2   (eBook)
https://doi.org/10.1007/978-1-0716-1213-2

This Humana imprint is published by the registered company Springer Science+Business Media, LLC, part of Springer Nature.
The registered company address is: 1 New York Plaza, New York, NY 10004, U.S.A.

Preface

The underlying pathology of MS is characterized by focal lesions and ongoing white and gray matter compartment pathology. Importantly, functional and structural reorganization occurs continuously during the disease course and plays an essential role for the long-term outcome. These cannot be precisely quantified in humans. Particularly at different disease stages processes of demyelination, remyelination, and axonal remodeling occur simultaneously and influence the disease course. An exact understanding of these distinct disease-related fingerprints could be achieved through the use of translational models (forward and reverse translation) of neuroinflammation. Unraveling complementary paths of tissue damage and restoration in humans and rodents could give important answers to inter-individual disease courses and build up an essential background for the improvement of actual therapeutic strategies and facilitate the development of future remedies. On the basis of recent results on the importance of MRI-derived characterization of tissue integrity and cerebral network fingerprints for neuroinflammation, we summed up current advances in the study of translational paths in central neuroinflammation, with a focus on ongoing pathophysiological processes and the transition between inflammatory stages on one side and progressive states with neurodegeneration on the other. We introduce the pathophysiological hallmarks of neuroinflammation from tissue damage to reorganization, bridging the gap between studies performed in mouse models (in vivo and ex vivo) and investigations of humans with MS (in vivo imaging and electrophysiological data). We put these processes into a conceptual framework of brain network dynamics addressing new developments of cerebral circuit characterizations and computational neuroscience for the description of the brain's transition through disease stages by the aid of neuroimaging and system electrophysiology. Finally, we link static and dynamic network dynamics related to neuroinflammation with clinical and functional outcome measures and actual therapeutic remedies to highlight important aspects for future studies and our conceptual understanding of MS as a chronic and progressive disease paving new ways to tackle neuroinflammation and neurodegeneration from the translational and clinical perspective.

Mainz, Germany
Düsseldorf, Germany

Sergiu Groppa
Sven G. Meuth

Contents

Contributors

MARIOS ANTONAKAKIS • *Institute for Biomagnetism and Biosignalanalysis, University of Münster, Münster, Germany*

MIRIAM BECKE • *Department of Clinical and Developmental Neuropsychology, University of Groningen, Groningen, The Netherlands*

STEFAN BITTNER • *Focus Program Translational Neurosciences (FTN) and Immunology (FZI), Rhine Main Neuroscience Network (rmn2), Department of Neurology, University Medical Center, Johannes Gutenberg University Mainz, Mainz, Germany*

MANUELA CERINA • *Department of Neurology with Institute of Translational Neurology, University of Münster, Münster, Germany*

VENKATA CHAITANYA CHIRUMAMILLA • *Movement Disorders and Neurostimulation, Biomedical Statistics and Multimodal Signal Processing Unit, Department of Neurology, Focus Program Translational Neuroscience (FTN), University Medical Center of the Johannes Gutenberg University, Mainz, Mainz, Germany*

DUMITRU CIOLAC • *Movement Disorders and Neurostimulation, Department of Neurology, Focus Program Translational Neuroscience (FTN), Rhine Main Neuroscience Network (rmn2), University Medical Center, Johannes Gutenberg University Mainz, Mainz, Germany; Laboratory of Neurobiology and Medical Genetics, Nicolae Testemitanu State University of Medicine and Pharmacy, Chisinau, Republic of Moldova; Department of Neurology, Institute of Emergency Medicine, Chisinau, Republic of Moldova*

ERIK ELLWARDT • *Focus Program Translational Neurosciences (FTN) and Immunology (FZI), Rhine Main Neuroscience Network (rmn2), Department of Neurology, University Medical Center of the Johannes Gutenberg University Mainz, Mainz, Germany*

CORNELIUS FABER • *Translational Imaging Center TRIC, Department of Clinical Radiology, University of Münster and University Hospital Münster, Münster, Germany*

LUCA FAZIO • *Department of Neurology with Institute of Translational Neurology, University of Münster, Münster, Germany*

MASSIMO FILIPPI • *Neuroimaging Research Unit, Division of Neuroscience, IRCCS San Raffaele Scientific Institute, Milan, Italy; Neurology Unit, Neurorehabilitation Unit, and Neurophysiology Service, IRCCS San Raffaele Scientific Institute, Milan, Italy; Vita-Salute San Raffaele University, Milan, Italy*

ANN-KATRIN FLECK • *Department of Neurology with Institute of Translational Neurology, University Hospital of Münster, Münster, Germany*

VINZENZ FLEISCHER • *Focus Program Translational Neurosciences (FTN) and Immunology (FZI), Rhine Main Neuroscience Network (rmn2), Department of Neurology, University Medical Center, Johannes Gutenberg University Mainz, Mainz, Germany*

FELIX M. GLASER • *Department of Neurology with Institute for Translational Neurology, University of Münster, Münster, Germany*

GABRIEL GONZALEZ-ESCAMILLA • *Movement Disorders and Neurostimulation, Department of Neurology, Focus Program Translational Neuroscience (FTN), Rhine Main Neuroscience Network (rmn2), University Medical Center, Johannes Gutenberg University Mainz, Mainz, Germany*

SERGIU GROPPA • *Neuroimaging and Neurostimulation, Department of Neurology, Focus Program Translational Neuroscience (FTN), Rhine-Main Neuroscience Network (rmn2), University Medical Center of the Johannes Gutenberg University Mainz, Mainz, Germany*

STANISLAV A. GROPPA • *Laboratory of Neurobiology and Medical Genetics, Nicolae Testemitanu State University of Medicine and Pharmacy, Chisinau, Republic of Moldova; Department of Neurology, Institute of Emergency Medicine, Chisinau, Republic of Moldova*

MATTHIAS GROTHE • *Department of Neurology, University Medicine Greifswald, Greifswald, Germany*

ASAD KHAN • *Institute for Biomagnetism and Biosignalanalysis, University of Münster, Münster, Germany*

LUISA KLOTZ • *Department of Neurology with Institute of Translational Neurology, University Hospital of Münster, Münster, Germany*

JULIA KRÄMER • *Department of Neurology with Institute of Translational Neurology, University Hospital Münster, Münster, Germany*

GURUMOORTHY KRISHNAMOORTHY • *Research Group Neuroinflammation and Mucosal Immunology, Max Planck Institute of Biochemistry, Martinsried, Germany*

HANS LASSMANN • *Center for Brain Research, Medical University of Vienna, Vienna, Austria*

MAREN LINDNER • *Department of Neurology with Institute of Translational Neurology, University Hospital of Münster, Münster, Germany*

DIRK LUCHTMAN • *Scientifica Ltd, Kingfisher Court, Uckfield, East Sussex, UK*

FELIX LUESSI • *Department of Neurology, Focus Program Translational Neuroscience (FTN), University Medical Center of the Johannes Gutenberg University, Mainz, Germany*

SVEN G. MEUTH • *Department of Neurology, University Hospital Düsseldorf, Düsseldorf, Germany*

MUTHURAMAN MUTHURAMAN • *Focus Program Translational Neurosciences (FTN) and Immunology (FZI), Rhine Main Neuroscience Network (rmn2), Department of Neurology, University Medical Center, Johannes Gutenberg University Mains, Mainz, Germany*

SHIN-YOUNG NA • *Research Group Neuroinflammation and Mucosal Immunology, Max Planck Institute of Biochemistry, Martinsried, Germany*

MARC PAWLITZKI • *Department of Neurology with Institute of Translational Neurology, University Hospital Münster, Münster, Germany*

MAREN PERSON • *Department of Neurology, University Medical Center of the Johannes Gutenberg University Mainz, Mainz, Germany*

MARIA CARLA PIASTRA • *Cognitive Neuroscience, Donders Institute for Brain, Cognition and Behaviour, Radboud University Nijmegen Medical Center, Nijmegen, The Netherlands*

PAOLO PREZIOSA • *Neuroimaging Research Unit, Division of Neuroscience, IRCCS San Raffaele Scientific Institute, Milan, Italy; Neurology Unit, IRCCS San Raffaele Scientific Institute, Milan, Italy*

ANGELA RADETZ • *Neuroimaging and Neurostimulation, Department of Neurology, Focus Program Translational Neuroscience (FTN), Rhine-Main Neuroscience Network (rmn2), University Medical Center of the Johannes Gutenberg University Mainz, Mainz, Germany*

MARIA A. ROCCA • *Neuroimaging Research Unit, Division of Neuroscience, IRCCS San Raffaele Scientific Institute, Milan, Italy; Neurology Unit, IRCCS San Raffaele Scientific Institute, Milan, Italy; Vita-Salute San Raffaele University, Milan, Italy*

LEONI ROLFES • *Department of Neurology with Institute of Translational Neurology, University Hospital Münster, Münster, Germany*

TOBIAS RUCK • *Department of Neurology, University of Düsseldorf, Düsseldorf, Germany*

MENNO M. SCHOONHEIM • *Department of Anatomy and Neurosciences, MS Center Amsterdam, Amsterdam UMC, Vrije Universiteit Amsterdam, Amsterdam Neuroscience, Amsterdam, The Netherlands*

TIMO UPHAUS • *Focus Program Translational Neurosciences (FTN) and Immunology (FZI), Rhine Main Neuroscience Network (rmn2), Department of Neurology, University Medical Center, Johannes Gutenberg University Mainz, Mainz, Germany*

PAOLA VALSASINA • *Neuroimaging Research Unit, Institute of Experimental Neurology, Division of Neuroscience, IRCCS San Raffaele Scientific Institute, Milan, Italy*

JOHANNES VORWERK • *Institute of Electrical and Biomedical Engineering, UMIT— University for Health Sciences, Medical Informatics and Technology, Hall in Tirol, Austria*

LYDIA WACHSMUTH • *Translational Imaging Center TRIC, Department of Clinical Radiology, University of Münster and University Hospital Münster, Münster, Germany*

CARSTEN H. WOLTERS • *Institute for Biomagnetism and Biosignalanalysis, University of Münster, Münster, Germany*

Part I

Pathophysiological Fingerprints of MS and Related Neuroimmunological Disorders

Chapter 1

Pathophysiological Bases of Autoimmune-Initiated/Mediated Neurodegeneration

Hans Lassmann

Abstract

Neurodegeneration in inflammatory conditions in the central nervous system follows a basic pathway, mediated by pro-inflammatory cytokines and macrophage and microglia activation, which leads to tissue injury through oxidative stress, mitochondrial injury, and ionic imbalance in glia, axons, and neurons. In inflammatory diseases of the brain and spinal cord, however, this downstream mechanism can be triggered and modified by a variety of different primary mechanisms of the adaptive immune system, engaging CD4$^+$ Th1 or Th17 cells, CD8$^+$ cytotoxic T-cells, tissue resident memory T-cells, B-cells, and autoantibodies. These different triggers result in distinct pathological entities and are reflected in different inflammatory diseases in humans. Their specific features are discussed in this chapter.

Key words Brain inflammation, Neurodegeneration, T-lymphocytes, B-lymphocytes, Antibodies

1 Introduction

Tissue damage in the central nervous system is a common sequela of brain inflammation. In addition, however, recent evidence suggests the involvement of inflammatory and immunological mechanisms in the pathogenesis of neurodegenerative diseases [1]. It is generally believed that the liberation of pro-inflammatory cytokines, either in the course of immune mediated inflammatory brain diseases or as a secondary consequence of central nervous system (CNS) tissue injury, plays a key role in the propagation of neurodegeneration, either directly or indirectly through the brain resident macrophages/microglia or astrocytes [2]. The basic principles of immune-mediated CNS injury have originally been identified and defined in models of autoimmune encephalomyelitis and similar mechanisms have then been identified in experimental models of neuro-metabolic or neuro-degenerative diseases. More recently, however, a much more complicated picture has emerged, which indicates disease-specific mechanisms of tissue injury in the brain and spinal cord,

Sergiu Groppa and Sven G. Meuth (eds.), *Translational Methods for Multiple Sclerosis Research*, Neuromethods, vol. 166, https://doi.org/10.1007/978-1-0716-1213-2_1, © Springer Science+Business Media, LLC, part of Springer Nature 2021

driven by various different cells and effector mechanisms of the immune system. These mechanisms, which are the topic of this review chapter, have emerged from a broad spectrum of experimental models and are gradually translated into the setting of human disease [3].

2   Basic Mechanisms of Immune-Mediated Tissue Injury in the CNS

The basic mechanisms of immune-mediated tissue injury have originally been elucidated in models of experimental autoimmune encephalomyelitis. These data showed that inflammation in the CNS is started by activated T-cells, which can enter the normal brain and spinal cord in the process of immune surveillance. When they encounter their specific antigen within the CNS, they get a second boost of activation and start the inflammatory process with the production of pro-inflammatory cytokines followed by the recruitment of other inflammatory cells, in particular of macrophages [4]. It is mainly the subset of activated macrophages, recruited from the circulation, which is responsible for the induction of clinical disease and the initiation and propagation of the tissue injury [5]. Although these models have originally been developed as models for demyelinating diseases with autoreactive T-cells directed against myelin antigen, it turned out later that the primary target for tissue damage is the axon [6]. Demyelination, which is present in most models too, in general, develops as a secondary consequence of axonal injury (secondary demyelination), while primary demyelination with preservation of axons is sparse. Axonal injury starts with focal disturbance of fast axonal transport, which can, in part, be reversible, but in many axons it results in complete transection. The prominent disturbance of fast axonal transport is related to oxidative injury and energy deficiency, partly due to mitochondrial dysfunction [6, 7]. It is suggested that microglia/macrophage activation leads to production of reactive oxygen and nitrogen species and oxidative damage [8], which is instrumental in mitochondrial injury, and this process is amplified by precipitation of fibrin at sites of blood–brain barrier injury [9]. A consequence of energy failure in axons is a cessation of fast axonal transport, leading to abnormal expression of axonal ion channels and ionic imbalance, disturbance of axonal membrane permeability, and intra-axonal calcium overload [10, 11]. Additional mechanisms of neurodegeneration involve cell stress, endoplasmic reticulum stress, and the activation of molecules involved in Wallerian degeneration [12]. Depending upon the intensity of the inflammatory stress these axonal changes may be reversible or may lead to focal areas of (mainly secondary) demyelination and tissue damage. This basic mechanism acts in all conditions of brain inflammation and is also a major mechanism of demyelination and neurodegeneration in

Microglia / Macrophage Activation

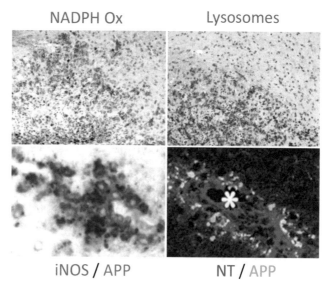

NADPH Ox   Lysosomes

iNOS / APP   NT / APP

Oxidative Injury Neurodegeneration

Fig. 1 Basic pathway of inflammatory tissue injury in brain inflammation is mediated through the activation of microglia or recruited macrophages, which produce reactive oxygen and nitrogen species; oxidative injury mainly affects cell processes, in particular axons [7]. NADPH Oxidase: nicotinamide adenine dinucleotide phosphate oxidase; iNOS: inducible nitric oxide synthase; APP: amyloid precursor protein (which accumulates in dystrophic axons due to disturbed axonal transport); NT: nitro-tyrosine

multiple sclerosis [13]. It is, however, modified by additional mechanisms, depending upon the specific disease conditions and immune mechanisms, involved in the inflammatory process (Fig. 1).

Oxidative injury with subsequent mitochondrial impairment has its most pronounced effects in cell processes, while the perinuclear cell body is more resistant [14]. This implies that in neurons damage of axons, dendrites and synapses prevails over global neuronal loss. Similarly, in oligodendrocytes this process results in distal oligodendrogliopathy, with primary degeneration of peri-axonal cell processes and subsequent demyelination. In astrocytes the polarization of the cell processes forming the perivascular and subpial glia limitans is dominantly affected, resulting in loss of aquaporins, connexins, and amino acid transporters.

Although this basic mechanism is seen in all conditions of inflammatory tissue damage in the brain and spinal cord, it is, as discussed below, triggered and amplified by disease-specific immune mechanisms (Fig. 2).

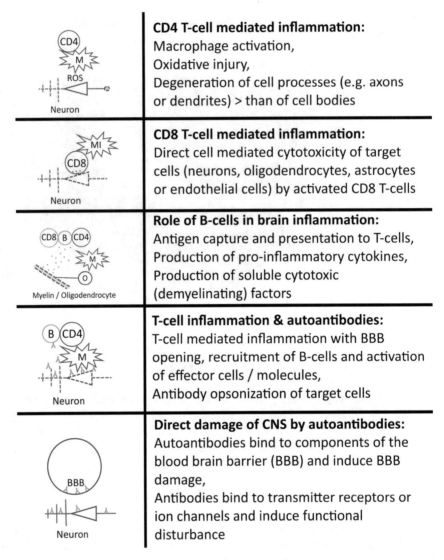

	CD4 T-cell mediated inflammation: Macrophage activation, Oxidative injury, Degeneration of cell processes (e.g. axons or dendrites) > than of cell bodies
	CD8 T-cell mediated inflammation: Direct cell mediated cytotoxicity of target cells (neurons, oligodendrocytes, astrocytes or endothelial cells) by activated CD8 T-cells
	Role of B-cells in brain inflammation: Antigen capture and presentation to T-cells, Production of pro-inflammatory cytokines, Production of soluble cytotoxic (demyelinating) factors
	T-cell inflammation & autoantibodies: T-cell mediated inflammation with BBB opening, recruitment of B-cells and activation of effector cells / molecules, Antibody opsonization of target cells
	Direct damage of CNS by autoantibodies: Autoantibodies bind to components of the blood brain barrier (BBB) and induce BBB damage, Antibodies bind to transmitter receptors or ion channels and induce functional disturbance

Fig. 2 Different pathways of inflammation in the brain and spinal cord and their pathological outcome. CD4: CD4+ T-cells; CD8: CD8+ T-cell; B: B-cell; M: macrophage; MI: microglia; ROS: reactive oxygen species; BBB: blood–brain barrier

3   CNS Inflammation Driven by MHC Class II Restricted CD4+ T-Lymphocytes

Most experimental models of autoimmune encephalomyelitis are mediated by MHC Class II restricted CD4+ T-cells, which may be polarized towards Th-1 T-cells (mainly in rat models, [15]) or Th-17 T-cells (dominant in mouse models; [16]). Brain inflammation can either be triggered by active immunization of the animals with (components of) brain tissue or by passive transfer of autoreactive T-cells. In addition, spontaneous disease is observed in T-cell receptor transgenic animals [3]. Encephalitogenic T-Cells can be directed against a large spectrum of different CNS autoantigens. The antigen specificity of the T-cells in part determines the location of the inflammatory lesions (e.g., spinal cord versus brain, white matter versus grey matter), but the nature of the lesions with respect to inflammation and tissue damage (neurodegeneration) is the same, irrespective whether the T-cells are directed against an antigen in oligodendrocytes, astrocytes, or neurons [3]. Inflammation and immune mediated tissue damage is mediated through the basic pattern of immune mediated tissue injury discussed above. Thus, the immune cell population, which drives tissue damage and neurodegeneration is the activated macrophage, recruited from the circulation. The role of microglia is less clear, and has been suggested to be involved in promotion or amelioration of tissue damage or having no effect, possibly depending upon their specific subpopulation present [8, 17]. In line with this observation it has been shown that cytokines involved in macrophage recruitment are instrumental in the induction of the disease [18] and that more prolonged and aggressive tissue damage is seen in NOD mice with an enhanced innate immunity response [19]. It is important to note that the disease in these EAE models is acute/subacute and self-limiting and that in animals with active sensitization the disease only remains active as long as the peripheral immune depot at the sensitization site is retained [20]. Overall, autoimmune encephalomyelitis mediated by CD4$^+$ T-cells is a good model for human acute disseminated encephalomyelitis (ADEM; [21]). Widespread primary demyelination, the hallmark of multiple sclerosis lesions, is absent and a disease course and pathology, which resembles progressive multiple sclerosis has so far not been convincingly shown in any of these models [3].

4 Direct Cellular Cytotoxicity through Autoreactive MHC Class I Restricted CD8$^+$ T-Cells

Although the induction of an MHC Class I restricted CD8$^+$ T-cell response against foreign intracellular (e.g., virus) antigen is a common feature, triggering of a CD8 T-cell mediated autoimmune inflammatory response in the brain by active sensitization has so far largely failed. The pathogenicity of CD8$^+$ T-cells in the CNS can at the moment be best studied by transgenic expression of a foreign antigen, such as influenza hemagglutinin [22] or ovalbumin [23], in the CNS and the passive transfer of activated CD8$^+$ T-cells directed against the respective antigen. With these models several important features of brain inflammation have been described. First, CD8$^+$ T-cells alone, without the help of other cells of the immune system, can trigger brain inflammation, although much higher cell numbers are required in comparison to those needed for transfer of CD4$^+$ T-cells. Much lower numbers of cells are necessary, when there is already a local inflammatory environment within the CNS, for instance due to the presence of foreign exogenous antigens [24]. Secondly, CD8$^+$ T-cells induce a very specific pathology in the CNS, depending upon the cellular expression of their cognate antigen, and this includes widespread inflammatory demyelination [22], astrocytopathy [25], degeneration of specific neuronal or axonal subtypes [26], or vasculitis [27]. Thus, these experimental models reflect characteristic features of different human disease, such as for example multiple sclerosis, paraneoplastic encephalomyelitis, narcolepsy, diabetes insipidus, or Susac syndrome. Thirdly, the pathology in these models is characterized by profound T-cell infiltration accompanied by microglia activation, but the recruitment and activation of circulating myeloid cells is sparse [25]. This implies that tissue injury is a highly specific reflection of the cellular antigen recognition of the T-cells, while macrophage mediated bystander injury of neurons and axons is mild or even absent. As in classical EAE models, in these models too, CNS inflammation is acute/subacute and self-limiting.

5   CD8$^+$ Tissue Resident Memory T-Cells May Drive Chronic CNS Inflammation, when Antigen Clearance Is Incomplete

The dominant leukocyte subset, which is present in the normal human CNS and in metabolic and neurodegenerative diseases, is the CD8$^+$ tissue resident effector memory cell [28]. In these conditions, these cells are dispersed within the CNS in low numbers and do not show signs of activation. Similarly, in multiple sclerosis patients tissue resident CD8$^+$ memory T-cells outnumber all other leukocyte subpopulations in the CNS [29] and are activated in a focally and temporally restricted manner [30]. It is unlikely that such tissue resident memory cells are directed against classical autoantigens, which are constantly present within the CNS.

Tissue resident memory cells have been characterized in experimental models of virus infection of solid tissues, including the brain [31]. These cells enter the brain during the phase of acute encephalitis and some of them then transform into tissue resident effector memory cells, when the infectious agent is cleared from the tissue. These T^{rm} cells persist in the tissue for long time as a potentially effective defense mechanism against reinfection with the same agent. When such reinfection occurs, they immediately react by activation and trigger a relapse or chronic progressive inflammation, with the aim to eliminate the infectious agent [31]. Whether such cells are present in the brain in experimental models of chronic autoimmune encephalomyelitis is currently unknown.

6   The Role of B-Cells in Brain Inflammation

B-cell depleting therapies in EAE animals suggested that there are B-cell dependent and B-cell independent inflammatory conditions in the brain [32]. B-cell directed therapy in acute EAE, induced by active sensitization with the MOG35-55 peptide in C57B6 mice or by passive transfer of MOG35-55 reactive T-cells, is ineffective or may even enhance disease, while in more chronic models, which are driven by a cooperation between encephalitogenic T-cells and demyelinating anti-MOG autoantibodies, the same therapies are beneficial. Since B-cell depletion with anti-CD20 antibodies has little effect on antibodies, the therapeutic effect in these models is mainly due to the reduction of antigen presentation by B-cells [33]. In addition, activated B-cells are potent source for pro-inflammatory cytokines and may also produce soluble toxic factors, that may induce demyelination in an antibody independent way [34, 35]. In the presence of a pathogenic autoantibody response it should be expected that a deletion of plasma cells should be beneficial. This is, however, not the case in many instances and may be related to the observations, that plasma cells, besides the antibody production, may also have immunoregulatory properties by the production interleukin 10 or transforming growth factor beta [36].

All these data are based on inflammatory disease processes of the central nervous system, which are induced by T-cells and then modified by B-cells or antibodies. Whether B-cells alone can induce brain inflammation is so far not clear and no experimental models is currently available for a B-cell driven inflammatory disease of the nervous system.

7   T-Cell Mediated Encephalomyelitis Modified by Pathogenic Autoantibodies

It has been shown more than 50 years ago, that the serum of animals in certain EAE models contains antibodies, which can trigger demyelination in organotypic tissue culture models in vitro or in vivo, after their local injection into the brain or spinal cord [37, 38]. However, the autoantibodies do not induce disease or tissue damage when present in the systemic circulation of naïve animals. They become highly pathogenic, when present in animals with CD4[+] T-cell mediated brain inflammation [39]. This has first been shown with antibodies against myelin oligodendrocyte glyco-protein (MOG), and has later been also found with autoantibodies against neuronal [40] or astrocytic antigens such as aquaporin 4 [41] Most importantly, these models can also be used to determine the pathogenic potential of human autoantibodies and were instrumental to define the pathogenesis neuromyelitis optica

(NMO), driven by aquaporin 4 reactive autoantibodies [42] or of the inflammatory demyelinating disease, associated with antibodies against MOG (MOGAD, [43]). The latter is particularly interesting, since it is an inflammatory demyelinating disease, which has originally been thought to be a variant of multiple sclerosis. When patients, however, were classified on the basis of the presence of anti-MOG antibodies, it turned out that the disease closely reflects autoimmune encephalitis, but has a clinical course, pathology, and response to therapy which is different from multiple sclerosis [44, 45].

8   Can Autoantibodies Alone Induce Brain Disease, Inflammation, or Tissue Damage?

Since the brain and spinal cord are shielded from the blood by the blood–brain barrier, only small amounts of circulating antibodies leak into the normal brain. These amounts are in general too small to pass the threshold for the induction of tissue damage. In addition, antibodies induce immunological tissue damage by the activation of complement or the interaction of activate macrophages. Both components are largely absent from the normal CNS tissue [46]. Thus, in order to induce damage circulating autoantibodies have to reach the CNS tissue at sites of blood–brain barrier injury and in an environment, which provides activated complement and/or activated macrophages. Such an environment is best provided in the context of T-cell mediated inflammation [39].

There are, however, two conditions where antibodies alone may trigger CNS disease. The first has been shown in an experimental condition induced by high systemic titers of high affinity autoantibodies against aquaporin 4. In this model the small amount of antibody, which passes the normal blood–brain barrier, binds to perivascular astrocytes and appears to be sufficient to induce profound blood–brain barrier disturbance [41]. However, even in these models the titer of circulating auto-antibody has to be much higher in comparison to that, necessary to induce disease in the presence of a T-cell mediated encephalomyelitis.

The second condition is seen in diseases, mediated by antibodies against neurotransmitter receptors or ion channels in the CNS [44, 46]. In these diseases even very small amounts of antibodies, which get access to the brains in spite of an intact blood–brain barrier, may induce disturbance of neuronal function in a similar way as pharmacological agonists or antagonists [47]. Due to the lack of immune mediated mechanisms of tissue injury in such a condition, there is no inflammation or tissue damage present, despite the very severe symptoms or deficits [44].

9   Conclusions

In this short review the principal mechanisms of immune-mediated tissue injury in the central nervous system are outlined. Experimental models have provided major contributions to their understanding and have also provided clues how to target them therapeutically. However, in the more complex situation of human disease it is rare that only a single mechanism drives inflammation or neurodegeneration. Nevertheless, the current data suggest that therapeutic interventions have to be different, depending upon the dominant mechanism of tissue injury.

References

1. Ransohoff RM (2016) How neuroinflammation contributes to neurodegeneration. Science 353:777–783
2. Becher B, Spath S, Goverman J (2017) Cytokine networks in neuroinflammation. Nat Rev Immunol 17:49–59
3. Lassmann H, Bradl M (2017) Multiple sclerosis: experimental models and reality. Acta Neuropathol 133:223–244
4. Flügel A, Berkowicz T, Ritter T et al (2001) Migratory activity and functional changes of green fluorescent effector cells before and during experimental autoimmune encephalomyelitis. Immunity 14:547–560
5. Wolf Y, Shemer A, Levy-Efrati L et al (2018) Microglial MHC class II is dispensable for experimental autoimmune encephalomyelitis and cuprizone-induced demyelination. Eur J Immunol 48:1308–1318
6. Nikic I, Merkler D, Sorbara C et al (2011) A reversible form of axon damage in experimental autoimmune encephalomyelitis and multiple sclerosis. Nat Med 17:495–499
7. Aboul-Enein F, Weiser P, Hoftberger R et al (2006) Transient axonal injury in the absence of demyelination: a correlate of clinical disease in acute experimental autoimmune encephalomyelitis. Acta Neuropathol 111:539–547
8. Mendiola AS, Ryu JK, Bardehle S et al (2020) Transcriptional profiling and therapeutic targeting of oxidative stress in neuroinflammation. Nat Immunol 21:513–524
9. Ryu JK, Rafalski VA, Meyer-Franke A et al (2018) Fibrin-targeting immunotherapy protects against neuroinflammation and neurodegeneration. Nat Immunol 19:1212–1223
10. Boscia F, De Rosa V, Cammarota M et al (2020) The Na+/Ca2+ exchangers in demyelinating diseases. Cell Calcium 85:102130
11. Friese MA, Schattling B, Fugger L (2014) Mechanisms of neurodegeneration and axonal dysfunction in multiple sclerosis. Nat Rev Neurol 10:225–238
12. Andhavarapu S, Mubariz F, Arvas M et al (2019) Interplay between ER stress and autophagy: a possible mechanism in multiple sclerosis pathology. Exp Mol Pathol 108:183–190
13. Mahad DH, Trapp BD, Lassmann H (2015) Pathological mechanisms in progressive multiple sclerosis. Lancet Neurol 14:183–193
14. Lassmann H, Van Horssen J (2016) Oxidative stress and its impact on neurons and glia in multiple sclerosis lesions. Biochim Biophys Acta 1862:506–510
15. Ben-Nun A, Wekerle H, Cohen IR (1981) The rapid isolation of clonable antigen-specific T lymphocyte lines capable of mediating autoimmune encephalomyelitis. Eur J Immunol 11:195–199
16. Korn T, Mitsdoerffer M, Croxford AL et al (2008) IL-6 controls Th17 immunity in vivo by inhibiting the conversion of conventional T cells into Foxp3+ regulatory T cells. Proc Natl Acad Sci 105:18460–18465
17. Wimmer I, Scharler C, Zrzavy T et al (2019) Microglia pre-activation and neurodegeneration precipitate neuroinflammation without exacerbating tissue injury in experimental autoimmune encephalomyelitis. Acta Neuropathol Commun 7:1–13
18. Becher B, Tugues S, Greter M (2016) GM-CSF: from growth factor to central mediator of tissue inflammation. Immunity 45:963–973
19. Basso AS, Frenkel D, Quintana FJ et al (2008) Reversal of axonal loss and disability in a mouse model of progressive multiple sclerosis. J Clin Invest 118:1532–1543

20. Tabira T, Itoyama Y, Kuroiwa Y (1984) Necessity of continuous antigenic stimulation by the locally retained antigens in chronic relapsing experimental allergic encephalomyelitis. J Neurol Sci 66:97–106

21. Alvord E Jr (1970) In: Vinken PJ, Bruyn GW (eds) Handbook of clinical neurology, vol 9. North Holland, Amsterdam

22. Saxena A, Bauer J, Scheikl T et al (2008) Cutting edge: multiple sclerosis-like lesions induced by effector CD8 T cells recognizing a sequestered antigen on oligodendrocytes. J Immunol 181:1617–1621

23. Na SY, Cao Y, Toben C et al (2008) Naive CD8 T-cells initiate spontaneous autoimmunity to a sequestered model antigen of the central nervous system. Brain 131:2353–2365

24. Na SY, Hermann A, Sanchez-Ruiz M et al (2012) Oligodendrocytes enforce immune tolerance of the uninfected brain by purging the peripheral repertoire of autoreactive CD8+ T cells. Immunity 37:134–146

25. Cabarrocas J, Bauer J, Piaggio E et al (2003) Effective and selective immune surveillance of the brain by MHC class I-restricted cytotoxic T lymphocytes. Eur J Immunol 33:1174–1182

26. Scheikl T, Pignolet B, Dalard C et al (2012) Cutting edge: neuronal recognition by CD8 T cells elicits central diabetes insipidus. J Immunol 188:4731–4735

27. Gross CC, Meyer C, Bhatia U et al (2019) CD8+ T cell-mediated endotheliopathy is a targetable mechanism of neuro-inflammation in Susac syndrome. Nat Commun 10:1–19

28. Smolders J, Heutinck KM, Fransen NL et al (2018) Tissue-resident memory T cells populate the human brain. Nat Commun 9:4593

29. Van Nierop GP, Van Luijn MM, Michels SS et al (2017) Phenotypic and functional characterization of T cells in white matter lesions of multiple sclerosis patients. Acta Neuropathol 134:383–401

30. Machado-Santos J, Saji E, Tröscher AR et al (2018) The compartmentalized inflammatory response in the multiple sclerosis brain is composed of tissue-resident CD8+ T lymphocytes and B cells. Brain 141:2066–2082

31. Steinbach K, Vincenti I, Merkler D (2018) Resident-memory T cells in tissue-restricted immune responses: for better or worse? Front Immunol 9:2827

32. Weber MS, Prod'homme T, Patarroyo JC et al (2010) B-cell activation influences T-cell polarization and outcome of anti-CD20 B-cell depletion in central nervous system autoimmunity. Ann Neurol 68:369–383

33. Hausler D, Hausser-Kinzel S, Feldmann L et al (2018) Functional characterization of reappearing B cells after anti-CD20 treatment of CNS autoimmune disease. Proc Natl Acad Sci U S A 115:9773–9778

34. Li R, Patterson KR, Bar-Or A (2018) Reassessing B cell contributions in multiple sclerosis. Nat Immunol 19:696–707

35. Lisak RP, Benjamins JA, Nedelkoska L et al (2012) Secretory products of multiple sclerosis B cells are cytotoxic to oligodendroglia in vitro. J Neuroimmunol 246:85–95

36. Fillatreau S (2018) Natural regulatory plasma cells. Curr Opin Immunol 55:62–66

37. Appel SH, Bornstein MB (1964) The application of tissue culture to the study of experimental allergic encephalomyelitis. II Serum factors responsible for demyelination. J Exp Med 119:303–312

38. Lassmann H, Kitz K, Wisniewski H (1981) In vivo effect of sera from animals with chronic relapsing experimental allergic encephalomyelitis on central and peripheral myelin. Acta Neuropathol 55:297–306

39. Linington C, Bradl M, Lassmann H et al (1988) Augmentation of demyelination in rat acute allergic encephalomyelitis by circulating mouse monoclonal antibodies directed against a myelin/oligodendrocyte glycoprotein. Am J Pathol 130:443–454

40. Mathey EK, Derfuss T, Storch MK et al (2007) Neurofascin as a novel target for autoantibody-mediated axonal injury. J Exp Med 204:2363–2372

41. Hillebrand S, Schanda K, Nigritinou M et al (2019) Circulating AQP4-specific auto-antibodies alone can induce neuromyelitis optica spectrum disorder in the rat. Acta Neuropathol 137:467–485

42. Bradl M, Misu T, Takahashi T et al (2009) Neuromyelitis optica: pathogenicity of patient immunoglobulin in vivo. Ann Neurol 66:630–643

43. Spadaro M, Winklmeier S, Beltran E et al (2018) Pathogenicity of human antibodies against myelin oligodendrocyte glycoprotein. Ann Neurol 84:315–328

44. Höftberger R, Lassmann H (2018) Immune-mediated disorders. In: Handbook of clinical neurology. Elsevier, pp 285–299

45. Jurynczyk M, Messina S, Woodhall MR et al (2017) Clinical presentation and prognosis in MOG-antibody disease: a UK study. Brain 140:3128–3138

46. Bradl M, Lassmann H (2016) Neurologic autoimmunity: mechanisms revealed by animal models. Handb Clin Neurol 133:121–143

47. Geis C, Planaguma J, Carreno M et al (2019) Autoimmune seizures and epilepsy. J Clin Invest 129:926–940

Chapter 2

Translational Animal Models for MS and Related Neuroimmunological Disorders

Felix M. Glaser and Tobias Ruck

Abstract

Experimental autoimmune encephalitis (EAE) is an important translational model in multiple sclerosis (MS) research. As no single model can mimic all aspects of this heterogeneous disease, different experimental approaches have been developed. In this chapter, we will present an overview of the different EAE models. We will describe the different clinical and histopathological aspects of MS which they address as well as their limitations. Furthermore we will focus on models mimicking aspects of primary/secondary progressive MS. Moreover, we will describe the use of EAE models in the evaluation of therapeutic approaches. Finally, we will provide an overview on experimental models on myositis, myasthenia gravis and peripheral neuropathies.

Key words Multiple sclerosis, EAE, Myositis, Myasthenia gravis, CIDP, Translational animal models

1 Introduction

Multiple sclerosis (MS) is an inflammatory demyelinating disease of the central nervous system (CNS) [1] characterized by a varying clinical course and different histopathological and immunological patterns. This variability reflects the complex interactions between the immune system and the CNS [2]. Experimental autoimmune encephalitis (EAE), the most common animal model for MS, is a scientific tool which has proven to be essential for the clarificaton of MS pathogenesis. In addition, it is available for the development of disease-modifying therapies (DMT). Recent transgenic models have provided deeper insights into, for example, soluble pro- and anti-inflammatory factors, T-cell mediated CNS damage, and effector cascades of innate immunity. However, it has to be emphasized that no single animal model represents all aspects of MS heterogeneity [1]. Therefore, a huge number of different models have been developed which complementarily contribute to a comprehensive knowledge of MS pathology. However, not only were animal models crucial in decoding MS pathology, Some of them also feature

Sergiu Groppa and Sven G. Meuth (eds.), *Translational Methods for Multiple Sclerosis Research*, Neuromethods, vol. 166, https://doi.org/10.1007/978-1-0716-1213-2_2, © Springer Science+Business Media, LLC, part of Springer Nature 2021

animal models with induced or spontaneous autoimmunity which permitting the exploration of diseases such as autoimmune myositis or autoimmune neuropathy. Nonetheless, usage of animal models has become a polarizing issue in public discourse, and the relevance of new approaches, for instance computer models in pharmacological testing, is increasing. We here present an overview of historic as well as current animal models of autoimmune diseases with a focus on experimental autoimmune encephalitis.

2   Methods

2.1   EAE: Active Immunization Model

The first implemented EAE model was the active immunization model. It requires an immunization of an animal with a CNS antigen and a strong adjuvant (e.g., Complete Freund's Adjuvant, CFA) [1]. In mice, additional administration of pertussis toxin is necessary to achieve a sufficient immune reaction. The reason for this is not fully understood, but data of Kamradt et al. (1991) [3] show that pertussis toxin interferes with the induction of Ag-induced peripheral T cell anergy and also enhances the permeability of the blood–brain barrier (BBB). Myelin basic protein (MBP) represents 30–40% of the CNS myelin protein and 5–15% of the peripheral myelin protein [4]. MBP was one of the first proteins shown to induce EAE [4]. It induces a relapsing-remitting EAE in SJL/J mice, a CD4$^+$-T-cell mediated disease characterized by a transient ascending limb paralysis [5]. Histopathological analysis of this model shows mononuclear cell infiltration, fibrin deposition in the brain and spinal cord, and adjacent areas of acute and chronic demyelination [5]. However, the most frequently used EAE model with an active immunization procedure uses MOG (myelin oligodendrocyte glycoprotein)$_{35-55}$ and CFA in mice [1, 6]. In C57BL/6J mice, immunized with MOG a chronic disease with ascending paralysis was observed [7], whereas PL/J mice developed an atypical chronic relapsing EAE [6]. The MOG$_{35-55}$ model has gained additional relevance by being favored in studies using transgenic approaches (see below; [8]). Despite the popularity of this model, there are important restrictions. It is a model for acute and chronic autoimmune encephalopathy with primary axonal injury mediated by MHC-class II restricted CD4$^+$ T-cells [1]. In contrast to MS, larger lesions are charactized by primary axonal damage with only demyelination, with primary demyelination being almost absent [1]. An induction of EAE in rats, guinea pigs, and primates using immunization with the recombinant extracellular domain of MOG, with myelin or with brain tissue solved in CFA features a closer pathophysiological and histopathological similarity to MS [1]. For example, rat strains possessing the isotypes and alleles RT1.BD in the MHC II region and RT1.C in the nonclassical MHC I region in common demonstrated considerable cortical demyelination following after immunization with MOG [1, 9]. However, although many aspects of MS are represented in this model, MOG antibodies are not present in the serum of MS patients. As a consequence these approaches are possibly a good model for MOG-mediated autoimmune syndromes [1]. Generally, the active EAE model is generated by immunization with self-peptides inducing CD4$^+$ T-cell activation [2]. CD8$^+$ T-cells however which outnumber CD4$^+$ T-cells in MS lesions seem to be underrepresented (*see also* Table 1). Therefore, other EAE models were developed, addressing these aspects of pathogenesis.

2.2 Adoptive Transfer Model

The data published by Ben-Nun et al. in 1981 [14] (see above) were an early example of adoptive T-cell transfer. Since then, its importance has increased as it is essential in order to investigate T-cell function and regulation in the context auf autoimmune neuroinflammation. In Lewis rats, clinical signs usually appear 3–4 days after adoptive transfer of MBP-specific T-cells reaching their maximum within 48 h. However, thereafter a rapid decline of clinical symptoms can be observed [8]. In 1991 Pender et al. demonstrated that in EAE apoptosis of myelin–/oligodendrocyte-specific T-cells occurs. It was hypothesized that this was an important mechanism of immunoregulation in EAE [15; reviewed by 8]. However, some crucial limitations of adoptive transfer EAE in rats have to be mentioned. These are the monophasic disease course and the minimal demyelination in the CNS, which is more similar to acute demyelinating encephalomyelitis (ADEM) [8] then MS. The adoptive transfer model proved that the T-cell arm of autoimmune reaction is crucial in the breakdown of the BBB, but not sufficient to induce demyelination at a larger scale in rats [reviewed by 8]. However, Zamvil et al. [16] demonstrated that adoptive transfer of MBP-specific T-cell clones in mice caused EAE with a relapsing paralysis. Histopathological analysis of these animals showed intense perivascular inflammation, demyelination and remyelination in the central nervous system [16]. Taken together, the adoptive transfer model is a convenient tool to study mechanisms of T-cell mediated inflammation (*see also* Table 1).

2.3 Transgenic Approaches

A crucial limitation of the EAE models described so far is the necessity of disease induction as a decisive difference to MS, which is a spontaneously occurring disease. In 1993 Goverman et al. [13] described an EAE model expressing a transgenic TCR (T-cell receptor) specific for a MHC class II-restricted myelin epitope. In this approach transgenic mice were generated expressing genes encoding a rearranged TCR specific for MBP. These transgenic animals developed EAE after immunization with MBP and adjuvant plus pertussis toxin as well as following an immunization with pertussis toxin alone [13]. Crossing of MBP-specific

Table 1
Overview of animal models for multiple sclerosis

Models	Induction	Characteristics	Comparison to human disease	Limitations
Active immunization model	– Immunization with CNS antigen (e.g., PLP [10, 11], MOG [1, 6], MBP [10, 12]) – Multiple different antigens possible (e.g., NF 155, NF 186, Neurofilament-M) [10]	– Highly reproducible inflammatory disease [1]; mainly CD4+ T-cell response [2] – Variable acute axonal injury, little permanent axonal loss or demyelination [1]	– Artificial disease induction; – Mainly affects spinal cord [2] – Relapsing–remitting (SJL, Biozzi) and chronic-progressive (C57BL/6) disease course	– Role of B-cells and CD8+ T-cells not adequately addressed [2] – Disease initiation usually highly artificial [2]; – Relative lack of brain involvement [7]
Adoptive transfer model (CD4+ and CD8+ T-cells)	– Immunization of donor animals, isolation of peripheral lymphoid cells, in vitro restimulation, adoptive transfer [2] – Adoptive transfer of CD4+ T cells or CD8+ T-cells [1]	– Homogenous and monophasic disease course, rapid onset, little demyelination [8] – Limited microglia activation, variable acute and little permanent axonal loss [1] – Adoptive CD8+ T-cell transfer: low macrophage recruitment, cytotoxic T-cell induced tissue damage [1]	– Monophasic disease course [8] – Artificial disease induction	– Encephalitogenic capacity of transferred T-cells not necessarily reflects the in vivo condition [2] – CD8+ T cell transfer difficult to handle, high intra-experimental variability [1]
Transgenic approaches	– Genetic modifications (e.g., transgenic T-cell receptor/humanized T-cell receptor) – Disease induction depending on model used (e.g., spontaneous, active immunization) – Mostly C57BL/6 mice used for transgenic approaches [8]	– High variability, depending on genetic modification. Study of neuroinflammation and immune system activation [2] – Specifically addresses role of specific immune molecules/neurotrophic cytokines [8] – Immune cells involved (depending on the approach): CD8, CD4, Th17 cells, monocytes, macrophages, B-cells, and Tregs [2]	– Mimic different aspects of MS depending on genetic modification. Some approaches with spontaneously occurring EAE [13]	– Depending on specific transgenic approach (e.g., extensive backcrossing [>10 times] on C57BL/6 background required) [8]

Theiler's murine encephalomyelitis virus (TMEV)	– Inoculation of mice with TMEV, also other virus models available (e.g., MHV/coronavirus) [1]	– Virus-induced inflammatory demyelinating disease [1] – Always chronic progressive disease course [2]	– Axonal damage in MS secondary to demyelination (outside-in model), TMEV encephalomyelitis: axonal damage precedes demyelination (inside-out model) [2]	– Only inducible in mice [2]
Toxic demyelination model	– Induction of focal CNS damage, for example, with cuprizone or lysolecithin [1, 2]	– Highly reproduceable process of demyelination [1] – Cuprizone induces oligodendroglial cell death, subsequent demyelination, profound activation of astrocytes and microglia [2] – Lysolecithin is an activator of phospholipase A2, induction of focal areas of rapid, highly reproducible demyelination [2]	Cuprizone-model represents histopathological features of MS, mainly pattern III [10] – Very efficient spontaneous remyelination permanent; remyelination failure, as seen in MS, only in models with prolonged cuprizone intoxication [1] – Lysolecithin-induced demyelination caused by direct toxic effect [2]	– Lysolecithin-induced demyelination not an immunological process [2] – Toxin-induced demyelination does not reflect MS disease, but an isolated demyelination and remyelination process [2]

transgenic mice with RAG-1-deficient mice created animals with T-cells expressing the transgenic TCR but no other lymphocytes. 14% of these H2^U T/R$^+$ and 100% of H2^U T/R$^-$ animals developed spontaneous EAE, demonstrating that MBP-specific CD4$^+$ T-cells can induce EAE in the absence of other lymphocytes and that lymphocytes which are present in T/R$^+$ but absent in T/R$^-$ animals have protective effects [17]. Another interesting aspect of this approach was the suggestion that the gut microbiome may be a critical determinant of CNS susceptibility in autoimmune disease [7]. TCR transgenic mice developed spontaneous EAE when housed in a nonsterile facility, but not when they were kept under germ-free conditions [13]. However, species-related differences in the molecular function of proteins are a limitation for EAE animal models [1]. While myelin proteins are highly conserved, antigen presentation is affected by this issue. For example, antigen recognition of T-cells depends on antigen presentation by major histocompatibility complexes (MHC) [1]. To address the problem of species-related differences in antigen presentation, humanized TCR transgenic mice along with the corresponding antigen presenting molecules have been developed [18; reviewed by 1]. Experiments with EAE induced in animals with humanized MHC class I or MHC class II proteins make it possible to investigate peptides with potential encephalitogenicity in humans [1]. Nonetheless, it has to be taken into consideration that encephalitogenicity also depends on antigen processing and cell migration processes. Therefore, an ideal model would require a humanization of all proteins involved in this process [1]. A main disadvantage of the transgenic model described so far is its over-representation of CD4$^+$ T-cell-mediated inflammation contrasting human MS pathology, which is characterized by CD8$^+$-T cell dominated inflammatory infiltrates (*see below*, e.g., [8], *see also* Table 1).

2.4 CD8$^+$ T Cell Model

CD8$^+$ T cells dominate the immune cell infiltrates in MS lesions and seem to play a significant role in MS pathogenesis [18]. The relevance of CD8$^+$ T cells in MS is supported by several lines of evidence. The (1) the extent of axonal damage correlates with the number of CD8$^+$ T cells and macrophages, (2) and neuroantigen specific CD8$^+$ but not CD4$^+$ T cells are enriched in the CNS of MS patients [7]. Adoptive transfer of MBP-specific CD8$^+$ T cells induced severe EAE with clinical characteristics and pathological patterns not seen in CD4$^+$ T cell induced EAE [19], resembling aspects of pattern III and IV lesions in MS patients [19; reviewed by 7]. In contrast, several subsets of CD8$^+$ T cells have been demonstrated to have a regulatory function in EAE. Antibody-induced depletion of CD8$^+$ T cells after active immunization enhanced the severity of EAE and abolished the need of pertussis toxin to induce EAE [20]. So far, the pathophysiological role of CD8$^+$-T cells in EAE seems to be diverse [7] and has not been comprehensively illuminated.

2.5 EAE Models of Progressive MS Forms

After active immunization or adoptive cell transfer, animals usually develop a monophasic disease. However, in MS approximately 50% of patients develop progressive disease courses either secondary relapsing remitting MS (secondary progressive MS, SPMS) or primary progressive MS (PPMS) [7] which is not represented in these models. Of note, the Biozzi ABH mouse recapitulates the clinical and pathological features of MS including relapsing-remitting episodes with immune-mediated demyelination and progressive disability with neurodegeneration [21]. The induction of immune tolerance with a $CD4^+$-T cell depleting, monoclonal antibody induces remyelination depending on timing of tolerization in the disease course [21]. Furthermore, an early tolerization showed to be partially neuroprotective [21], and also successfully eliminated further clinical relapses [21, 22]. However, it was not able to suppress ongoing neuronal degeneration and associated gliosis [21]. This mimics a change in the pathogenic mechanisms hypothesized to occur when patients with MS transition from relapsing remitting (RR)- to SP-MS [7] and shows a partial uncoupling of inflammation and CNS neurodegeneration [21]. In SPMS, inflammation in the CNS decreases, while brain atrophy and axonal loss steadily dominate the pathologic process, not responding to immunosuppressive agents [7]. Therefore, early inflammation might induce neuroaxonal damage triggering a potentially self-sustaining neurodegeneration [23]. The processes involved in this transition from acute/subacute inflammation to chronic neurodegeneration seem to be variegated and have not been fully delineated so far. It has been suggested that chronic inflammation leads to a production auf reactive oxygen species (ROS) and reactive nitrogen species (NOS), thereby promoting mitochondrial injury [23]. Mitochondrial injury causes energy deficiency, thus inhibiting neuroaxonal function which is highly energy demanding [23]. The function of initially infiltrating immune cells as well as of resident cells like microglia and macrophages has not yet been fully understood, and some results describing the function of distinct cell lines such as microglia in chronic neuroinflammation are partially contradictory. An attempt to further decipher the process of transition from inflammation to neurodegeneration as well as the function of distinct cell lines is to investigate the effect of cytokines. For example, Vogelaar et al. demonstrated that Interleukin 4 produced by T helper 2 (T_H2) cells, reverses disease progression in chronic EAE by direct neuronal signalling via the IRS1-PI3K-PKC pathway, thereby inducing cytoskeletal remodelling and axonal repair [24]. They used different mouse models to mimic the disease course of SP-MS and PP-MS. To mimic the SP-MS disease course, myelin oligodendrocyte glycoprotein (MOG_{35-55})-induced EAE in C57BL/6 mice with certain modifications was used. The mice received methylprednisolone (intraperitoneally) for 5 days during disease peak. The withdrawal of methylprednisolone then caused a progressive deterioration in the chronic phase. To reproduce clinical aspects of SPMS in this model, experiments were initiated in a more chronic stage (14 days after the initial peak of the disease). TCR1640 mice expressing a transgenic MOG-reactive T cell receptor develop spontaneous EAE with a chronic progressive disease course to mimicing features of PP-MS [24]. Additional interesting aspects result from histopathological studies, as progressive MS is histopathologically characterized by B-cell follicle-like structures in the inflamed meninges, increased subpial cortical demyelination, and cortical atrophy with reduced density of pyramidal neurons in layers III and V of the motor cortex. Moreover, glia limitans damage with astrocyte loss and an increased density of activated microglia has been described [25]. Based on the observation of microglia activation in MS, EAE was induced in nonobese diabetic (NOD) mice characterized by a genetic background prone to activation of innate immune responses [1]. Immunization of NOD mice with MOG_{35-55} peptide lead to a relapsing remitting disease course, evident from day 20 to day 70, which then switches into a chronic progressive stage [26]. Nonetheless, meningeal inflammation followed by subpial demyelination as described in progressive MS has not yet been adequately addressed in any EAE model. Taken together, many of the pathophysiological und histopathological aspects that have been described above and which are obviously relevant in the pathophysiology of chronic progressive MS are currently not adequately addressed in EAE. This emphasizes the actual limitations of this model.

2.6 Theiler's Murine Encephalomyelitis Virus (TMEV)

Viral infection can induce demyelination in mice. The best studied virus is picornavirus, for example, Theiler's murine encephalomyelitis virus (TMEV) which is a nonenveloped, positive sense, single-stranded RNA virus [2]. Two virus strains have to be distinguished. The DA strain infects the gray matter of the CNS during acute phase and the white matter during the following chronic phase of the disease, leading to demyelination. The GDVII strain causes an acute fatal polioencephalomyelitis without demyelination [2, 27]. In susceptible mice such as SJL/J [27] the disease is—unlike EAE—always chronic progressive [2]. Axonal injury occurs approximately 1 week after infection and after 2–3 weeks perivascular infiltration of microglia/macrophages—remarkably in the absence of T-cells—can be detected. The localization of axonal injury corresponds with the subsequent demyelination in the chronic phase of the disease [27]. In this chronic phase, TMEV infects glial cells and macrophages inducing immune-mediated demyelination with oligodendrocyte apoptosis and axonal degeneration [28; also reviewed by 2]. Notably, axonal damage in MS occurs secondary to demyelination (outside-in model), whereas in TMEV-induced encephalomyelitis axonal damage precedes demyelination (inside-out model) [2]. Nevertheless, TMEV-induced encephalomyelitis is a convenient tool to investigate demyelination in mice (*see also* Table 1).

2.7 Models of Toxic Demyelination

A pathological hallmark of MS is demyelination which distinguishes MS from other CNS diseases [1]. Additional possibilities to induce demyelination in mice are approaches using toxic agents such as cuprizone (*see also* Table 1). Cuprizone bis(cyclohexylidene hydrazide), a copper-chelating agent, induces a highly reproducible demyelination in distinct brain regions, especially in the corpus callosum [29]. This demyelination is an early observable event, without damage to CNS cell types other than oligodendrocytes [30] (also reviewed by [1]). Most models apply cuprizone in C57BL/6 mice for 4 weeks, which induces demyelination followed by rapid and extensive remyelination, but cuprizone can also be used in different mouse strains and other animal models [1]. While cuprizone intoxication is inducing demyelination, oligodendrocyte progenitor cells (OPCs) are proliferating and differentiating to remyelinate axons [31]. As the Akt/mTOR pathway is involved in the regulation of myelination, co-administration of cuprizone and mTOR-inhibitor rapamycin produces more intense demyelination and provides a longer time frame to investigate remyelination than treatment with cuprizone alone [31]. The distinct localization and time course of the lesions induced by cuprizone are suited to define the molecular mechanisms promoting demyelination and remyelination [1]. Another example for toxic induced demyelination is the use of lysolecithin (lysophosphatidylcholine). Intracerebral injection of lysolecithin produces focal demyelination, followed by spontaneous progressive remyelination [32]. Lysolecithin is an activator of phospholipase A2 and induces demyelination following injection into the spinal cord of several animal species [2]. Lysolecithin triggers rapid demyelination without relevant damage to adjacent cells and axons; it is not immune-mediated since it occurs even in immune-deficient mice [2]. Generally, toxin-induced demyelination offers an opportunity to study processes of demyelination and remyelination without representing other aspects of pathogenesis of MS.

2.8 Further Experimental Autoimmune Diseases

Experimental animal models have not only been used in the delineation of MS, but also in other autoimmune disease. As we focus on MS models, we refer the reader to the below cited, detailed reviews on these topics, but provide a short overview of important existing animal models.

The role of animal models in myositis has been reviewed in detail by Afzali et al. (2017). As our aim is to convey a short overview, we relegate this review for further details. Inflammatory immune myopathies (IIM) are a group of rare heterogenic muscle diseases with largely unclear pathogenesis [33]. To further delineate pathophysiological processes and may develop more specific therapeutic approaches, different animal models have been implemented. Interestingly, there are models of spontaneous myositis in dogs known as canine masticatory muscles myositis (CMMM) and a general form that symmetrically affects the extremities called canine polymyositis (CPM), representing aspects of dermatomyositis and polymyositis, respectively [33, 34]. However, the implementation of experimental settings for dogs remains difficult [33], and more standardized approaches are necessary. Therefore, for example, infectious animal models using coxsackievirus B1 (CVB) or alphaviruses such as Ross River Virus (RRV) or Chikungunya Virus (CHIKV) [33, 35–37] have been used. For example, BALB/c, Swiss C3H [37], and COH mice develop rapidly progressing myositis of the proximal hind limbs with mortality rates of up to 30% [33]. Another approach are immunological animal models, also known as experimental autoimmune myositis (EAM). EAM can be induced by partially purified or purified (>95%) Myosin B and C-protein—emulsified in CFA followed by simultaneous i.p. injections of Pertussis toxin—respectively [33, 38]. Myosin-B-based models have been used in rodents, rabbits, rats, and mice [33]. Improved purification of myosin (>95%) led to a more severe, self-limiting disease course. By using mice and especially the C57BL/6 mice strain, transgenic approaches were integrated in the EAM model. For example, interleukin 6 deficient mice ($Il\text{-}6^{-/-}$) turned out to be completely resistant to EAM [33, 39]. This demonstrates the importance of transgenic approaches in the delineation of EAM pathology. Another approach mentioned above is C-protein–induced myositis (CIM), originally using C-protein purified from skeletal muscle, later using recombinant C-Protein produced in *E. coli*, of which C protein fragment 2 showed the highest myositogenic potential [33]. For example, CIM is induced in C57BL/6 mice with almost 100% incidence and a self-limiting course [33, 40]. Histopathological analysis in this model revealed a disease maximum at day 14 with an enrichment of macrophages, CD8+ and CD4+ T cells and full recovery after day 21 [33]. CIM as well as myosin-B-based models have also been used in the evaluation of IIM therapies such as prednisolone or immunoglobulins [33]. A more recent model is the use of *Caenorhabditis elegans* (*C. elegans*), thereby partially replacing rodent experiments. As an example, a transgenic *C. elegans* model has been described expressing human Aβ under the control of a muscle-specific gene, showing immune reaction to thioflavin S and amyloid-β linked with oxidative stress and activation of stress responses [33, 41].

Conclusively, animal models including EAM have turned out to be crucial in IIM research, however further models have been developed for the investigation of other muscle diseases such as myasthenia gravis (MG).MG is a disease characterized by a chronic, fatigable weakness of vulnerable muscles mainly caused by antibodies against acetylcholine receptor (AChR) destroying the postsynaptic membrane and degrading receptors [42]. However, also other antibodies, for example antibodies directed against muscle-specific tyrosine kinase (MuSK), can be involved in MG pathogenesis, especially in patients without AChR-antibodies [43]. Active immunization of different animal species (e.g., rat, mouse, rhesus monkey, guinea pig, frog) using purified AChR from electric eels (*Electrophorus electricus*), *Torpedo californica* (Pacific electric ray), mammalian muscles, or recombinant or synthetic AChR fragments generates a model of MG called experimental autoimmune myasthenia gravis (EAMG) [42, 44]. Nonetheless, mainly mice (e.g., susceptible strains C67BL/6, SJL) are used for EAMG experiments due to the high rate of animals developing clinical symptoms [43]. The EAMG model has been used to describe mechanisms involved in the pathophysiological process of MG including antigen presentation, T-cell function, and selection and proliferation of autoreactive B-cells, and also in therapy evaluation [44]. During immunization, usually adjuvants (CFA, CFA with Bordetella pertussis toxin, incomplete Freund's adjuvant with additional heat-killed *Mycobacterium tuberculosis*) are also administered [44]. However, only approximately 1% of the produced antibodies cross-react with human AChR and thus induce EAMG. Disease severity, moreover, depends on the animal strain, the age of the animals used, and the amount of AChR used to induce disease [44]. Moreover, passive transfer models, and also MuSK-induced EAMG models are available [43], and EAMG has been extensively used in the evaluation of new therapeutic approaches [43].

Experimental animal models in autoimmune neuropathies further expand the spectrum of models for peripheral neuroimmunological disorders. A comprehensive review by Soliven [45] (2014) describes the different available approaches. Peripheral neuroimmunological disorders comprise various diseases caused by cell-mediated or autoantibody-mediated immune attack on the peripheral nervous system (PNS) [45] such as Guillain-Barré syndrome (GBS) and chronic inflammatory demyelinating polyradiculoneuropathy (CIDP). Molecular mimicry between PNS gangliosides and surface lipo-oligosaccharides is suggested to be involved in the pathogenesis of GBS and chronic dysimmune neuropathies [45]. Immunization of animals with peripheral nerve homogenate or with PNS myelin proteins (e.g., P0, P2, PMP22) induces a disease called experimental autoimmune neuropathy (EAN), a disease model which has contributed to the understanding of immune mechanisms involved in pathogenesis of immune neuropathies [45, 46]. For example, it was demonstrated that macrophages are the main effector cells by stripping myelin and producing proinflammatory cytokines, as protection against EAN is induced by depletion of macrophages or blocking macrophage function [45, 47].

In NOD mice, elimination of a costimulatory molecule B7-2 (CD86) triggers a spontaneous autoimmune polyneuropathy (SAP) at 6–7 months of age [45, 48]. This interesting model mimics different characteristics of CIDP such as disease course and histological and electrophysiological aspects, and experimental findings which, for example, delineate the role of the B7-1/B7-2–CD28/CTLA4 pathway have contributed to a further understanding of pathogenesis [45, 48]. Of note, various additional models mimicking aspects of, for example, multifocal motor neuropathy (MMN), anti-MAG neuropathy, or paraneoplastic neuropathies exist [45].

3   Alternatives to EAE, New Approaches

Although their value has been proven over the last decades, animal experiments have become a controversial issue in the public debate. Besides those ethical concerns, animal experiments show a number of disadvantages such as high cost and the need of skilled manpower [49]. Moreover, the reliability of animal models has been generally questioned as more than 30% of promising medications have failed due to toxicity in humans despite demonstrating safety in animal models [50]. Therefore, a strategy comprising reduction, refinement, and replacement ("3 Rs") has been developed. Reduction comprises methods that minimize the number of animals used per experiment [51]. For example, cell culture can be used to screen the compounds at early stages. Accordingly, human hepatocyte cultures are used to investigate drug metabolism and elimination from the body. Inclusion of such methods in study design is suitable to screen compounds in preliminary stages and to reduce the number of experiments requiring the use of animals [49]. Refinement comprises methods that minimize pain, suffering, distress, or lasting harm that may be experienced by the animals [51]. Stress and discomfort, for example, may lead to an imbalance of hormonal levels of animals leading to a reduced validity and reproducibility of results and refinement may counteract these factors [49]. Replacement means to avoid or replace the use of animals in experimental settings where they would have otherwise been used [51]. An interesting tool that has been implemented are computer models. One example is a software known as Computer Aided Drug Design (CADD). It is used to predict the receptor binding site for a potential drug molecule and identifies probable binding sites. Consequently, testing of unsuitable pharmacological agents with no or adverse biological activity can be avoided reducing and partly replacing animal experiments [49]. Another example for computer models is the structure–activity relationship (SARs) models which predicts biological activity of a drug candidate based on the presence of chemical moieties attached to the parent compound

[49]. Today, even software exists that can simulate an entire organism like *Mycoplasma genitalium* [50]. In summary, reduction, refinement, and replacement strategies actually cannot fully replace animal models, but they are an important addition and can enhance the quality of animal experiments.

4 Conclusion

Despite many limitations, experimental autoimmune diseases like experimental autoimmune encephalomyelitis (EAE), experimental autoimmune neuropathy (EAN), and experimental autoimmune myositis (EAM) have proven to be decisive tools in the investigation of autoimmune diseases, their pathogenetic mechanisms and etiology, and also in the evaluation of new therapeutic approaches. It has to be emphasized that no single experimental model can mimic all aspects of heterogeneous diseases such as MS. However, different approaches like virus-induced disease, active immunization, adoptive transfer, or transgenic models offer an opportunity to address different aspects of autoimmune diseases. The choice of the model significantly affects the attainable results. Therefore, detailed knowledge of the used animal models is an inevitable prerequisite for study design. New computer-based models are able to replace certain aspects; however, for mechanistical studies, basic research will continue to rely on animal models.

References

1. Lassmann H, Bradl M (2017) Multiple sclerosis: experimental models and reality. Acta Neuropathol 133:223–244
2. Procaccini C, De Rosa V, Pucino V et al (2015) Animal models of multiple sclerosis. Eur J Pharmacol 759:182–191
3. Kamradt T, Soloway PD, Perkins DL et al (1991) Pertussis toxin prevents the induction of peripheral T cell anergy and enhances the T cell response to an encephalitogenic peptide of myelin basic protein. J Immunol 147:3296–3302
4. Glatigny S, Bettelli E (2018) Experimental autoimmune encephalomyelitis (EAE) as animal models of multiple sclerosis (MS). Cold Spring Harb Perspect Med 8(11):a028977. https://doi.org/10.1101/cshperspect. a028977
5. Miller SD, Karpus WJ, Davidson TS (2010) Experimental autoimmune encephalomyelitis in the mouse. Curr Protoc Immunol Chapter 15:Unit 15.1. https://doi.org/10. 1002/0471142735.im1501s88
6. Mendel I, Kerlero de Rosbo N, Ben-Nun A (1995) A myelin oligodendrocyte glycoprotein peptide induces typical chronic experimental autoimmune encephalomyelitis in H-2b mice: fine specificity and T cell receptor V beta expression of encephalitogenic T cells. Eur J Immunol 25:1951–1959
7. Simmons SB, Pierson ER, Lee SY et al (2013) Modeling the heterogeneity of multiple sclerosis in animals. Trends Immunol 34:410–422
8. Gold R, Linington C, Lassmann H (2006) Understanding pathogenesis and therapy of multiple sclerosis via animal models: 70 years of merits and culprits in experimental autoimmune encephalomyelitis research. Brain 129:1953–1971
9. Storch MK, Bauer J, Linington C et al (2006) Cortical demyelination can be modeled in specific rat models of autoimmune encephalomyelitis and is major histocompatibility complex (MHC) haplotype-related. J Neuropathol Exp Neurol 65:1137–1142

10. Mix E, Meyer-Rienecker H, Hartung HP (2010) Animal models of multiple sclerosis - potentials and limitations. Prog Neurobiol 92:386–404

11. Tuohy VK, Lu ZJ, Sobel RA et al (1988) A synthetic peptide from myelin proteolipid protein induces experimental allergic encephalomyelitis. J Immunol 141:1126–1130

12. Einstein ER, Robertson DM, Dicaprio JM et al (1962) The isolation from bovine spinal cord of a homogeneous protein with encephalitogenic activity. J Neurochem 9:353–361

13. Goverman J, Woods A, Larson L et al (1993) Transgenic mice that express a myelin basic protein-specific T cell receptor develop spontaneous autoimmunity. Cell 72:551–560

14. Ben-Nun A, Wekerle H, Cohen IR (1981) The rapid isolation of clonable antigenspecific, T lymphocyte lines capable of mediating autoimmune encephalomyelitis. Eur J Immunol 11:195–199

15. Pender MP, Nguyen KB, McCombe PA et al (1991) Apoptosis in the nervous system in experimental allergic encephalomyelitis. J Neurol Sci 104:81–87

16. Zamvil S, Nelson P, Trotter J et al (1985) T-cell clones specific for myelin basic protein induce chronic relapsing paralysis and demyelination. Nature 317:355–35850

17. Lafaille JJ, Nagashima K, Katsuki M et al (1994) High incidence of spontaneous autoimmune encephalomyelitis in immunodeficient anti-myelin basic protein T cell receptor transgenic mice. Cell 78:399–408

18. Ben-Nun A, Kaushansky N, Kawakami N et al (2014) From classic to spontaneous and humanized models of multiple sclerosis: impact on understanding pathogenesis and drug development. J Autoimmun 54:33–50

19. Huseby ES, Liggitt D, Brabb T et al (2001) A pathogenic role for myelin-specific CD8(+) T cells in a model for multiple sclerosis. J Exp Med 194:669–676

20. Montero E, Nussbaum G, Kaye JF et al (2004) Regulation of experimental autoimmune encephalomyelitis by CD4+, CD25+ and CD8+ T cells: analysis using depleting antibodies. J Autoimmun 23:1–7

21. Hampton DW, Serio A, Pryce G et al (2013) Neurodegeneration progresses despite complete elimination of clinical relapses in a mouse model of multiple sclerosis. Acta Neuropathol Commun 1:84

22. Pryce G, O'Neill JK, Croxford JL et al (2005) Autoimmune tolerance eliminates relapses but fails to halt progression in a model of multiple sclerosis. J Neuroimmunol 165:41–52

23. Dendrou CA, Fugger L, Friese MA (2015) Immunopathology of multiple sclerosis. Nat Rev Immunol 15:545–558

24. Vogelaar CF, Mandal S, Lerch S et al (2018) Fast direct neuronal signaling via the IL-4 receptor as therapeutic target in neuroinflammation. Sci Transl Med. https://doi.org/10.1126/scitranslmed.aao2304

25. Magliozzi R, Howell OW, Reeves C et al (2010) A gradient of neuronal loss and meningeal inflammation in multiple sclerosis. Ann Neurol 68:477–493

26. Levy H, Assaf Y, Frenkel D (2010) Characterization of brain lesions in a mouse model of progressive multiple sclerosis. Exp Neurol 226:148–158

27. Tsunoda I, Kuang LQ, Libbey JE et al (2003) Axonal injury heralds virus-induced demyelination. Am J Pathol 162:1259–1269

28. Tsunoda I, Fujinami RS (2010) Neuropathogenesis of Theiler's murine encephalomyelitis virus infection, an animal model for multiple sclerosis. J Neuroimmune Pharmacol 5:355–369

29. Zhen W, Liu A, Lu J et al (2017) An alternative cuprizone-induced demyelination and remyelination mouse model. ASN Neuro 9 (4):1759091417725174. https://doi.org/10.1177/1759091417725174

30. Matsushima GK, Morell P (2001) The neurotoxicant, cuprizone, as a model to study demyelination and remyelination in the central nervous system. Brain Pathol 11:107–116

31. Sachs HH, Bercury KK, Popescu DC et al (2014) A new model of cuprizone-mediated demyelination/remyelination. ASN Neuro. https://doi.org/10.1177/1759091414551955

32. Zhang M, Hugon G, Bouillot C et al (2019) Evaluation of myelin radiotracers in the lysolecithin rat model of focal demyelination: beware of pitfalls! Contrast Media Mol Imaging 2019:9294586. https://doi.org/10.1155/2019/9294586

33. Afzali AM, Ruck T, Wiendl H et al (2017) Animal models in idiopathic inflammatory myopathies: how to overcome a translational roadblock? Autoimmun Rev 16:478–494

34. Dalakas MC (2004) From canine to man: on antibodies, macrophages, and dendritic cells in inflammatory myopathies. Muscle Nerve 29:753–7556

35. Morrison TE, Oko L, Montgomery SA et al (2011) A mouse model of chikungunya virus-induced musculoskeletal inflammatory disease: evidence of arthritis, tenosynovitis, myositis, and persistence. Am J Pathol 178:32–40

36. Morrison TE, Whitmore AC, Shabman RS et al (2006) Characterization of Ross River virus tropism and virus-induced inflammation in a mouse model of viral arthritis and myositis. J Virol 80:737–749

37. Strongwater SL, Dorovini-Zis K, Ball RD et al (1984) A murine model of polymyositis induced by coxsackievirus B1 (Tucson strain). Arthritis Rheum 27:433–442

38. Kohyama K, Matsumoto Y (1999) C-protein in the skeletal muscle induces severe autoimmune polymyositis in Lewis rats. J Neuroimmunol 98:130–135

39. Scuderi F, Mannella F, Marino M et al (2006) IL-6-deficient mice show impaired inflammatory response in a model of myosin-induced experimental myositis. J Neuroimmunol 176:9–11540

40. Sugihara T, Okiyama N, Suzuki M (2010) Definitive engagement of cytotoxic CD8 T cells in C protein-induced myositis, a murine model of polymyositis. Arthritis Rheum 62:3088–3092.45

41. Link CD (1995) Expression of human beta-amyloid peptide in transgenic Caenorhabditis elegans. Proc Natl Acad Sci U S A 92:9368–9372

42. Fuchs S, Aricha R, Reuveni D et al (2014) Experimental autoimmune myasthenia gravis (EAMG): from immunochemical characterization to therapeutic approaches. J Autoimmun 54:51–59

43. Mantegazza R, Cordiglieri C, Consonni A et al (2016) Animal models of myasthenia gravis: utility and limitations. Int J Gen Med 9:53–64

44. Losen M, Martinez-Martinez P, Molenaar PC et al (2015) Standardization of the experimental autoimmune myasthenia gravis (EAMG) model by immunization of rats with Torpedo californica acetylcholine receptors—recommendations for methods and experimental designs. Exp Neurol 270:8–28

45. Soliven B (2014) Animal models of autoimmune neuropathy. ILAR J 54:282–290

46. Csurhes PA, Sullivan AA, Green K et al (2005) T cell reactivity to P0, P2, PMP-22, and myelin basic protein in patients with Guillain-Barre syndrome and chronic inflammatory demyelinating polyradiculoneuropathy. J Neurol Neurosurg Psychiatry 76:1431–1439

47. Hartung HP, Schäfer B, Heininger K et al (1988) The role of macrophages and eicosanoids in the pathogenesis of experimental allergic neuritis. Serial clinical, electrophysiological, biochemical and morphological observations. Brain 111:1039–1059

48. Kim HJ, Jung CG, Jensen MA et al (2008) Targeting of myelin protein zero in a spontaneous autoimmune polyneuropathy. J Immunol 181:8753–8760

49. Doke SK, Dhawale SC (2015) Alternatives to animal testing: a review. Saudi Pharm J 23:223–229

50. Rosania K (2013) Synthetic research tools as alternatives to animal models. Lab Anim (NY) 42:189–190

51. Graham ML, Prescott MJ (2015) The multifactorial role of the 3Rs in shifting the harm-benefit analysis in animal models of disease. Eur J Pharmacol 759:19–29

Chapter 3

White Matter Pathology

Angela Radetz and Sergiu Groppa

Abstract

The central nervous system of patients with multiple sclerosis (MS) is affected by diverse pathological processes. Whereas focal tissue injury is partly detectable with conventional magnetic resonance imaging (MRI), widespread diffuse damage is also present in the normal-appearing white matter (NAWM) and can be quantified using advanced MRI and positron emission tomography (PET) techniques. In this chapter, we first address imaging methods for description of pathological processes in focal lesions and the NAWM. Next, we describe imaging techniques that reflect inflammation, demyelination and remyelination and axonal damage and loss as the central pathologies observed in MS. Lastly, we present novel methods based on diffusion-weighted imaging to quantify microstructural integrity. This chapter aims at providing an overview of imaging sequences, acquisition strategies, and models that can be chosen in dependence of the tissue type and pathological process researchers plan to noninvasively investigate.

Key words Multiple sclerosis, Lesions, Normal-appearing white matter, Inflammation, Demyelination, Remyelination, Axonal damage, Axonal loss, Diffusion-weighted imaging

1 Introduction

White matter pathology in multiple sclerosis (MS) is mainly characterized by focal lesions. Widespread diffuse tissue damage has also been, however, detected in white matter that appears macroscopically normal with conventional and advanced magnetic resonance imaging (MRI) methods [1]. Even if tissue damage is detectable with conventional MRI techniques, the underlying pathological processes cannot be inferred from the acquired images. Further, it is not always clear, whether tissue has been damaged irreversibly or if recovery is still possible. The following subchapters point out methodological tools for the characterization of white matter pathology through imaging methods as MRI and positron emission tomography (PET). Imaging techniques are presented that aid to dissolve pathological processes occurring in focal lesions and normal-appearing white matter.

Sergiu Groppa and Sven G. Meuth (eds.), *Translational Methods for Multiple Sclerosis Research*, Neuromethods, vol. 166, https://doi.org/10.1007/978-1-0716-1213-2_3, © Springer Science+Business Media, LLC, part of Springer Nature 2021

2   Focal Lesions

Lesions in the white matter were among the first prominent markers of MS pathology found during autopsies in the nineteenth century as well as in the twentieth century using imaging methods [2, 3]. MRI provides robust means to characterize focal pathology and is part of the routine diagnostic methods. Lesions manifest as focal, ovoid areas of hyperintensity in T2-weighted or proton density-weighted sequences and, depending on the exact location, should be at least 3 mm in axial length for MS diagnosis [4] (Fig. 1a). MS lesions show a widespread asymmetric distribution in the central nervous system, with predilection sites in the periventricular and juxtacortical white matter, corpus callosum, and infratentorial regions [4], which is partly explained by the distribution of venules [5].

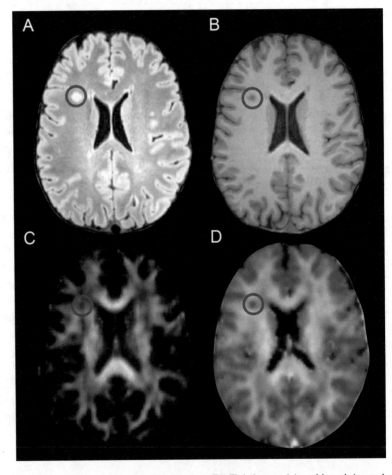

Fig. 1 Lesion appearing hyperintense in a T2-Flair image (**a**) and hypointense in a T1-weighted image (**b**). Diffusion-weighted images reveal a decrease in fractional anisotropy (**c**), caused by a decreased neurite density (**d**)

Focal lesions are commonly classified as active, chronic active, or chronic in literature, yet without any exact and unique definition [6]. While *activity* usually describes inflammatory, demyelinating, and/or neurodegenerative events that manifest as gadolinium or other contrast agent enhancement, *chronic* refers to inactive and chronic active lesions or lesions without enhancement [6,7]. Several attempts have been made for a more specific lesion classification [8]. For diagnosis of MS, however, criteria of dissemination in space and time are paramount [9]. Lesions need to be detected in an MS-typical region (periventricular, juxtacortical, infratentorial, or spinal cord), with either gadolinium-enhancing and non-enhancing lesions in one scan, or a new lesion at follow-up. Yet MRI signal abnormalities originating from vascular or other neurological diseases including migraine, neuromyelitis optica spectrum disorders, and cerebral small vessel disease might overlap with MS-typical findings [10]. Sensitivity and specificity for differential diagnosis of MS was increased by considering the occurrence of perivenular lesions, termed the *central vein sign* [11–13], that best can be visualized with 7 T MRI, but also at 1.5 T and 3 T [12, 14]. Hence, studies are conducted in order to potentially establish guidelines for applying the central vein sign in the clinical context.

Filippi et al. [15] devised a recent practical guideline with a focus on describing most suitable sequences for capturing typical MS MRI findings. While T2-Flair sequences are most sensitive for detection of periventricular lesions and differentiation from the perivascular space, confirming evidence might be attained using a second acquisition, such as T2-, PD-, or T1-weighted sequences [15, 16]. Periventricular lesions often have their main axis perpendicular to the lateral ventricular walls and are termed *Dawson's fingers*. Juxtacortical lesions, that is, lesions in direct contact with the cortex, as well as infratentorial lesions, comprising the brainstem, cerebellar peduncles, and the cerebellum, are best detected using a T2-Flair sequence [15, 17]. Focal lesions frequently span both the cortex and adjacent subcortical white matter. While T2-Flair sequences are capable of detecting such lesions, grey matter lesions are even better localized using double inversion recovery (DIR), phase-sensitive inversion recovery (PSIR), or T1-weighted MPRAGE sequences [4]. Regarding lesion detection in the posterior fossa and spinal cord, PD- and T2-weighted imaging displays better sensitivity for lesion detection [5].

Gadolinium-enhancing lesions together with non-enhancing lesions observed at one time point can support the criteria for dissemination in time [4]. Evolving lesions initially are typically characterized by a disruption of the blood–brain barrier, most likely accompanied by an inflammatory response and demyelination [16, 18, 19]. Most lesions exhibit a nodular shape and up to one third a ring-like enhancement, whereby the latter are larger in size and persist longer [20]. Ring-enhancing lesions have been suggested to represent the most destructive lesions [21, 22] and are linked to pronounced macrophage infiltration [23].

A further MS-specific lesion characteristic that has not been observed in neuromyelitis optica or cerebrovascular diseases [24, 25] is a hypointense rim in some white matter lesions observed with susceptibility-based MRI at 7 T [26]. Among other factors as oxidative stress and disruption of tissue microstructural organization, the rim mainly reflects iron accumulation within macrophages/activated microglia that contain myelin degradation products, T cells and axonal transections and could therefore predict inflammation and tissue damage [4, 26]). Rim lesions typically demonstrate slow tissue loss and are hence categorized as slowly evolving/expanding or smoldering lesions [15].

T2-hyperintensity of some lesions might be accompanied by hypointensity in T1-weighted images [27]. A longitudinal study observed a probability of 56% for black holes to become chronic [28], which is associated with marked demyelination, axonal loss, and increased disability [29, 30].

Although lesions can be detected very sensitively with imaging methods, specificity of underlying pathological processes is low. These might include gliosis, demyelination or remyelination, axonal loss, inflammation as well as edema [23]. This non-specificity partly explains the low correlation between findings in MR images and clinical disability of the patients, known as the clinical–radiologic paradox [31].

3   Normal-Appearing White Matter

White matter tissue measured with conventional MRI that does not contain any observable damage in MS is typically referred to as normal-appearing white matter (NAWM) [1, 32]. Tissue damage in MS appears not only in focal lesions, but also diffusely in the NAWM. In a postmortem study, 72% of the histologically investigated tissue was abnormal, where macroscopically analyzed NAWM changes included gliosis, demyelination and axonal loss [32]. These ex-vivo findings underline the importance of investigating tissue that initially appears normal for predictions of further disease progression. Applying magnetization transfer imaging in vivo, a significant decrease in magnetization transfer ratios (MTR) in NAWM of later enhancing lesions was observed, demonstrating changes before visible lesions appear [33]. These changes potentially include demyelination, inflammation, increased water content, and axonal density reduction [34–37]. A more pathologically specific method, proton magnetic resonance spectroscopy imaging (^{1}H-MRSI), revealed a decrease in N-acetyl-aspartate and N-acetyl-aspartyl-glutamate, and an increase in myoinositol. Such findings are suggestive of axonal density reduction and an increase in the number of glial cells [38]. In diffusion tensor imaging (DTI) studies, an increased mean diffusivity (MD) and decreased fractional anisotropy (FA) in NAWM of MS patients compared to healthy controls were detected. Here, increased diffusivity is caused by a loss of structural barriers, and an altered organization of barriers causes a more pronounced isotropic water diffusion profile [39–41]. Not only MTR but also DTI studies showed tissue abnormalities before lesion evolution. While FA values decreased in NAWM compared to white matter of healthy controls, apparent diffusion coefficient (ADC) was higher before plaque formation. Although abnormalities appear to be widespread in the brain, they occur more pronounced in affinity to lesioned regions [42] and in predilection sites of lesions [40, 43]. Integrating the findings on abnormalities in the NAWM, various underlying pathological processes appear to occur, such that the appropriate nonconventional imaging tools need to be chosen depending on the specific evaluation of tissue changes.

4   Imaging of Pathologies

4.1   Inflammation

Lesions characterized by inflammatory processes resulting from a permeable blood–brain barrier are conventionally visualized with T1-weighted sequences. Subsequently to intravenous administration of a paramagnetic contrast agent, typically gadolinium, lesions appear hyperintense in the T1-weighted images [3, 44, 45]. Enhancement is proportional to the number of infiltrating inflammatory cells, mainly macrophages, in lesioned tissue [46, 47]. Although demyelination is observed in lesions as well, it does not cause gadolinium enhancement, making this observation unique for inflammatory processes [48]. Lesion enhancement observed in the early stages of MS typically persists for a few days to weeks, on average 3 weeks [20]. Also re-enhancement of lesions has been observed with varying frequencies of occurrence [22, 49]. The intensity and size of enhancement depends on the concentration of gadolinium, permeability of the blood–brain barrier, and the volume of the leakage space [47, 50]. While using a triple dose of gadolinium increases sensitivity of smaller lesion detection, delaying the acquisition has minor advantages for detection of larger lesions [51–53]. A significant increase in sensitivity for the detection of enhancing lesions has also been reported by applying magnetization transfer pulses to T1-weighted imaging in conjunction with gadolinium administration [53, 54]. However, gadolinium enhancement is neither informative about the specific immune system cells crossing the disrupted blood–brain barrier nor about the extent and duration of inflammation [55].

Proton magnetic resonance spectroscopy (^{1}H-MRS) allowed for observing increases in glutamate and choline in areas prior to lesion formation that could be explained by macrophage/microglial activation [56, 57].

In a more recent approach, susceptibility MRI at 7 Tesla was used to characterize the T2*-weighted signal magnitude and phase in order to infer inflammatory activity [58]. Here, an increased permeability of the blood–brain barrier was accompanied by a paramagnetic rim on phase, reflecting iron accumulation in proinflammatory activated macrophages/microglia [26, 58].

In contrast to the described MR methods, PET detects the involvement of specific cells and hence can complement conventional MRI with selective immunological characterizations [59–62]. Here, mostly microglial activation was quantified with the aid of 18-kDa translocator protein (TSPO) radioligands, as TSPO is predominantly expressed on membranes of activated, but not surveying microglia and therefore strongly indicates both focal and diffuse inflammatory activity [62]. A further promising step towards neuroinflammation characterization via PET appears to be adenosine receptor imaging [63, 64].

4.2 Demyelination and Remyelination

Demyelination, that is, loss of the myelin sheath that surrounds axons, is a characteristic pathological process of MS. Also remyelination, the process of myelin sheath restoration, is observed in MS [34]. In order to monitor evolution of pathological processes including demyelination and remyelination and the potential effect of therapeutic interventions, imaging methods that track myelin integrity are of great importance. With conventional imaging, hypointensity of T1-weighted lesions is indicative of demyelination, yet not exclusively, since it also captures further pathology as axonal loss [65]. Nonconventional quantitative imaging is an important means to directly quantify absolute tissue parameters such as relaxation times, proton density, and magnetization transfer, and is thereby not prone to technical sources of signal variability [66, 67]. It allows for reproducible interindividual and intraindividual, as well as longitudinal comparisons independent of the scanning site [68]. Magnetization transfer imaging (MTI) is based on interactions between bounded protons in macromolecules and water [69]. The semiquantitative magnetization transfer ratio (MTR) quantifies these interactions, where a low MTR is indicative of reduced magnetization exchange between macromolecules and the surrounding water, reflecting structural damage [69, 70]. In vivo and postmortem MS studies observed decreased MTR in demyelinating lesions; however, it needs to be considered that MTR is also dependent on inflammation, water content, and axonal density [34–37, 71]. The apparent transverse relaxation time (T2*) and the reciprocal transverse relaxation rate (R2*) have been linked to myelin as well, but also to iron, fiber orientation and water content [72–74]. Stüber et al. [74] concluded from their postmortem MRI and histology investigation, that longitudinal relaxation time (T1) maps reflect myelin more specifically than do T2*-based images. In vivo studies showed decreases of R1 (1/T1) and R2* in the normal-appearing white matter (NAWM) of MS patients compared to healthy controls [75].

Variants of MRI sequences have been developed that more specifically allow investigating myelin integrity through the estimation of myelin water fraction (MWF), that is, the signal component captured from water within the myelin sheaths compared to that in other tissue compartments [76, 77]. MWF based on T2* relaxation was significantly reduced in lesions compared to normal-appearing white matter of MS patients, and lower in NAWM compared to that in healthy control subjects [78]. However, it needs to be considered that MWF is likely affected by edema from inflammatory activity [79]. Further restrictions of MWF techniques are the long acquisition time, low contrast-to-noise-ratio, and computationally expensive processing [80, 81]. An improvement in scanning time was made by implementation of the gradient and spin echo technique (*GRASE*) [80] and the multicomponent driven equilibrium single pulse observation of T1 and T2 (*mcDESPOT*) that are promising for application in clinics [82–84].

The ratio of T1- and T2-weighted images (T1w/T2w) has been proposed as an alternative proxy for microstructural and especially myelin imaging by using conventional imaging methods and keeping scanning time minimal [81]. T1w/T2w was significantly lower in the normal-appearing white matter of MS patients compared to healthy controls, and ratios correlated with disability scores [85].

A more technically demanding tool, proton magnetic resonance spectroscopy imaging (^{1}H-MRSI), is a further promising method to investigate demyelination and remyelination by imaging cerebral metabolites such as choline [86, 87]. Lastly, myelin integrity can be estimated with diffusion-weighted imaging that measures motion of water molecules. Here, motion is restricted along fiber tracts, such that a diffusion tensor can be described for every voxel's signal [88]. Lower fractional anisotropy and higher mean diffusivity are characteristic for demyelination, which, however, are also affected by axonal loss [89].

Using PET and histological staining methods, Stankoff et al. [90] demonstrated high specificity of the PET marker thioflavin-T derivative   2-(40-methylaminophenyl)-6-hydroxybenzothiazole (PIB) to myelin in postmortem MS brains that can be used for in vivo assessments of demyelination and remyelination [91].

Lastly, which method should be used for imaging depends on the expected amount of specificity for myelin, the need of sensitivity to differences in tissue type, and the availability of acquisition and analysis time [92].

4.3 Axonal damage and Loss

Axonal damage and loss mainly manifest as hypointense lesions in T1-weighted images, termed *black holes* [23, 29]. Hypointense T1-weighted lesion volume exhibits the strongest relationship with disability compared to other conventional imaging methods [93]. Yet extracellular edema also have been shown to contribute to T1-weighted hypointensities [23]. In brain regions including the brain stem and the optic nerves, *black holes* are difficult to detect [94, 95]. In order to dissolve tissue pathology more specifically, nonconventional imaging techniques are more adequate [94]. Magnetization transfer ratio correlates, also in normal-appearing white matter, strongly with axonal density but shows a weaker association to myelin density [71]. Proton magnetic resonance spectroscopy imaging (^{1}H-MRSI) studies have revealed N-acetyl-aspartate (NAA) as a specific marker of axonal damage and loss that appears largely independent of inflammation [96–98] and is specific to neurons and their processes [99, 100]. Here, lower NAA mainly reflects a loss of axonal volume or density or alterations in axonal metabolism [98]. NAA has often been reported relative to Creatine (Cr) as a marker of axonal damage and loss, as Cr appears relatively stable in all brain cells and chronic conditions [101, 102]. While conventional methods are insensitive to the detection of diffuse widespread axonal damage and loss, MRS can detect this tissue abnormality also in normal-appearing white matter [102]. As NAA decreases precede brain atrophy, it appears to be an early marker of severe tissue destruction relevant for treatment monitoring [103, 104].

5 Novel Diffusion-Based Methods for Imaging of Microstructural Integrity

Diffusion-weighted imaging has proven to be a promising method for imaging of microstructural integrity. Diffusion tensor imaging (DTI) comprises the most widely used techniques that are based on eigenvectors and eigenvalues of a diffusion tensor fitted in each brain voxel. Computed DTI parameters include fractional anisotropy (FA), mean diffusivity (MD), axial diffusivity and radial diffusivity. Typically, increases in MD and decreases in FA are observed in focal lesions compared to NAWM [40]. MD increases have been observed in NAWM even before lesion formation, suggesting a high sensitivity to emerging pathological processes [105]. A drawback of DTI is, however, that the applied Gaussian models are oversimplified when considering complex diffusion profiles as occurring in brain tissue [106]. In the case of crossing fibers, the estimated diffusion tensor of a voxel appears isotropic, although the underlying diffusion might be highly anisotropic in multiple orientations [107]. Importantly, a proportion of up to 90% of crossing fibers has been estimated in the human brain [108]. For better estimation of those, advanced acquisition protocols based on high angular resolution diffusion-weighted imaging (HARDI) have been developed, that require a large number of diffusion-weighting gradient directions, typically minimally 40 [109]. Using HARDI protocols, estimates of crossing fibers can be reconstructed directly using various compartment models that make assumptions about the tissue structure. With multi-tensor fitting, two or more populations of fiber orientations are assumed [110, 111]. In the ball-and-stick model, axons are represented as parallel cylinders and extracellular dispersion as an isotropic diffusion tensor [112, 113]. Composite hindered and restricted models of diffusion (CHARMED) model a hindered extra-axonal region with a diffusion tensor together with a restricted diffusion model of the intra-axonal compartment, thereby constituting a biologically more plausible representation [114, 115]. A drawback is the requirement of acquiring multiple nonzero b-values (multi-shell), resulting in longer scanning times. Further, these models necessitate prior assumptions about the number of fiber populations. Spherical deconvolution, on the contrary, includes a continuous representation of fiber orientations, thereby modeling the distribution of the relative amounts of each fiber population over a sphere, called the fiber orientation distribution [116, 117].

Diffusion basis spectrum imaging (DBSI) incorporates multiple anisotropic diffusion tensors that reflect crossing (un)-myelinated axons [118]. A spectrum of isotropic diffusion tensors represents cells, subcellular structures, and edema [119, 120]. Neurite orientation dispersion and density imaging (NODDI) allows for estimating neurite density and orientation dispersion and the isotropic volume fraction, yet does not resolve crossing fiber angles [121]. In Fig. 1, the same MS patient's brain imaged with different modalities is depicted. A lesion appearing hyperintense in the T2-Flair, and hypointense in the T1-weighted image, is highlighted. Here, diffusion-weighted imaging reveals a decrease in fractional anisotropy (FA), which indicates the restriction of water flow. NODDI reveals more specifically that the decrease in FA can be attributed to a decrease in neurite density.

Characterizing the diffusion process itself is a further method for disentangling crossing fibers, dissimilarly to the model-based approaches. Numerous q-space based techniques have been developed based on a theory relating the diffusion signal to the displacement distribution function [122, 123]. A major drawback that these methods have in common is, however, the increase in scanning time. Techniques based on q-space theory include diffusion kurtosis imaging (DKI) [124], diffusion spectrum imaging (DSI) [125, 126], and q-ball MRI [127].

Numerous toolboxes are available that allow for estimating microstructural properties, including FMRIB's Software Library (FSL, https://fsl.fmrib.ox.ac.uk/fsl/fslwiki/), the NODDI Matlab Toolbox (https://www.nitrc.org/projects/noddi_toolbox), AMICO (https://github.com/daducci/AMICO), DSI

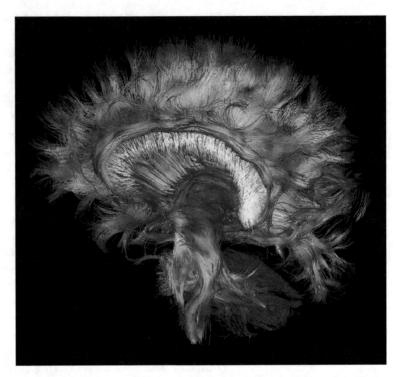

Fig. 2 Probabilistic tractography image based on multi-tensor fitting, color-coded by fractional anisotropy values. Tracts were reconstructed using Slicer 4 [129]

Studio (http://dsi-studio.labsolver.org/), CAMINO (http://cam ino.cs.ucl.ac.uk/index.php?n=Main.HomePage), SLICER (https://www.slicer.org/), and PANDA (https://www.nitrc.org/projects/panda/) amongst many others.

In order to investigate white matter trajectories that connect brain regions, tractography algorithms have been developed to reconstruct fiber bundles (Fig. 2) [128].

DTI has been widely used for tractography in order to infer whether adjacent voxels are part of the same tract. However, it is limited due to the crossing fiber problem, such that several other of the aforementioned models are recommended in estimating the fiber orientation distribution function (fODF) [107]. Two major classes of tractography are available. In deterministic tractography algorithms, the inferred orientation is defined by the major eigenvector of the diffusion tensor, such that only a single estimate is considered. On the contrary, probabilistic approaches provide a probabilistic estimation of white matter fiber tracts [107]. FMRIB's Diffusion Toolbox (FDT), which is part of FSL, is an exemplary software that is frequently used for the analysis of diffusion data (https://fsl.fmrib.ox.ac.uk/fsl/fslwiki/FDT). Subsequently to preprocessing of the diffusion-weighted images, the toolbox uses the algorithm implemented in *BEDPOSTX* (Bayesian Estimation of

Diffusion Parameters Obtained using Sampling Techniques, X for crossing fibers) for the generation of distributions on diffusion parameters [112, 113, 130–132]. Here, crossing fibers can, for example, be modeled based on the ball-and-stick model [112, 113]. *PROBTRACKX* then repetitively samples from the fODFs to each time compute a streamline on the samples, resulting in a probabilistic streamline [113, 130, 133]. Both deterministic and probabilistic tractography have been employed in multiple sclerosis research to evaluate the integrity of white matter networks [134–136].

Diffusion parameters as FA have further been used in voxelwise analyses to compare disease-related changes in the white matter [137]. Due to registration problems of FA maps to a standard space, an approach termed tract-based spatial statistics (TBSS) has been developed that allows statistical testing of diffusion parameters with minimized registration confounders [138]. Any other diffusion parameter than FA could equally be applied using this method. In MS patients compared to healthy controls, lower FA values were observed in the white matter, and decreases in FA were related to cognitive impairment [139, 140].

The multitude of acquisition paradigms, diffusion and compartment models, and analyses strategies shows that the field of diffusion-weighted imaging has steadily evolved in the last decades. The continuous development of these methods ensures appropriate prognosis for disease progression, as well as treatment monitoring in MS.

References

1. Miller DH, Thompson AJ, Filippi M (2003) Magnetic resonance studies of abnormalities in the normal appearing white matter and grey matter in multiple sclerosis. J Neurol 250 (12):1407–1419. https://doi.org/10.1007/s00415-003-0243-9

2. Carswell R (1838) Pathological anatomy: illustrations of the elementary forms of disease. Longman, Orme, Brown, Green and Longman, London

3. McDonald W, Barnes D (1989) Lessons from magnetic resonance imaging in multiple sclerosis. Trends Neurosci 12(10):376–379

4. Filippi M, Preziosa P, Banwell BL, Barkhof F, Ciccarelli O, De Stefano N et al (2019) Assessment of lesions on magnetic resonance imaging in multiple sclerosis: practical guidelines. Brain 142(7):1858–1875. https://doi.org/10.1093/brain/awz144

5. Trip SA, Miller DH (2005) Imaging in multiple sclerosis. J Neurol Neurosurg Psychiatry 76(Suppl 3):iii11–iiii8. https://doi.org/10.1136/jnnp.2005.073213

6. Kuhlmann T, Ludwin S, Prat A, Antel J, Bruck W, Lassmann H (2017) An updated histological classification system for multiple sclerosis lesions. Acta Neuropathol 133 (1):13–24. https://doi.org/10.1007/s00401-016-1653-y

7. Miller DH, Barkhof F, Nauta JJP (1993) Gadolinium enhancement increases the sensitivity of MRI in detecting disease activity in multiple sclerosis. Brain 116(5):1077–1094

8. van der Valk P, de Groot C (2000) Staging of multiple sclerosis (MS) lesions: pathology of the time frame of MS. Neuropathol Appl Neurobiol 26(1):2–10

9. Thompson AJ, Banwell BL, Barkhof F, Carroll WM, Coetzee T, Comi G et al (2018) Diagnosis of multiple sclerosis: 2017 revisions of the McDonald criteria. Lancet Neurol 17 (2):162–173. https://doi.org/10.1016/s1474-4422(17)30470-2

10. Geraldes R, Ciccarelli O, Barkhof F, De Stefano N, Enzinger C, Filippi M et al (2018) The current role of MRI in

differentiating multiple sclerosis from its imaging mimics. Nat Rev Neurol 14(4):199

11. Tallantyre EC, Morgan PS, Dixon JE, Al-Radaideh A, Brookes MJ, Evangelou N et al (2009) A comparison of 3T and 7T in the detection of small parenchymal veins within MS lesions. Investig Radiol 44 (9):491–494

12. Maggi P, Absinta M, Grammatico M, Vuolo L, Emmi G, Carlucci G et al (2018) Central vein sign differentiates multiple sclerosis from central nervous system inflammatory vasculopathies. Ann Neurol 83 (2):283–294

13. Sinnecker T, Clarke MA, Meier D, Enzinger C, Calabrese M, De Stefano N et al (2019) Evaluation of the central vein sign as a diagnostic imaging biomarker in multiple sclerosis. JAMA Neurol 76(12):1446–1456

14. Sati P, Oh J, Constable RT, Evangelou N, Guttmann CR, Henry RG et al (2016) The central vein sign and its clinical evaluation for the diagnosis of multiple sclerosis: a consensus statement from the North American imaging in multiple sclerosis cooperative. Nat Rev Neurol 12(12):714

15. Filippi M, Brück W, Chard DT, Fazekas F, Geurts JJG, Enzinger C et al (2019) Association between pathological and MRI findings in multiple sclerosis. Lancet Neurol 18 (2).198–210. https://doi.org/10.1016/s1474-4422(18)30451-4

16. Miller DH, Grossman RI, Reingold SC, McFarland HF (1998) The role of magnetic resonance techniques in understanding and managing multiple sclerosis. Brain 121 (1):3–24

17. Bakshi R, Ariyaratana S, Benedict RH, Jacobs L (2001) Fluid-attenuated inversion recovery magnetic resonance imaging detects cortical and juxtacortical multiple sclerosis lesions. Arch Neurol 58(5):742–748

18. Miller DH, Rudge P, Johnson G, Kendall BE, Macmanus DG, Moseley IF et al (1988) Serial gadolinium enhanced magnetic resonance imaging in multiple sclerosis. Brain 111 (4):927–939

19. Prineas J, Connell F (1979) Remyelination in multiple sclerosis. Ann Neurol 5(1):22–31

20. Cotton F, Weiner HL, Jolesz FA, Guttmann CR (2003) MRI contrast uptake in new lesions in relapsing-remitting MS followed at weekly intervals. Neurology 60(4):640–646

21. Morgen K, Jeffries N, Stone R, Martin R, Richert N, Frank J et al (2001) Ring-enhancement in multiple sclerosis: marker of disease severity. Mult Scler 7(3):167

22. van Waesberghe JHTM, van Walderveen MAA, Castelijns JA, Scheltens P (1998) Lycklama à Nijeholt GJ, Polman CH, et al. patterns of lesion development in multiple sclerosis: longitudinal observations with T1-weighted spin-echo and magnetization transfer MR. AJNR Am J Neuroradiol 19 (4):675–683

23. Brück W, Bitsch A, Kolenda H, Brück Y, Stiefel M, Lassmann H (1997) Inflammatory central nervous system demyelination: correlation of magnetic resonance imaging findings with lesion pathology. Ann Neurol 42 (5):783–793

24. Chawla S, Kister I, Wuerfel J, Brisset J-C, Liu S, Sinnecker T et al (2016) Iron and non-iron-related characteristics of multiple sclerosis and neuromyelitis optica lesions at 7T MRI. AJNR Am J Neuroradiol 37 (7):1223–1230

25. Kilsdonk ID, Wattjes MP, Lopez-Soriano A, Kuijer JP, de Jong MC, de Graaf WL et al (2014) Improved differentiation between MS and vascular brain lesions using FLAIR* at 7 Tesla. Eur J Radiol 24(4):841–849

26. Dal-Bianco A, Grabner G, Kronnerwetter C, Weber M, Hoftberger R, Berger T et al (2017) Slow expansion of multiple sclerosis iron rim lesions: pathology and 7 T magnetic resonance imaging. Acta Neuropathol 133 (1):25–42. https://doi.org/10.1007/s00401-016-1636-z

27. Sahraian MA, Eshaghi A (2010) Role of MRI in diagnosis and treatment of multiple sclerosis. Clin Neurol Neurosurg 112(7):609–615. https://doi.org/10.1016/j.clineuro.2010.03.022

28. Bagnato F, Jeffries N, Richert ND, Stone RD, Ohayon JM, McFarland HF et al (2003) Evolution of T1 black holes in patients with multiple sclerosis imaged monthly for 4 years. Brain 126(8):1782–1789

29. Truyen L, Van Waesberghe J, Van Walderveen M, Van Oosten B, Polman C, Hommes O et al (1996) Accumulation of hypointense lesions ("black holes") on T1 spin-echo MRI correlates with disease progression in multiple sclerosis. Neurology 47 (6):1469–1476

30. Bitsch A, Kuhlmann T, Stadelmann C, Lassmann H, Lucchinetti C, Brück W (2001) A longitudinal MRI study of histopathologically defined hypointense multiple sclerosis lesions. Ann Neurol 49(6):793–796

31. Barkhof F (1999) MRI in multiple sclerosis: correlation with expanded disability status scale (EDSS). Mult Scler 5(4):283–286

32. Allen IV, McKeown SR (1979) A histological, histochemical and biochemical study of the macroscopically normal white matter in multiple sclerosis. J Neurol Sci 41(1):81–91

33. Filippi M, Rocca MA, Martino G, Horsfield MA, Comi G (1998) Magnetization transfer changes in the normal appering white matter precede the appearance of enhancing lesions in patients with multiple sclerosis. Ann Neurol 43(6):809–814

34. Mallik S, Samson RS, Wheeler-Kingshott CA, Miller DH (2014) Imaging outcomes for trials of remyelination in multiple sclerosis. J Neurol Neurosurg Psychiatry 85 (12):1396–1404. https://doi.org/10.1136/jnnp-2014-307650

35. Schmierer K, Scaravilli F, Altmann DR, Barker GJ, Miller DH (2004) Magnetization transfer ratio and myelin in postmortem multiple sclerosis brain. Ann Neurol 56(3):407–415. https://doi.org/10.1002/ana.20202

36. Vavasour IM, Laule C, Li DK, Traboulsee AL, MacKay AL (2011) Is the magnetization transfer ratio a marker for myelin in multiple sclerosis? J Magn Reson Imaging 33 (3):713–718. https://doi.org/10.1002/jmri.22441

37. Chen JT, Collins DL, Atkins HL, Freedman MS, Arnold DL, Canadian MS/BMT Study Group (2008) Magnetization transfer ratio evolution with demyelination and remyelination in multiple sclerosis lesions. Ann Neurol 63(2):254–262. https://doi.org/10.1002/ana.21302

38. Chard D, Griffin C, McLean M, Kapeller P, Kapoor R, Thompson A et al (2002) Brain metabolite changes in cortical grey and normal-appearing white matter in clinically early relapsing–remitting multiple sclerosis. Brain 125(10):2342–2352

39. Ceccarelli A, Rocca MA, Falini A, Tortorella P, Pagani E, Rodegher M et al (2007) Normal-appearing white and grey matter damage in MS. a volumetric and diffusion tensor MRI study at 3.0 tesla. J Neurol 254(4):513–518. https://doi.org/10.1007/s00415-006-0408-4

40. Rovaris M, Gass A, Bammer R, Hickman S, Ciccarelli O, Miller D et al (2005) Diffusion MRI in multiple sclerosis. Neurology 65 (10):1526–1532

41. Werring D, Clark C, Barker G, Thompson A, Miller D (1999) Diffusion tensor imaging of lesions and normal-appearing white matter in multiple sclerosis. Neurology 52(8):1626

42. Guo AC, MacFall JR, Provenzale JM (2002) Multiple sclerosis: diffusion tensor MR imaging for evaluation of normal-appearing white matter. Radiology 222(3):729–736

43. Filippi M, Iannucci G, Cercignani M, Rocca MA, Pratesi A, Comi G (2000) A quantitative study of water diffusion in multiple sclerosis lesions and normal-appearing white matter using echo-planar imaging. Arch Neurol 57 (7):1017–1021

44. Katz D, Taubenberger JK, Cannella B, McFarlin DE, Raine CS, McFarland HF (1993) Correlation between magnetic resonance imaging findings and lesion development in chronic, active multiple sclerosis. Ann Neurol 34(5):661–669

45. Filippi M (2000) Enhanced magnetic resonance imaging in multiple sclerosis. Mult Scler 6(5):320–326

46. Nesbit GM, Forbes GS, Scheithauer BW, Okazaki H, Rodriguez M (1991) Multiple sclerosis: histopathologic and MR and/or CT correlation in 37 cases at biopsy and three cases at autopsy. Radiology 180 (2):467–474

47. Martino G, Adorini L, Rieckmann P, Hillert J, Kallmann B, Comi G et al (2002) Inflammation in multiple sclerosis: the good, the bad, and the complex. Lancet Neurol 1 (8):499–509. https://doi.org/10.1016/s1474-4422(02)00223-5

48. Dousset V, Brochet B, Vital A, Gross C, Benazzouz A, Boullerne A et al (1995) Lysolecithin-induced demyelination in primates: preliminary in vivo study with MR and magnetization transfer. AJNR Am J Neuroradiol 16(2):225–231

49. Ciccarelli O, Giugni E, Paolillo A, Mainero C, Gasperini C, Bastianello S et al (1999) Magnetic resonance outcome of new enhancing lesions in patients with relapsing-remitting multiple sclerosis. Eur J Neurol 6(4):455–459

50. Tofts PS, Kermode AG (1991) Measurement of the blood-brain barrier permeability and leakage space using dynamic MR imaging. 1. Fundamental concepts. Magn Reson Med 17(2):357–367

51. Filippi M, Yousry T, Campi A, Kandziora C, Colombo B, Voltz R et al (1996) Comparison of triple dose versus standard dose gadolinium-DTPA for detection of MRI enhancing lesions in patients with MS. Neurology 46(2):379–384

52. Filippi M, Rovaris M, Capra R, Gasperini C, Yousry TA, Sormani MP et al (1998) A multicentre longitudinal study comparing the sensitivity of monthly MRI after standard and triple dose gadolinium-DTPA for monitoring disease activity in multiple sclerosis.

Implications for phase II clinical trials. Brain 121(10):2011–2020

53. van Waesberghe JHTM, Castelijns JA, Roser W, Silver NC, Yousry T, Lycklama à Nijeholt GJ et al (1997) Single-dose gadolinium with magnetization transfer versus triple-dose gadolinium in the MR detection of multiple sclerosis lesions. AJNR Am J Neuroradiol 18(7):1279–1285

54. Silver N, Good C, Barker G, MacManus D, Thompson A, Moseley I et al (1997) Sensitivity of contrast enhanced MRI in multiple sclerosis. Effects of gadolinium dose, magnetization transfer contrast and delayed imaging. Brain 120(7):1149–1161

55. Charil A, Filippi M (2007) Inflammatory demyelination and neurodegeneration in early multiple sclerosis. J Neurol Sci 259 (1–2):7–15. https://doi.org/10.1016/j.jns.2006.08.017

56. Klauser AM, Wiebenga OT, Eijlers AJ, Schoonheim MM, Uitdehaag BM, Barkhof F et al (2018) Metabolites predict lesion formation and severity in relapsing-remitting multiple sclerosis. Mult Scler 24(4):491–500

57. Tartaglia M, Narayanan S, De Stefano N, Arnaoutelis R, Antel S, Francis S et al (2002) Choline is increased in pre-lesional normal appearing white matter in multiple sclerosis. J Neurol 249(10):1382–1390

58. Absinta M, Sati P, Gaitan MI, Maggi P, Cortese IC, Filippi M et al (2013) Seven-tesla phase imaging of acute multiple sclerosis lesions: a new window into the inflammatory process. Ann Neurol 74(5):669–678. https://doi.org/10.1002/ana.23959

59. Rissanen E, Tuisku J, Rokka J, Paavilainen T, Parkkola R, Rinne JO et al (2014) In vivo detection of diffuse inflammation in secondary progressive multiple sclerosis using PET imaging and the radioligand (1)(1)C-PK11195. J Nucl Med 55(6):939–944. https://doi.org/10.2967/jnumed.113.131698

60. Ratchford JN, Endres CJ, Hammoud DA, Pomper MG, Shiee N, McGready J et al (2012) Decreased microglial activation in MS patients treated with glatiramer acetate. J Neurol 259(6):1199–1205. https://doi.org/10.1007/s00415-011-6337-x

61. Airas L, Rissanen E, Rinne JO (2015) Imaging neuroinflammation in multiple sclerosis using TSPO-PET. Clin Transl Imaging 3:461–473. https://doi.org/10.1007/s40336-015-0147-6

62. Airas L, Rissanen E, Rinne JO (2017) Imaging of microglial activation in MS using PET:

research use and potential future clinical application. Mult Scler 23(4):496–504

63. Rissanen E, Virta JR, Paavilainen T, Tuisku J, Helin S, Luoto P et al (2013) Adenosine A2A receptors in secondary progressive multiple sclerosis: a [11C] TMSX brain PET study. J Cereb Blood Flow Metab 33(9):1394–1401

64. Barret O, Hannestad J, Vala C, Alagille D, Tavares A, Laruelle M et al (2015) Characterization in humans of 18F-MNI-444, a PET radiotracer for brain adenosine 2A receptors. J Nucl Med 56(4):586–591

65. van Walderveen M, Kamphorst W, Scheltens P, van Waesberghe J, Ravid R, Valk J et al (1998) Histopathologic correlate of hypointense lesions on T1-weighted spin-echo MRI in multiple sclerosis. Neurology 50(5):1282–1288

66. Gracien RM, Reitz SC, Hof SM, Fleischer V, Droby A, Wahl M et al (2017) Longitudinal quantitative MRI assessment of cortical damage in multiple sclerosis: a pilot study. J Magn Reson Imaging 46(5):1485–1490. https://doi.org/10.1002/jmri.25685

67. Reitz SC, Hof S-M, Fleischer V, Brodski A, Gröger A, Gracien R-M et al (2017) Multiparametric quantitative MRI of normal appearing white matter in multiple sclerosis, and the effect of disease activity on T2. Brain Imaging Behav 11(3):744–753

68. Tofts PS (2003) Concepts: measurement and MR. Quantitative MRI of the brain: measuring changes caused by disease. Wiley, Chichester, pp 3–16

69. Wolff SD, Balaban RS (1994) Magnetization transfer imaging: practical aspects and clinical applications. Radiology 192(3):593–599

70. Filippi M (2015) MRI measures of neurodegeneration in multiple sclerosis: implications for disability, disease monitoring, and treatment. J Neurol 262(1):1–6. https://doi.org/10.1007/s00415-014-7340-9

71. van Waesberghe JHTM, Kamphors W, De Groot CJA, Van Walderveen MAA, Castelijns JA, Ravid R et al (1999) Axonal loss in multiple sclerosis lesions: magnetic resonance imaging insights into substrates of disability. Ann Neurol 46(5):747–754

72. Bagnato F, Hametner S, Boyd E, Endmayr V, Shi Y, Ikonomidou V et al (2018) Untangling the R2* contrast in multiple sclerosis: a combined MRI-histology study at 7.0 Tesla. PLoS One 13(3):e0193839

73. Cohen-Adad J (2014) What can we learn from T2* maps of the cortex? NeuroImage 93:189–200

74. Stüber C, Morawski M, Schäfer A, Labadie C, Wähnert M, Leuze C et al (2014) Myelin and iron concentration in the human brain: a quantitative study of MRI contrast. NeuroImage 93:95–106

75. Lommers E, Simon J, Reuter G, Delrue G, Dive D, Degueldre C et al (2019) Multiparameter MRI quantification of microstructural tissue alterations in multiple sclerosis. Neuroimage Clin 23:101879

76. Nguyen TD, Wisnieff C, Cooper MA, Kumar D, Raj A, Spincemaille P et al (2012) T2 prep three-dimensional spiral imaging with efficient whole brain coverage for myelin water quantification at 1.5 tesla. Magn Reson Med 67(3):614–621. https://doi.org/10.1002/mrm.24128

77. Mackay A, Whittall K, Adler J, Li D, Paty D, Graeb D (1994) In vivo visualization of myelin water in brain by magnetic resonance. Magn Reson Med 31(6):673–677

78. Laule C, Vavasour IM, Moore GR, Oger J, Li DK, Paty DW et al (2004) Water content and myelin water fraction in multiple sclerosis. A T2 relaxation study. J Neurol 251 (3):284–293. https://doi.org/10.1007/s00415-004-0306-6

79. Vavasour IM, Laule C, Li DK, Oger J, Moore GR, Traboulsee A et al (2009) Longitudinal changes in myelin water fraction in two MS patients with active disease. J Neurol Sci 276 (1–2):49–53. https://doi.org/10.1016/j.jns.2008.08.022

80. Prasloski T, Rauscher A, MacKay AL, Hodgson M, Vavasour IM, Laule C et al (2012) Rapid whole cerebrum myelin water imaging using a 3D GRASE sequence. NeuroImage 63(1):533–539. https://doi.org/10.1016/j.neuroimage.2012.06.064

81. Uddin MN, Figley TD, Solar KG, Shatil AS, Figley CR (2019) Comparisons between multi-component myelin water fraction, T1w/T2w ratio, and diffusion tensor imaging measures in healthy human brain structures. Sci Rep 9(1):1–17

82. Deoni SC, Rutt BK, Jones DK (2007) Investigating the effect of exchange and multicomponent T(1) relaxation on the short repetition time spoiled steady-state signal and the DESPOT1 T(1) quantification method. J Magn Reson Imaging 25(3):570–578. https://doi.org/10.1002/jmri.20836

83. Kitzler HH, Su J, Zeineh M, Harper-Little C, Leung A, Kremenchutzky M et al (2012) Deficient MWF mapping in multiple sclerosis using 3D whole-brain multi-component relaxation MRI. NeuroImage 59

(3):2670–2677. https://doi.org/10.1016/j.neuroimage.2011.08.052

84. Deoni SC, Rutt BK, Jones DK (2008) Investigating exchange and multicomponent relaxation in fully-balanced steady-state free precession imaging. J Magn Reson Imaging 27(6):1421–1429

85. Beer A, Biberacher V, Schmidt P, Righart R, Buck D, Berthele A et al (2016) Tissue damage within normal appearing white matter in early multiple sclerosis: assessment by the ratio of T1-and T2-weighted MR image intensity. J Neurol 263(8):1495–1502

86. De Stefano N, Filippi M (2007) MR spectroscopy in multiple sclerosis. J Neuroimaging 17 (Suppl 1):31S–35S. https://doi.org/10.1111/j.1552-6569.2007.00134.x

87. Rovira A, Alonso J (2013) 1H magnetic resonance spectroscopy in multiple sclerosis and related disorders. Neuroimaging Clin N Am 23(3):459–474. https://doi.org/10.1016/j.nic.2013.03.005

88. Basser PJ, Mattiello J, LeBihan D (1994) MR diffusion tensor spectroscopy and imaging. Biophys J 66(1):259–267

89. Schmierer K, Wheeler-Kingshott CA, Boulby PA, Scaravilli F, Altmann DR, Barker GJ et al (2007) Diffusion tensor imaging of post mortem multiple sclerosis brain. NeuroImage 35 (2):467–477. https://doi.org/10.1016/j.neuroimage.2006.12.010

90. Stankoff B, Freeman L, Aigrot MS, Chardain A, Dolle F, Williams A et al (2011) Imaging central nervous system myelin by positron emission tomography in multiple sclerosis using [methyl-(1)(1)C]-2-(4'-methylaminophenyl)-6-hydroxybenzothiazole. Ann Neurol 69 (4):673–680. https://doi.org/10.1002/ana.22320

91. Bodini B, Veronese M, García-Lorenzo D, Battaglini M, Poirion E, Chardain A et al (2016) Dynamic imaging of individual remyelination profiles in multiple sclerosis. Ann Neurol 79(5):726–738

92. O'Muircheartaigh J, Vavasour I, Ljungberg E, Li DKB, Rauscher A, Levesque V et al (2019) Quantitative neuroimaging measures of myelin in the healthy brain and in multiple sclerosis. Hum Brain Mapp 40 (7):2104–2116. https://doi.org/10.1002/hbm.24510

93. Zivadinov R, Leist TP (2005) Clinical-magnetic resonance imaging correlations in multiple sclerosis. J Neuroimaging 15 (4 Suppl):10S–21S. https://doi.org/10.1177/1051228405283291

94. Filippi M, Grossman RI (2002) MRI techniques to monitor MS evolution: the present and the future. Neurology 58(8):1147–1153

95. Gass A, Filippi M, Rodegher M, Schwartz A, Comi G, Hennerici M (1998) Characteristics of chronic MS lesions in the cerebrum, brainstem, spinal cord, and optic nerve on T1—weighted MRI. Neurology 50(2):548–550

96. Grossman RI, Lenkinski R, Ramer K, Gonzalez-Scarano F, Cohen J (1992) MR proton spectroscopy in multiple sclerosis. AJNR Am J Neuroradiol 13(6):1535–1543

97. Wolinsky JS, Narayana PA (2002) Magnetic resonance spectroscopy in multiple sclerosis: window into the diseased brain. Curr Opin Neurol 15(3):247–251

98. Filippi M, Bozzali M, Rovaris M, Gonen O, Kesavadas C, Ghezzi A et al (2003) Evidence for widespread axonal damage at the earliest clinical stage of multiple sclerosis. Brain 126 (2):433–437

99. Simmons M, Frondoza C, Coyle J (1991) Immunocytochemical localization of N-acetyl-aspartate with monoclonal antibodies. Neuroscience 45(1):37–45

100. Lin A, Ross BD, Harris K, Wong W (2005) Efficacy of proton magnetic resonance spectroscopy in neurological diagnosis and neurotherapeutic decision making. NeuroRx 2 (2):197–214

101. De Stefano N, Narayanan S, Francis GS, Arnaoutelis R, Tartaglia MC, Antel JP et al (2001) Evidence of axonal damage in the early stages of multiple sclerosis and its relevance to disability. Arch Neurol 58(1):65–70

102. Fu L, Matthews P, De Stefano N, Worsley K, Narayanan S, Francis G et al (1998) Imaging axonal damage of normal-appearing white matter in multiple sclerosis. Brain 121 (1):103–113

103. Narayana PA (2005) Magnetic resonance spectroscopy in the monitoring of multiple sclerosis. J Neuroimaging 15:46S–57S

104. Ge Y, Gonen O, Inglese M, Babb J, Markowitz C, Grossman R (2004) Neuronal cell injury precedes brain atrophy in multiple sclerosis. Neurology 62(4):624–627

105. Rocca M, Cercignani M, Iannucci G, Comi G, Filippi M (2000) Weekly diffusion-weighted imaging of normal-appearing white matter in MS. Neurology 55(6):882–884

106. Farquharson S, Tournier JD, Calamante F, Fabinyi G, Schneider-Kolsky M, Jackson GD et al (2013) White matter fiber tractography: why we need to move beyond DTI. J Neurosurg 118(6):1367–1377. https://doi.org/10.3171/2013.2.JNS121294

107. Tournier JD, Mori S, Leemans A (2011) Diffusion tensor imaging and beyond. Magn Reson Med 65(6):1532–1556. https://doi.org/10.1002/mrm.22924

108. Jeurissen B, Leemans A, Tournier JD, Jones DK, Sijbers J (2013) Investigating the prevalence of complex fiber configurations in white matter tissue with diffusion magnetic resonance imaging. Hum Brain Mapp 34 (11):2747–2766. https://doi.org/10.1002/hbm.22099

109. Farquharson S, Tournier J-D (2016) High angular resolution diffusion imaging. Diffusion Tensor Imaging. Springer, New York, NY, pp 383–406

110. Tuch DS, Reese TG, Wiegell MR, Makris N, Belliveau JW, Wedeen VJ (2002) High angular resolution diffusion imaging reveals intravoxel white matter fiber heterogeneity. Magn Reson Med 48(4):577–582. https://doi.org/10.1002/mrm.10268

111. Parker GJ, Alexander DC (2003) Probabilistic Monte Carlo based mapping of cerebral connections utilising whole-brain crossing fibre information. In: Biennial international conference on information processing in medical imaging. Springer, pp. 684–695

112. Jbabdi S, Sotiropoulos SN, Savio AM, Grana M, Behrens TE (2012) Model-based analysis of multishell diffusion MR data for tractography: how to get over fitting problems. Magn Reson Med 68(6):1846–1855. https://doi.org/10.1002/mrm.24204

113. Behrens TE, Woolrich MW, Jenkinson M, Johansen-Berg H, Nunes RG, Clare S et al (2003) Characterization and propagation of uncertainty in diffusion-weighted MR imaging. Magn Reson Med 50(5):1077–1088. https://doi.org/10.1002/mrm.10609

114. Assaf Y, Freidlin RZ, Rohde GK, Basser PJ (2004) New modeling and experimental framework to characterize hindered and restricted water diffusion in brain white matter. Magn Reson Med 52(5):965–978. https://doi.org/10.1002/mrm.20274

115. Assaf Y, Blumenfeld-Katzir T, Yovel Y, Basser PJ (2008) AxCaliber: a method for measuring axon diameter distribution from diffusion MRI. Magn Reson Med 59(6):1347–1354. https://doi.org/10.1002/mrm.21577

116. Tournier JD, Calamante F, Connelly A (2007) Robust determination of the fibre orientation distribution in diffusion MRI: non-negativity constrained super-resolved spherical deconvolution. NeuroImage 35 (4):1459–1472

117. Kaden E, Knösche TR, Anwander A (2007) Parametric spherical deconvolution: inferring anatomical connectivity using diffusion MR imaging. NeuroImage 37(2):474–488

118. Wang Y, Sun P, Wang Q, Trinkaus K, Schmidt RE, Naismith RT et al (2015) Differentiation and quantification of inflammation, demyelination and axon injury or loss in multiple sclerosis. Brain 138(Pt 5):1223–1238. https://doi.org/10.1093/brain/awv046

119. Wang Y, Wang Q, Haldar JP, Yeh FC, Xie M, Sun P et al (2011) Quantification of increased cellularity during inflammatory demyelination. Brain 134(12):3590–3601. https://doi.org/10.1093/brain/awr307

120. Anderson AW (2005) Measurement of fiber orientation distributions using high angular resolution diffusion imaging. Magn Reson Imaging 54(5):1194–1206

121. Zhang H, Schneider T, Wheeler-Kingshott CA, Alexander DC (2012) NODDI: practical in vivo neurite orientation dispersion and density imaging of the human brain. NeuroImage 61(4):1000–1016. https://doi.org/10.1016/j.neuroimage.2012.03.072

122. Callaghan PT, Coy A, MacGowan D, Packer KJ, Zelaya FO (1991) Diffraction-like effects in NMR diffusion studies of fluids in porous solids. Nature 351(6326):467–469

123. King MD, Houseman J, Roussel SA, Van Bruggen N, Williams SR, Gadian DG (1994) Q-space imaging of the brain. Magn Reson Imaging 32(6):707–713

124. Jensen JH, Helpern JA, Ramani A, Lu H, Kaczynski K (2005) Diffusional kurtosis imaging: the quantification of non-gaussian water diffusion by means of magnetic resonance imaging. Magn Reson Med 53 (6):1432–1440. https://doi.org/10.1002/mrm.20508

125. Wedeen VJ, Hagmann P, Tseng WY, Reese TG, Weisskoff RM (2005) Mapping complex tissue architecture with diffusion spectrum magnetic resonance imaging. Magn Reson Med 54(6):1377–1386. https://doi.org/10.1002/mrm.20642

126. Wedeen VJ, Wang RP, Schmahmann JD, Benner T, Tseng WY, Dai G et al (2008) Diffusion spectrum magnetic resonance imaging (DSI) tractography of crossing fibers. NeuroImage 41(4):1267–1277. https://doi.org/10.1016/j.neuroimage.2008.03.036

127. Tuch DS (2004) Q-ball imaging. Magn Reson Med 52(6):1358–1372. https://doi.org/10.1002/mrm.20279

128. Jeurissen B, Descoteaux M, Mori S, Leemans A (2019) Diffusion MRI fiber tractography of the brain. NMR Biomed 32(4):e3785

129. Fedorov A, Beichel R, Kalpathy-Cramer J, Finet J, Fillion-Robin J-C, Pujol S et al (2012) 3D Slicer as an image computing platform for the quantitative imaging network. Magn Reson Imaging 30(9):1323–1341

130. Behrens TE, Berg HJ, Jbabdi S, Rushworth MF, Woolrich MW (2007) Probabilistic diffusion tractography with multiple fibre orientations: what can we gain? NeuroImage 34 (1):144–155. https://doi.org/10.1016/j.neuroimage.2006.09.018

131. Sotiropoulos SN, Hernández-Fernández M, Vu AT, Andersson JL, Moeller S, Yacoub E et al (2016) Fusion in diffusion MRI for improved fibre orientation estimation: an application to the 3T and 7T data of the Human Connectome Project. NeuroImage 134:396–409

132. Hernández M, Guerrero GD, Cecilia JM, García JM, Inuggi A, Jbabdi S et al (2013) Accelerating fibre orientation estimation from diffusion weighted magnetic resonance imaging using GPUs. PLoS One 8(4):e61892

133. Hernandez-Fernandez M, Reguly I, Jbabdi S, Giles M, Smith S, Sotiropoulos SN (2019) Using GPUs to accelerate computational diffusion MRI: from microstructure estimation to tractography and connectomes. NeuroImage 188:598–615

134. Radetz A, Koirala N, Krämer J, Johnen A, Fleischer V, Gonzalez-Escamilla G et al (2020) Gray matter integrity predicts white matter network reorganization in multiple sclerosis. Hum Brain Mapp 41(4):917–927. https://doi.org/10.1002/hbm.24849

135. Shu N, Liu Y, Li K, Duan Y, Wang J, Yu C et al (2011) Diffusion tensor tractography reveals disrupted topological efficiency in white matter structural networks in multiple sclerosis. Cereb Cortex 21(11):2565–2577. https://doi.org/10.1093/cercor/bhr039

136. Fleischer V, Gröger A, Koirala N, Droby A, Muthuraman M, Kolber P et al (2017) Increased structural white and grey matter network connectivity compensates for functional decline in early multiple sclerosis. Mult Scler 23(3):432–441. https://doi.org/10.1177/1352458516651503

137. Pierpaoli C, Basser PJ (1996) Toward a quantitative assessment of diffusion anisotropy. Magn Reson Imaging 36(6):893–906

138. Smith SM, Jenkinson M, Johansen-Berg H, Rueckert D, Nichols TE, Mackay CE et al (2006) Tract-based spatial statistics: voxelwise analysis of multi-subject diffusion data. NeuroImage 31(4):1487–1505. https://doi.org/10.1016/j.neuroimage.2006.02.024

139. Dineen RA, Vilisaar J, Hlinka J, Bradshaw CM, Morgan PS, Constantinescu CS et al (2009) Disconnection as a mechanism for cognitive dysfunction in multiple sclerosis. Brain 132(Pt 1):239–249. https://doi.org/10.1093/brain/awn275

140. Roosendaal SD, Geurts JJ, Vrenken H, Hulst HE, Cover KS, Castelijns JA et al (2009) Regional DTI differences in multiple sclerosis patients. NeuroImage 44(4):1397–1403. https://doi.org/10.1016/j.neuroimage.2008.10.026

Chapter 4

Pathophysiology of Grey Matter Affection in MS

Gabriel Gonzalez-Escamilla and Dumitru Ciolac

Abstract

There is a striking relationship between cognitive performance, such as executive function, attention and motor processing, and grey matter (GM) surface- and voxel-based morphometric measures, as derived from magnetic resonance imaging (MRI). In addition, loss of GM has shown to be a reliable index of atrophy in neurological diseases, including multiple sclerosis (MS). Therefore, morphometric measures are highly valuable tools to noninvasively study brain pathology.

Among different MRI morphometric measures, cortical thinning has gained great importance for characterizing neurodegeneration in MS. While measures of dendrite density and myelin content assess further processes of GM pathology. Hence, MRI is a unique and versatile, noninvasive method for computer-aided lesion detection and brain-wide evaluation of the pathogenic neurodegenerative process in MS.

This chapter provides an overview of quantitative image analysis methods used to investigate GM pathology in MS and how to derive them, while describing the potential of inferring microstructural changes based on the microscopic and mesoscopic measurements obtained from MRI acquisitions.

Key words Structural MRI, Grey matter, Cerebral cortex, Image processing, Microstructure

1 Introduction

1.1 Grey Matter Pathology in Multiple Sclerosis

Since the nineteenth-century preliminary studies about multiple sclerosis (MS) reported sclerotic plaques at the border between the grey matter (GM) and white matter (WM) [69], or in the cerebral cortex [26], together with descriptions of possible underlying pathophysiological processes [71]. However, the potential role of GM pathology was initially largely ignored due to both suboptimal histological staining techniques that limited the study of the cortical GM and the inability to study these processes in vivo.

Development of more advanced histopathological and magnetic resonance imaging (MRI) techniques subsequently led to more insights into GM pathology in MS. Cortical pathology from MRI was initially reported in up to 70–80% of MS patients [11, 60] and was found to be extensive, involving both demyelination (lesions) and neuroaxonal loss (known as neurodegeneration).

Sergiu Groppa and Sven G. Meuth (eds.), *Translational Methods for Multiple Sclerosis Research*, Neuromethods, vol. 166, https://doi.org/10.1007/978-1-0716-1213-2_4, © Springer Science+Business Media, LLC, part of Springer Nature 2021

Today, neurodegeneration of the GM is recognized as the major cause of permanent neurological disability in individuals with MS [31, 65]. GM damage occurs from the earliest stages of MS [23] and can be detected in individuals with very low brain WM lesion burden [24, 68, 74]. GM atrophy may precede WM impaired connectivity [62], and can occur independently of brain WM lesions [78]. Findings from postmortem and MRI studies support cortical neuronal loss, atrophy, and demyelination in the absence of cerebral WM demyelination [17, 22, 78].

Proposed mechanisms of neurodegeneration are inflammatory-mediated demyelination and altered remyelination [32, 66], reduced number of cortical synapses [37, 47] and axonal density (axonal damage/axonal repair) [27, 78], disrupted neural excitability [29, 53], toxic ion accumulations [6, 45, 73], and insufficient restoration of the network functionality [16, 35, 72].

1.2 General Assets for Studying GM Properties

Thanks to the wide use of voxel-based morphometry (VBM) [4, 41, 85], volume is considered a robust metric for studying brain integrity and morphometry. Volume can be estimated either through the use of voxel-wise partial volume effects using a volume-based representation of the brain, such as in VBM, or in a vertex-wise manner as in surface-based morphometry (SBM), in which volume can be measured as the amount of tissue present between the surface placed at the site of the pia mater, and the surface at the interface between GM and WM [83]. VBM has been shown to exhibit comparable accuracy to manual measures [36, 82], therefore providing confidence in the biological validity of the VBM approach. Nevertheless, the exact underlying microstructural properties of volume remain still unknown [2]. Volume is thought to result from a combination of microstructural properties of the GM. Volume measurement may be, then, almost ubiquitous in studies of brain disease, whereas surface-derived cortical thickness has been used as a key morphometric measure in characterizing MS GM pathology in vivo [12, 76]. Notably, it is very likely that not all factors affecting GM tissue integrity (e.g., demyelination, inflammation, axonal and cell damage, impaired excitability, altered gene expression, or protein alteration and aggregation) will be exclusively reflected by thickness or area measures. For this reason, volume is a way to assess the effects of nonspecific factors on the cortex. The use of different morphometric measures is, hence, of great utility to comprehensively investigate disease-related GM integrity.

Despite whole-brain GM volume, computed as the average value across the brain GM tissue, has been largely used as common marker of GM damage, the strength of the correlation between this whole-brain value and clinical deterioration in different MS stages is only moderate. This suggests that averaged values over the whole brain fail to capture the spatial variability of GM integrity loss.

Therefore, in this chapter we focus on non-average measures. In general, we will discuss a wide variety of volume-based and surface-based morphometric measures that enable investigators to assess local differences and alterations in tissue meso- and micro-structure, with high regional specificity across the brain. We focus mainly on structural MRI, diffusion tensor imaging (DTI), and imaging proxies of myelin content given that these techniques are the most commonly used modalities in clinical settings.

2 Methods for Assessing Mesoscopic GM Integrity

The architectural topology of the brain cortex can be seen in different dimensions: its surface area, folding pattern (i.e. gyri and sulci), and thickness. Here, we discuss how to derive these measurements using MRI data, how they relate to each other, and how they depict processes of GM pathology.

2.1 Common Preprocessing Steps for Structural MRI Data

The standard preprocessing workflow starts with "motion correction and averaging" if more than one MRI of the same subject from the same session exist; "intensity correction" also called "intensity normalization"; removal of non-brain tissue, known as "skull stripping"; and classification of the brain tissue into GM, WM, and cerebrospinal fluid (CSF) compartments, in a process known as "segmentation." The main differences between the available tools for MRI preprocessing are the algorithms used to perform these single steps and thereby derive the final images. Similarly, for each morphometric measure to be analyzed the further steps also differ between different software packages.

2.2 Volume-Based Analyses

In this particular case segmentation is followed by "spatial normalization," whereby the individual tissue segments of interest (usually GM) are brought into a common space via registration to a standard stereotactic atlas (also called a template) to ensure one-to-one correspondence across locations in different brains. The spatial normalization changes the volume of the tissue segments locally (whereas some regions expand, others contract). For this reason, improved registration algorithms, such as high-dimensional diffeomorphic anatomical registration through exponentiated lie algebra (DARTEL) [3] from SPM (Statistical Parametric Mapping software, http://www.fil.ion.ucl.ac.uk/spm/) or the nonlinear diffeomorphic normalization from advanced normalization tools (ANTS; http://stnava.github.io/ANTs/) [5, 80], are commonly used leaving only very small differences between the template and individual images and, thus, across individual images. When using DARTEL to normalize the data, the original anatomical differences are coded in matrices called the "deformation fields." Via this deformation information and by applying a "modulation" (i.e., multiplying the normalized GM tissue segments with the Jacobian determinant from the deformation matrix), the induced volume changes are corrected and the original local volumes are preserved (even in the new space). Although some controversy exists with respect to modulation, the recommendation is to use it as part of the processing protocol [63]. Regardless of whether modulation is implemented or not, the normalized tissue segments are then convoluted with a Gaussian function, which is commonly referred to as "spatial smoothing." Spatial smoothing ensures that the random errors have a Gaussian distribution (this is a prerequisite for parametric tests), compensates for small inaccuracies in spatial normalization (even the most advanced normalization algorithms do not yield a perfect voxel-to-voxel correspondence) and determines the spatial scale at which effects are most sensitively detected in order to discriminate true effects from random noise. Common sizes for the smoothing Gaussian kernel are six to ten mm^3 full-width at half maximum (FWHM). The spatially normalized and smoothed GM tissue segments then constitute the input for the voxel-wise statistical analyses.

2.3 Surface-Based Morphometry

Surface implies the derivation of morphometric measures from geometric models of the cortical surface.

The most widely used software allowing the computation of surface-based morphometric measures are listed in Table 1 and is described in detail below.

Table 1
Comparison of surface reconstruction methods and their respective cortical thickness estimation

Software	Surface model	Surface creation method	Between subject correspondence	Cortical thickness estimation
FreeSurfer	Pial and white boundaries	Tissue boundary polygonal tesselation	Spherical surface registration to the FreeSurfer reference atlas	Average of the distance from each white surface vertex to their corresponding closest point on the pial surface
CAT12	Central surface	Laplacian map and partial volume effect probabilistic classification	Adapted DARTEL algorithm to work with spherical maps using a multi-grid approach	Projection-based: project the local maxima from white matter to grey matter voxels by using a neighbor relationship
CIVET	Grey and white matter surfaces	Marching-cubes and CLASP	Nonlinear registration of the sulcal geodesic depth map with an average sulcal depth sphere surface	Distance between the original white matter and grey matter surfaces

DARTEL—diffeomorphic anatomical registration through exponentiated lie algebra; CLASP—constrained Laplacian anatomic segmentation using proximity

2.3.1 FreeSurfer

FreeSurfer (http://freesurfer.net) is a freely available open source suite and one of the most widely used software for processing and analyzing human brain MR images. The "recon-all" preprocessing stream performs the common processing pipeline (see Sect. 2.1 above). Here, after tissue segmentation, delineation and tessellation of the GM-CSF boundary (referred to as pial surface) and GM-WM boundary (referred to as white surface) is performed, followed by topological surface correction, and parcellation/labeling with surface spatial normalization using a spherical registration algorithm [33]. All stages of cortical reconstruction are commonly performed in a semi-automated fashion, meaning that check and manual correction at different steps serve to ensure the quality of the preprocessing. The resulting cortical surface is composed of vertices, which in turn form triangular faces (Fig. 1).

FreeSurfer allows for the computation of diverse cortical morphometric measures, such as thickness, area, or volume, across the vertices of the cortical surface (Fig. 1).

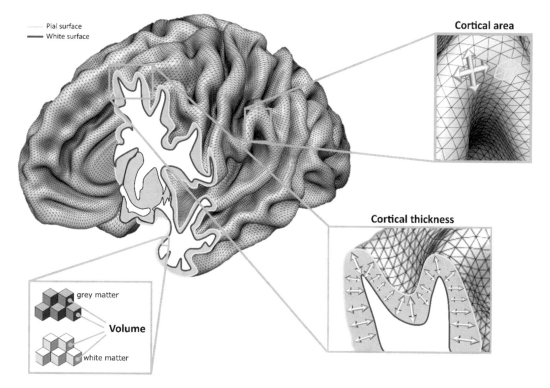

Fig. 1 Cortical surfaces from a structural MRI scan. First, boundaries between the grey and white matter tissue and between the grey matter and cerebrospinal fluid are detected to determine the white and pial surfaces, respectively. The reconstructed surfaces consist of vertices that are connected by faces forming triangles, from which measures describing local and widespread grey matter morphology are computed: cortical thickness is estimated as the average distance (in mm) between the white and pial surfaces; cortical surface area is calculated by summing up the areas of the triangles of the white surface (in mm^2); cortical volume is calculated as the volume contained between the white and pial surfaces (in mm^3)

In FreeSurfer, the thickness at each vertex is calculated as the average of the distance between each white surface vertex and their corresponding closest point on the pial surface and the distance from the corresponding pial vertex to the closest point on the white surface [34].

For each triangular face across the cortex, the area is calculated as $|u \times v|/2$, where $u = a - c$; $v = b - c$; $\times$ represents the cross product; the bars $|\ |$ represent the vector norm; and a, b, and c are three neighboring vertices forming the triangle. Then the area per face (i.e., the face-wise area) is converted to vertex-wise area by assigning to each vertex one-third of the sum of the areas of all faces that have that vertex in common [84].

Once the surface area and thickness are known the conventional method for computing surface-based volume consists of multiplying the area at each vertex by the thickness at that vertex [81, 83]. However, this procedure underestimates the volume in gyri due to gaps between non-matching vertices and overestimates volume in sulci due to overlapping vertex matches [83]. Therefore, in FreeSurfer volumes are computed using the three vertices that define a face on the white surface and the three matching vertices on the pial surface, thus defining an oblique truncated triangular pyramid, which in turn is subdivided into three tetrahedra, which do not overlap or leave gaps. The volumes of these tetrahedra are computed analytically, summed, and assigned to each face of the surface representation (for further details see AM Winkler et al. [83]). Conversion from face-wise to vertex-wise volume is performed in the same manner as for face-wise area.

FreeSurfer's procedure is impervious to variations in MRI sequence parameters, and generates models accurate enough to measure cortical morphometry reliably [33]. For this reason, it has been used in several different imaging studies characterizing MS pathology in vivo [41].

2.3.2 Alternative Tools to Derive Surface-Based Structural Morphometrics

A further tool, the computational anatomy toolbox (CAT12: http://www.neuro.uni-jena.de/cat/) for SPM allows both VBM and SBM analyses. After the typical VBM preprocessing, CAT12 offers a fast and easy-to-use alternative approach for cortical thickness estimations without the extensive surface reconstructions [21], but instead applies surface topology correction, which accounts for topological defects using spherical harmonics [86]. Furthermore, spherical mapping is applied to reparameterize the surface mesh into a common coordinate system [87], while spherical registration adapts the volume-based diffeomorphic DARTEL algorithm [3] to the surface. This volume-based approach uses projection-based thickness (PBT) [21], where a projection scheme is used, using the information of blurred sulci to create a correct cortical thickness map.

Another preprocessing software used for automated morphometric and volumetric analyses of MRI data is the corticometry analysis tools (CIVET; http://www.bic.mni.mcgill.ca/Ser vicesSoftware/CIVET). With this specific software, the MRI processing starts with radio-frequency nonuniformity artifact correction, followed by stereotaxic space transformation using a 9 to 12 parameter linear registration [19], brain mask creation, tissue classification into WM, GM, or CSF using an advanced neural net classifier [89], inner and outer cortical surface extraction via Marching-cubes [56] and constrained Laplacian anatomic segmentation using proximity (CLASP) algorithms, respectively [51]; intersubject spatial normalization to a template model is done using nonlinear registration of cortical surfaces. Cortical thickness is then computed by evaluating the distance between the inner WM and outer GM surfaces transformed back to native space in the MR images [55].

3 MRI-Derived Markers of GM Microstructure

Coming back to the architectural topology of the brain cortex, a further key aspect at a microscopic scale is its layered structure (Fig. 2 left), with each layer being identified by its cellular composition and density of neurons, which varies across the cortex. Although interregional morphological differences, as derived from MRI (e.g., cortical thickness, area, and volume), have been successfully used as biomarkers of pathological changes occurring in MS, such studies do not precisely relate microstructural alterations to clinical or neurobiological processes.

3.1 Diffusion-Derived Metrics

DTI, also known as diffusion MRI (dMRI), is based on the directionality (anisotropy and orientation) as well as the magnitude of water diffusion inside, outside, around, and through cellular structures [9, 10]. The diffusion model assumes that the motion of water molecules has a dominant tissue orientation per voxel within single water compartments (i.e., GM, WM, or CSF). This means that water will diffuse more rapidly in the direction aligned with the internal structure of cell tissue and more slowly as it moves perpendicular to it, depending on the local microarchitecture, including axons, dendrites and cell bodies. At each voxel, a matrix called the diffusion tensor is created containing the three orthogonal axes of diffusion (V1, V2, and V3) along the x, y, and z axes (Fig. 2 middle boxes). Then, diffusion parameters (or scalars) can be derived by fitting a linear regression to the tensor matrix; common programs for this task are FMRIB Software Library (FSL; http://www.fmrib.ox.ac.uk/fsl) [48], MRtrix (https://www.mrtrix.org/) [77], the MATLAB (The MathWorks, Inc., Natick, MA) based exploreDTI

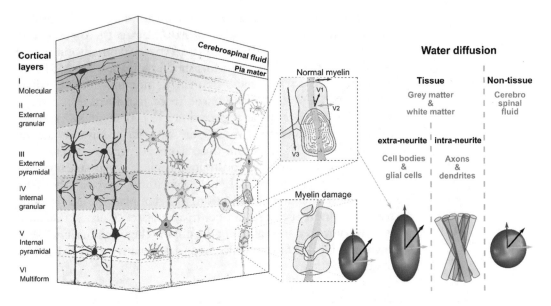

Fig. 2 Schematic representation of the cortical layers (I to VI) in relation to the models of water diffusion used for both traditional diffusion tensor imaging (DTI) and neurite orientation dispersion and density imaging (NODDI). In DTI, a diffusion tensor models three orthogonal axes of diffusion (V1, V2, and V3), from which diffusion scalars, such as fractional anisotropy and mean diffusivity, can be estimated. NODDI expands the diffusion model according to three compartments: hindered diffusion in the extra-neurite compartment, restricted diffusion in the intra-neurite compartment, and free diffusion in cerebrospinal fluid. From this model, parameter maps representing neurite density and orientation dispersion index can be estimated

(http://www.exploredti.com/) [54], or DTI studio software (Johns Hopkins School of Medicine, www.mristudio.org) [49], among others.

Diffusion imaging scalars within the GM can be used for inferring the microstructural tissue changes related to plasticity and disease. For example, fractional anisotropy (FA) and mean diffusivity (MD) are consider to be related to changes in myelination in brain tissue. Although not much is known about GM DTI measures in MS patients, increased FA and reduced MD within GM regions, correlating with disease severity scores [13, 15], have been suggested to reflect progressive GM degeneration [43, 44], such as the loss of dendrites or axonal degeneration due to fiber transections by remote focal MS lesions [18, 43]. Despite these reports, DTI parameters within the GM have not yet been shown to be specific to microstructural features of axons and dendrites (collectively known as neurites) [30] and are often sensitive to partial volume effects from tissue compartments other than neurites (i.e., CSF) [61].

Newer models to infer microstructural GM features from diffusion MRI, namely neurite orientation dispersion and density imaging (NODDI) [88], have been proposed to overcome the limitations of DTI and to provide an improved description of the microstructural morphology of dendrites and axons in the GM [39]. NODDI works by modeling dMRI signals as a combination of three tissue compartments: neurites, extra-neurites, and CSF (Fig. 2 most right part), each with different parameters of diffusion motion, enabling in vivo estimation of a neurite density index (NDI) and an orientation dispersion index (ODI). NDI describes axons and dendrites, modeling the amount of neurites within voxels. ODI, obtained by fitting a Watson distribution to a zero-radius cylinder orientation distribution [88], models neurite orientation variability, ranging from 0 (axon/dendrites all parallel) to 1 (axon/dendrites isotropically randomly oriented). dMRI acquisitions for NODDI require multiple b-values (e.g., $b = 1000$ and 2000 s/mm^2) and a higher number of diffusion gradient directions (e.g., > 64 directions over two b-shell) compared to DTI [38, 88]. The NODDI parameters can be obtained from the diffusion data on a voxel-by-voxel basis in MATLAB with the NODDI toolbox (http://nitrc.org/projects/noddi_toolbox) or using accelerated microstructure imaging via the convex optimization (https://github.com/daducci/AMICO/) toolbox [20], which is a linear implementation of the NODDI model.

The cortical distribution of NDI has a remarkably similar distribution and is closely related to cortical myelin, as estimated from the T1w/T2w ratio, in both somatosensory and motor areas among others [39], while ODI is associated with cortical organization of radial/horizontal fibers [58, 79]. This suggests that NODDI provides valuable information regarding decreased axonal density and secondary fiber degeneration that is related and complementary to GM myeloarchitecture. Moreover, current studies in MS patients have shown that NODDI parameters match histological data [42, 75], both depicting lower neurite dispersion in demyelinated lesions [40, 42]. Therefore, advocating NODDI as a strong potential marker for assessing microscopic GM damage in MS.

In MS, primary morphological changes have recently been detected at the level of the dendritic spines and branches [50]. These findings suggest that mapping neurite morphology in MS is of high interest, as it can potentially provide novel insights into pathophysiology and better explain motor, sensory, and cognitive deficits often encountered in the disease. The large overlap between NODDI and DTI findings [75] demonstrates that noninvasive markers offer potential for overcoming the radiological–clinical paradox [8], referring to the poor association between the extent of MRI-detected abnormalities and the clinical status and rate of development of disability in MS patients.

3.2 Additional Markers of GM Pathology

Cortical demyelination is a functionally relevant aspect of tissue damage in MS and is associated with cognitive deficits [60, 67] and with a higher risk for transition from the relapsing–remitting to the secondary progressive phase of MS [14]. Newer MRI techniques have increased the level of sensitivity and specificity to focal and diffuse GM pathology in MS [30]. For example, relaxometry and magnetization transfer (MT) imaging provide measures of biophysical parameters that show different sensitivity to the amount of free water (i.e., CSF), the amount of water bound to nonaqueous molecules (i.e., lipids and proteins in myelin and glial cells), and the amount of paramagnetic substances (i.e., iron), for review see [46]. The premise of MT is that hydrogens attached to nonaqueous molecules have a much broader range of magnetic resonance frequencies compared to hydrogens attached to water and preferentially exchange magnetization with water molecules [28]. Accordingly, decreases in MT are considered sensitive to the extent of tissue demyelination and have been extensively studied in MS [7, 25, 57]. It is worth noting that MT is not myelin specific but may be influenced by axonal density and edema [70]. A more recently proposed method for indirect quantification of myelin content divides the standard structural MRI (i.e., T1w contrast) by the inverse contrast image (i.e., T2w contrast). This image is known as the T1w/T2w ratio and takes advantage of the fact that the myelin content is tightly linked with the signal intensity in both T1w and T2w but in opposite directions [40]. These images present high test—retest reliability [1] for differentiating between myelinated and demyelinated cortex in MS patients [59] and when comparing across MS clinical subtypes and healthy controls [64]. These studies have shown markedly decreased myelin content in the temporal, frontal and cingulate cortices in MS, correlating with dendrite density obtained from postmortem histopathology [64]. Moreover, T1w/T2w ratio and MT are strongly associated with myelin status identified via postmortem tissue staining [30].

Additional imaging techniques assessing the neuroinflammatory aspects of MS pathology, including further imaging modalities, such as positron emission tomography (PET), are discussed in Chap. 5

---

4   Conclusion

The imaging techniques discussed here may be further implemented for direct translational studies investigating MS pathology in the GM, see Krämer et al. [52] for a more detailed discussion. The proper combination of different MRI-derived measures is key to improve the sensitivity and specificity for the in vivo characterization of the consequences of GM pathology in MS patients [41]. Thus, they are essential assets to achieve better disease diagnosis, to allow for single-subject predictions and to monitor disease progression.

References

1. Arshad M, Stanley JA, Raz N (2017) Test-retest reliability and concurrent validity of in vivo myelin content indices: myelin water fraction and calibrated T1 w/T2 w image ratio. Hum Brain Mapp 38:1780–1790

2. Ashburner J (2009) Computational anatomy with the SPM software. Magn Reson Imaging 27:1163–1174

3. Ashburner J (2007) A fast diffeomorphic image registration algorithm. NeuroImage 38:95–113

4. Ashburner J, Friston KJ (2000) Voxel-based morphometry--the methods. NeuroImage 11:805–821

5. Avants BB, Tustison NJ, Song G et al (2011) A reproducible evaluation of ANTs similarity metric performance in brain image registration. NeuroImage 54:2033–2044

6. Baecher-Allan C, Kaskow BJ, Weiner HL (2018) Multiple sclerosis: mechanisms and immunotherapy. Neuron 97:742–768

7. Bagnato F, Franco G, Ye F et al (2019) Selective inversion recovery quantitative magnetization transfer imaging: toward a 3 T clinical application in multiple sclerosis. Mult Scler 26(4):457–467

8. Barkhof F, Calabresi PA, Miller DH et al (2009) Imaging outcomes for neuroprotection and repair in multiple sclerosis trials. Nat Rev Neurol 5:256–266

9. Basser PJ, Mattiello J, Lebihan D (1994) Estimation of the effective self-diffusion tensor from the NMR spin echo. J Magn Reson B 103:247–254

10. Basser PJ, Mattiello J, Lebihan D (1994) MR diffusion tensor spectroscopy and imaging. Biophys J 66:259–267

11. Calabrese M, Agosta F, Rinaldi F et al (2009) Cortical lesions and atrophy associated with cognitive impairment in relapsing-remitting multiple sclerosis. Arch Neurol 66:1144–1150

12. Calabrese M, Magliozzi R, Ciccarelli O et al (2015) Exploring the origins of grey matter damage in multiple sclerosis. Nat Rev Neurosci 16:147–158

13. Calabrese M, Rinaldi F, Seppi D et al (2011) Cortical diffusion-tensor imaging abnormalities in multiple sclerosis: a 3-year longitudinal study. Radiology 261:891–898

14. Calabrese M, Romualdi C, Poretto V et al (2013) The changing clinical course of multiple sclerosis: a matter of gray matter. Ann Neurol 74:76–83

15. Cavallari M, Ceccarelli A, Wang GY et al (2014) Microstructural changes in the striatum and their impact on motor and neuropsychological performance in patients with multiple sclerosis. PLoS One 9:e101199

16. Cerqueira JJ, Compston DS, Geraldes R et al (2018) Time matters in multiple sclerosis: can early treatment and long-term follow-up ensure everyone benefits from the latest advances in multiple sclerosis? J Neurol Neurosurg Psychiatry 89:844–850

17. Chard DT, Griffin CM, Rashid W et al (2004) Progressive grey matter atrophy in clinically early relapsing-remitting multiple sclerosis. Mult Scler 10:387–391

18. Ciccarelli O, Werring D, Wheeler-Kingshott C et al (2001) Investigation of MS normal-appearing brain using diffusion tensor MRI with clinical correlations. Neurology 56:926–933

19. Collins DL, Neelin P, Peters TM et al (1994) Automatic 3D intersubject registration of MR volumetric data in standardized Talairach space. J Comput Assist Tomogr 18:192–205

20. Daducci A, Canales-Rodriguez EJ, Zhang H et al (2015) Accelerated microstructure imaging via convex optimization (AMICO) from diffusion MRI data. NeuroImage 105:32–44

21. Dahnke R, Yotter RA, Gaser C (2013) Cortical thickness and central surface estimation. NeuroImage 65:336–348

22. Dalton CM, Chard DT, Davies GR et al (2004) Early development of multiple sclerosis is associated with progressive grey matter atrophy in patients presenting with clinically isolated syndromes. Brain 127:1101–1107

23. De Stefano N, Giorgio A, Battaglini M et al (2010) Assessing brain atrophy rates in a large population of untreated multiple sclerosis subtypes. Neurology 74:1868–1876

24. De Stefano N, Matthews PM, Filippi M et al (2003) Evidence of early cortical atrophy in MS: relevance to white matter changes and disability. Neurology 60:1157–1162

25. Derakhshan M, Caramanos Z, Narayanan S et al (2014) Surface-based analysis reveals regions of reduced cortical magnetization transfer ratio in patients with multiple sclerosis: a proposed method for imaging subpial demyelination. Hum Brain Mapp 35:3402–3413

26. Dinkler M (1904) Zur Kasuistik der multiplen Herdsklerose des Gehirns und Ruckenmarks. Deuts Zeits f Nervenheilk 26:233–247

27. Dutta R, Chen J, Ohno N et al (2017) Axonal loss and Neurodegeneration in multiple sclerosis. Neurodegeneration 238–247

28. Duval T, Stikov N, Cohen-Adad J (2016) Modeling white matter microstructure. Funct Neurol 31:217–228

29. Ellwardt E, Pramanik G, Luchtman D et al (2018) Maladaptive cortical hyperactivity upon recovery from experimental autoimmune encephalomyelitis. Nat Neurosci 21:1392

30. Enzinger C, Barkhof F, Ciccarelli O et al (2015) Nonconventional MRI and microstructural cerebral changes in multiple sclerosis. Nat Rev Neurol 11:676–686

31. Eshaghi A, Prados F, Brownlee WJ et al (2018) Deep gray matter volume loss drives disability worsening in multiple sclerosis. Ann Neurol 83:210–222

32. Filippi M, Bar-Or A, Piehl F et al (2018) Multiple sclerosis. Nat Rev Dis Primers 4:43

33. Fischl B (2012) FreeSurfer. NeuroImage 62:774–781

34. Fischl B, Dale AM (2000) Measuring the thickness of the human cerebral cortex from magnetic resonance images. Proc Natl Acad Sci U S A 97:11050–11055

35. Fleischer V, Radetz A, Ciolac D et al (2019) Graph theoretical framework of brain networks in multiple sclerosis: a review of concepts. Neuroscience 403:35–53

36. Focke NK, Trost S, Paulus W et al (2014) Do manual and voxel-based morphometry measure the same? A proof of concept study. Front Psych 5:39

37. Friese MA (2016) Widespread synaptic loss in multiple sclerosis. Brain 139:2–4

38. Fukutomi H, Glasser MF, Murata K et al (2019) Diffusion tensor model links to neurite orientation dispersion and density imaging at high b-value in cerebral cortical gray matter. Sci Rep 9:12246

39. Fukutomi H, Glasser MF, Zhang H et al (2018) Neurite imaging reveals microstructural variations in human cerebral cortical gray matter. NeuroImage 182:488–499

40. Glasser MF, Van Essen DC (2011) Mapping human cortical areas in vivo based on myelin content as revealed by T1- and T2-weighted MRI. J Neurosci 31:11597–11616

41. Gonzalez-Escamilla G, Ciolac D, De Santis S, et al (2020) Gray matter network reorganization in multiple sclerosis from 7-Tesla and 3-Tesla MRI data. Ann Clin Transl Neurol 7:543–553

42. Good CD, Johnsrude I, Ashburner J et al (2001) Cerebral asymmetry and the effects of sex and handedness on brain structure: a voxel-based morphometric analysis of 465 normal adult human brains. NeuroImage 14:685–700

43. Grussu F, Schneider T, Tur C et al (2017) Neurite dispersion: a new marker of multiple sclerosis spinal cord pathology? Ann Clin Transl Neurol 4:663–679

44. Hannoun S, Durand-Dubief F, Confavreux C et al (2012) Diffusion tensor-MRI evidence for extra-axonal neuronal degeneration in caudate and thalamic nuclei of patients with multiple sclerosis. AJNR Am J Neuroradiol 33:1363–1368

45. Hasan KM, Halphen C, Kamali A et al (2009) Caudate nuclei volume, diffusion tensor metrics, and T(2) relaxation in healthy adults and relapsing-remitting multiple sclerosis patients: implications for understanding gray matter degeneration. J Magn Reson Imaging 29:70–77

46. Heidker RM, Emerson MR, Levine SM (2017) Metabolic pathways as possible therapeutic targets for progressive multiple sclerosis. Neural Regen Res 12:1262–1267

47. Helms G (2015) Tissue properties from quantitative MRI. In: Toga AW (ed) Brain mapping: an encyclopedic reference, vol 1. Elsevier, San Diego, CA, pp 287–294

48. Henstridge CM, Tzioras M, Paolicelli RC (2019) Glial contribution to excitatory and inhibitory synapse loss in Neurodegeneration. Front Cell Neurosci 13:63

49. Jenkinson M, Beckmann CF, Behrens TE et al (2012) FSL. NeuroImage 62:782–790

50. Jiang H, Van Zijl PC, Kim J et al (2006) DtiStudio: resource program for diffusion tensor computation and fiber bundle tracking. Comput Methods Prog Biomed 81:106–116

51. Jurgens T, Jafari M, Kreutzfeldt M et al (2016) Reconstruction of single cortical projection neurons reveals primary spine loss in multiple sclerosis. Brain 139:39–46

52. Kim JS, Singh V, Lee JK et al (2005) Automated 3-D extraction and evaluation of the inner and outer cortical surfaces using a Laplacian map and partial volume effect classification. NeuroImage 27:210–221

53. Krämer J, Brück W, Zipp F et al (2019) Imaging in mice and men: pathophysiological insights into multiple sclerosis from conventional and advanced MRI techniques. Prog Neurobiol 182:101663

54. Ksiazek-Winiarek DJ, Szpakowski P, Glabinski A (2015) Neural plasticity in multiple sclerosis: the functional and molecular background. Neural Plast 2015:307175

55. Leemans A, Jeurissen B, Sijbers J et al (2009) ExploreDTI: a graphical toolbox for processing, analyzing, and visualizing diffusion MR data. Proc Intl Soc Mag Reson Med 17:3537

56. Lerch JP, Evans AC (2005) Cortical thickness analysis examined through power analysis and a population simulation. NeuroImage 24:163–173

57. Lorensen WE, Cline HE (1987) Marching cubes: a high resolution 3D surface construction algorithm. SIGGRAPH Comput Graph 21:163–169

58. Mckeithan LJ, Lyttle BD, Box BA et al (2019) 7T quantitative magnetization transfer (qMT) of cortical gray matter in multiple sclerosis correlates with cognitive impairment. NeuroImage 203:116190

59. Mollink J, Kleinnijenhuis M, Cappellen Van Walsum AV et al (2017) Evaluating fibre orientation dispersion in white matter: comparison of diffusion MRI, histology and polarized light imaging. NeuroImage 157:561–574

60. Nakamura K, Chen JT, Ontaneda D et al (2017) T1-/T2-weighted ratio differs in demyelinated cortex in multiple sclerosis. Ann Neurol 82:635–639

61. Nelson F, Datta S, Garcia N et al (2011) Intracortical lesions by 3T magnetic resonance imaging and correlation with cognitive impairment in multiple sclerosis. Mult Scler 17:1122–1129

62. Pierpaoli C, Basser PJ (1996) Toward a quantitative assessment of diffusion anisotropy. Magn Reson Med 36:893–906

63. Radetz A, Koirala N, Kraemer J et al (2020) Gray matter integrity predicts white matter network reorganization in multiple sclerosis. Hum Brain Mapp 41:917–927

64. Radua J, Canales-Rodriguez EJ, Pomarol-Clotet E et al (2014) Validity of modulation and optimal settings for advanced voxel-based morphometry. NeuroImage 86:81–90

65. Righart R, Biberacher V, Jonkman LE et al (2017) Cortical pathology in multiple sclerosis detected by the T1/T2-weighted ratio from routine magnetic resonance imaging. Ann Neurol 82:519–529

66. Rocca MA, Sormani MP, Rovaris M et al (2017) Long-term disability progression in primary progressive multiple sclerosis: a 15-year study. Brain 140:2814–2819

67. Rodriguez EG, Wegner C, Kreutzfeldt M et al (2014) Oligodendroglia in cortical multiple sclerosis lesions decrease with disease progression, but regenerate after repeated experimental demyelination. Acta Neuropathol 128:231–246

68. Roosendaal SD, Moraal B, Pouwels PJ et al (2009) Accumulation of cortical lesions in MS: relation with cognitive impairment. Mult Scler 15:708–714

69. Sailer M, Fischl B, Salat D et al (2003) Focal thinning of the cerebral cortex in multiple sclerosis. Brain 126:1734–1744

70. Sander M (1898) Hirnrindenbefunde bei multipler Sklerose. Eur Neurol 4:427–436

71. Schmierer K, Tozer DJ, Scaravilli F et al (2007) Quantitative magnetization transfer imaging in postmortem multiple sclerosis brain. J Magn Reson Imaging 26:41–51

72. Schob F (1907) Ein Beitrag zur pathologischen Anatomie der multiplen Sklerose. Eur Neurol 22:62–87

73. Schoonheim MM, Meijer KA, Geurts JJ (2015) Network collapse and cognitive impairment in multiple sclerosis. Front Neurol 6:82

74. Schumacher AM, Mahler C, Kerschensteiner M (2017) Pathology and pathogenesis of progressive multiple sclerosis: concepts and controversies. Aktuel Neurol 44:476–488

75. Shiee N, Bazin PL, Zackowski KM et al (2012) Revisiting brain atrophy and its relationship to disability in multiple sclerosis. PLoS One 7: e37049

76. Spano B, Giulietti G, Pisani V et al (2018) Disruption of neurite morphology parallels MS progression. Neurol Neuroimmunol Neuroinflamm 5:e502

77. Steenwijk MD, Geurts JJ, Daams M et al (2016) Cortical atrophy patterns in multiple sclerosis are non-random and clinically relevant. Brain 139:115–126

78. Tournier J-D, Smith R, Raffelt D et al (2019) MRtrix3: a fast, flexible and open software framework for medical image processing and visualisation. NeuroImage 202:116137

79. Trapp BD, Vignos M, Dudman J et al (2018) Cortical neuronal densities and cerebral white matter demyelination in multiple sclerosis: a retrospective study. Lancet Neurol 17:870–884

80. Triarhou LC (2008) The 107 cortical cytoarchitectonic areas of Constantin Von Economo and Georg N. Koskinas in the Adult human brain: excerpt from: "Atlas of Cytoarchitectonics of the Adult Human Cerebral Cortex", Authors, Von Economo, C. (Vienna), Koskinas, GN (Athens). Karger

81. Tustison NJ, Avants BB (2013) Explicit B-spline regularization in diffeomorphic image registration. Front Neuroinform 7:39

82. Van Essen DC (2005) A population-average, landmark- and surface-based (PALS) atlas of

human cerebral cortex. NeuroImage 28:635–662

83. Wang WY, Yu JT, Liu Y et al (2015) Voxel-based meta-analysis of grey matter changes in Alzheimer's disease. Transl Neurodegener 4:6

84. Winkler AM, Greve DN, Bjuland KJ et al (2018) Joint analysis of cortical area and thickness as a replacement for the analysis of the volume of the cerebral cortex. Cereb Cortex 28:738–749

85. Winkler AM, Sabuncu MR, Yeo BT et al (2012) Measuring and comparing brain cortical surface area and other areal quantities. NeuroImage 61:1428–1443

86. Winkler AM, Webster MA, Brooks JC et al (2016) Non-parametric combination and related permutation tests for neuroimaging. Hum Brain Mapp 37:1486–1511

87. Yotter RA, Dahnke R, Thompson PM et al (2011) Topological correction of brain surface meshes using spherical harmonics. Hum Brain Mapp 32:1109–1124

88. Yotter RA, Thompson PM, Gaser C (2011) Algorithms to improve the reparameterization of spherical mappings of brain surface meshes. J Neuroimaging 21:e134–e147

89. Zhang H, Schneider T, Wheeler-Kingshott CA et al (2012) NODDI: practical in vivo neurite orientation dispersion and density imaging of the human brain. NeuroImage 61:1000–1016

90. Zijdenbos AP, Forghani R, Evans AC (2002) Automatic "pipeline" analysis of 3-D MRI data for clinical trials: application to multiple sclerosis. IEEE Trans Med Imaging 21:1280–1291

Chapter 5

Translational Characterization of the Glia Role in Multiple Sclerosis

Dumitru Ciolac, Stanislav A. Groppa, and Gabriel Gonzalez-Escamilla

Abstract

Glial cells are important players in the pathogenesis of multiple sclerosis (MS)-induced injury to the central nervous system. Characterization of their roles is a crucial milestone to advance therapeutic strategies for MS patients. The focus of this chapter is to provide an overview of the methods used to study glial cells in preclinical models of MS that can be translated into the field of human MS. Firstly, microanatomy and physiology of glial cells in the healthy brain are introduced. Then, main immunohistological, electrophysiological, structural, and molecular imaging techniques to study glial cells are described. Additionally, the function of each glial cell type in the development and restoration of MS-induced damage is highlighted. Finally, evidence from preclinical studies on glial cells as potential targets to be approached in order to design novel therapies for MS patients is discussed.

Key words Glial cells, Multiple sclerosis, Immunohistochemistry, Electrophysiology, Neuroimaging, Translational research

1 Introduction

1.1 Morphology and Physiology of Glial Cells

Glia represent a large group of morphologically different cells of nervous tissue that functionally orchestrate essential neuron-orientated activities, ranging from physical/nutrient support and maintenance of homeostasis to myelin production and immune surveillance. There are four types of glial cells within the central nervous system (CNS): astrocytes, oligodendrocytes, microglial cells, and ependymal cells.

1.1.1 Astrocytes

Astrocytes are morphologically heterogeneous cells, comprising the largest class of neuroglial cells in the adult brain (Fig. 1). Two types of astrocytes can be identified: *protoplasmic astrocytes*, found mainly within the grey matter (GM), are characterized by numerous branching processes that coat synapses and blood vessels, and *fibrillary astrocytes*, more common in the white matter (WM) with long and thin processes that come in contact with Ranvier nodes as

Sergiu Groppa and Sven G. Meuth (eds.), *Translational Methods for Multiple Sclerosis Research*, Neuromethods, vol. 166, https://doi.org/10.1007/978-1-0716-1213-2_5, © Springer Science+Business Media, LLC, part of Springer Nature 2021

Glial cells

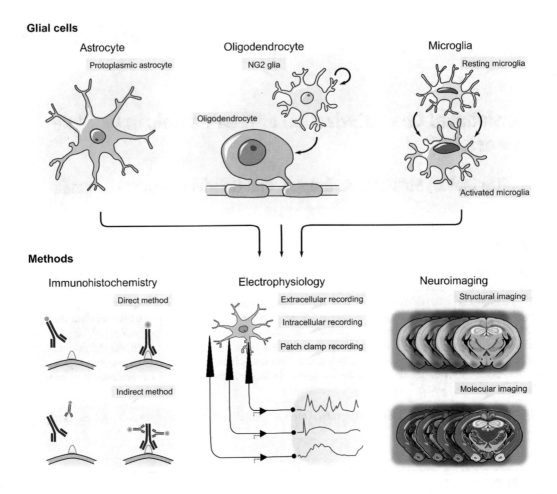

Fig. 1 Types of glial cells in the central nervous system and main methods used to study their structure and function in cell cultures, brain slices, and experimental animal models of multiple sclerosis (modified from [1, 3])

well as with blood vessels [1]. Both cell types contain glial fibrillary acidic protein (GFAP), whose expression is commonly used as specific marker for astrocyte detection. Astrocytes display various surface receptors to neurotransmitters: adrenergic, serotonin, histamine, acetylcholine, glycine, cannabinoid, glutamate, purinergic (P2X, P2Y), and γ-aminobutiric acid (GABA) receptors. Astrocyte ion channels: potassium, calcium, sodium, chloride, and aquaporins. Astrocyte functions: providing structural scaffold and guidance for migrating neural precursors and regulation of synaptogenesis during brain development; modulation of synaptic transmission and synaptic plasticity; formation of blood–brain barrier (BBB) and neurovascular units; trophic and metabolic support (glucose, lactate) for neurons; regulation of neurotransmitter homeostasis (e.g., glutamate, adenosine, and GABA); regulation of brain microcirculation by secretion of constricting and dilating factors;

participation in the synthesis of allopregnanolone and dehydroepi-androsterone that modulate neural excitability and facilitate myeli-nation; maintenance of extracellular ionic milieu of neurons by controlling homeostasis of potassium, chloride, sodium, and cal-cium; regulation of water homeostasis and extracellular pH; consti-tutive low-level secretion of anti-inflammatory cytokines such as transforming growth factor beta (TGF-β) and interleukin (IL)-10; involvement in immune responses and reactive astrogliosis [2].

1.1.2   Oligodendrocytes

Oligodendrocytes represent small cells with short processes that differentiate from oligodendrocyte progenitor cells (OPCs) and are primarily involved in the production and maintenance of myelin sheath in the CNS (Fig. 1) [1]. Cytoplasmic processes of a single oligodendrocyte can myelinate one or several axons (<30) at the same time. Myelin sheath represents a lipid-rich multi-lamellated wrapping around the axon. The proteins that build up the myelin are myelin basic protein (MBP), proteolipid protein (PLP), myelin associated glycoprotein (MAG), myelin oligodendrocyte glycopro-tein (MOG), and oligodendrocyte myelin glycoprotein (OMgp). General structure of a myelinated fiber within the CNS: axons are covered by concentric layers of myelin with no cytoplasm at Ranvier nodes, this structure imposing a more rapid conduction of action potentials, called saltatory conduction. According to their mor-phology myelinating oligodendrocytes are classified into *type I, II, III,* and *IV oligodendrocytes*. Oligodendrocytes express following types of receptors: glutamate (ionotropic, metabotropic), puriner-gic, GABA, glycine, and acetylcholine receptors. Oligodendrocyte ion channels: outwardly/inward rectifying potassium channels, voltage-operated calcium and sodium, and chloride channels.

1.1.3   NG2 Glia

NG2-glia (neuron/glia antigen 2) are cells with a highly branched morphology that are recognized by their expression of NG2 and platelet-derived growth factor receptor alpha (PDGFRα), and are considered as precursors of oligodendrocytes (Fig. 1). These cells are able to form synapses with neurons and display spontaneous synaptic electrical activity. Neurotransmitter receptors: glutamate and GABA receptors. NG2 glia ion channels: potassium, sodium, and calcium channels. Functions of NG2 glia: generation and maintenance of the oligodendrocyte lineage under physiological and pathological conditions; maintenance of the microglia homeo-static state; neurotrophic function and neuronal survival; synapto-genesis and synaptic plasticity [4].

1.1.4   Microglia

Microglia are the smallest of the neuroglial cells and primarily are scavenger cells (Fig. 1) that possess a central role in immune and inflammatory responses within the CNS. Microglial cells are derived from yolk-sac primitive myeloid progenitor cells, enter early into the developing CNS, where they proliferate and migrate to colonize the CNS [1]. Morphologically microglia are classified into *resting* phenotype, in healthy tissue, and *activated* phenotype, in pathological conditions. The latter can acquire one of two functional states, *M1* or *M2*. M1 microglia produce pro-inflammatory cytokines like tumor necrosis factor alpha (TNF-α), IL-1β, and IL-6, while M2 microglia promote tissue repair and neuronal survival [5]. Neurotransmitter receptors expressed by microglia: glutamate, purinergic, adrenergic, GABA, serotonin and acetylcholine receptors. Receptors for other active substances: bradykinin, histamine, endothelin, angiotensin II, glucocorticoid, cannabinoid, and opioid receptors. Chemokine receptors: CCR1, CCR2 and CCR5, CXCR1, CXCR3, and CCR3; TNF-α receptors: TNFR1 and TNFR2; IL receptors: IL-1R1, IL-1R2, IL-2R, IL-4R, IL-10R, IL-13R, and IL-15R. Microglia ion channels: potassium, calcium, sodium, chloride, and proton channels. Functions of microglia: phagocytosis of apoptotic cells and pathogens, removal of cellular debris from the sites of injury; promotion of embryonic synaptogenesis and neurogenesis by secretion of growth factors; control of postnatal synaptogenesis and neurogenesis; modulation of morphogenesis and remodeling of neural network circuitries of adult brain; modulation of synaptic transmission and plasticity [6].

2   Methods Used to Characterize the Role of Glia in Multiple Sclerosis

2.1   Histology and Microscopy Techniques

2.1.1   Immunohistochemistry

The principle of immunohistochemistry is based on the recognition by an antibody and binding to a specific surface-associated or intracellular protein (antigen). Two variations of this technique exist: direct and indirect immunohistochemistry [1]. In the former, a primary antibody coupled directly with a fluorochrome or chromogenic substrate binds to the antigen. In the latter, a secondary antibody recognizing and attaching to the primary antibody is used, thus an amplified signal is obtained (Fig. 1) [1]. Antibodies to GFAP, S100β-protein (a calcium binding protein), excitatory amino acid transporter 1 (EAAT1/GLAST-1), EAAT2/GLT-1, and glutamine synthetase are used as specific stains to identify astrocytes in brain sections and tissue cultures. Usually, astrocytes are labeled with anti-GFAP antibodies, staining their main processes. Depending on the brain region, the expression level of GFAP varies significantly between the astrocytes. Perivascular astrocytes are labeled with antibodies against S100β-protein. Antibodies to MBP, PLP, Olig 1/2/3, oligodendrocyte specific protein (OSP), and 2′,3′-cyclic nucleotide 3′-phosphodiesterase (CNPase) are used with varying staining specificity for detection of mature oligodendrocytes and myelin sheaths. For example, MBP immunolabeling is found only in the compacted myelin sheaths, whereas PLP is labeled within oligodendrocyte somata and myelin. The oligodendrocyte differentiation can be tracked in vitro by immunolabeling different stage-specific markers: NG2, A2B5 and PDGFRα for OPC, and O4-sulfatide for pro-oligodendrocytes (late OPC). Double immunolabeling with MBP antibody and Rip antibody allows the three-dimensional visualization of mature oligodendrocyte morphology. Microglial cells can be immunolabeled with following antibodies, which recognize and other tissue macrophages at the same time: antibodies against leukocyte common antigen (CD45/LCA), CD68 (macrosialin), $\alpha_M\beta_2$ integrin (CD11b/18), ionized calcium-binding adaptor molecule 1 (Iba1), glucose transporter 5 (GLUT5), CD163 (scavenger receptor M130, ED2), or F4/80 antigen with different expression level depending on the functional state of microglia. For example, GLUT5 is a marker of resting and activated microglia; the expression of some molecules (e.g., CD11b or Iba1) is enhanced during microglia activation, while others (e.g., major histocompatibility complex (MHC) class II antigens) are expressed only by activated microglia [7]. After immunostaining, glial cells are visualized by light, fluorescence, or confocal microscopy.

2.2 Electro-physiology Techniques

Except neurons, electrophysiological properties of astrocytes can be studied in vitro and in vivo (Fig. 1). Astrocytes possess a negative resting membrane potential of −80 to −90 mV due to the high permeability of the cell membrane for potassium. Depending on the electrode placement—outside or inside the cell or in close proximity to the cell membrane, three techniques can be distinguished—*extracellular, intracellular*, and *patch clamp recording*, respectively. Electric signals of the astrocytes are usually recorded in vitro cell cultures. A standard set of equipment required for experiments: microelectrodes, headstage (with micromanipulator and microdrive), microscope, amplifier, and a computer with software for data acquisition and analysis. Depending on the nature of the experiment, other set of instruments can be used: stereotaxic/microsurgery equipment, Faraday cage, staining and pharmacological manipulation equipment, and so on. Types of microelectrodes used: (a) glass micropipettes filled with different (e.g., sodium chloride) electrolyte solutions, (b) metal electrodes (gold, steel, platinum-iridium, tungsten, etc.) and carbon-fiber electrodes. Glass micropipettes are used for patch clamp recordings, while metal electrodes are more used for extracellular and intracellular recordings. The headstage holds the microelectrode within the tissue preparation from one side and connects the microelectrode to the amplifier from another side. By controlling the micromanipulator, a precise positioning of the microelectrode within the tissue preparation is achieved and by adjusting the microdrive, the desired depth of the microelectrode inside the tissue can be obtained. To visualize the macroscopic features of the tissue preparation for an extracellular electrophysiological set-up, a low-power microscope is used. A high magnification power microscope is used to visualize individual cells while performing intracellular or patch clamp experiments. The weak electrical signals from the recording and from the reference microelectrodes are enhanced and compared by the amplifier. Afterwards, the digitized signals are received by the computer for subsequent analysis. The built-in software packages allow for real-time data analysis, modulation of recording thresholds and timing the delivery of electrical stimuli. By employing electrophysiological recordings, the alterations in voltage- and time-dependent properties of the astrocytes and neuron–astrocyte crosstalk related to MS can be studied.

2.2.1 Extracellular Recording

Within extracellular setup, the recording and the reference electrodes are placed outside the astrocytes and at different sites of the extracellular space. The advantage of this technique: relatively simple to perform and record the activity of multiple cells is possible at the same time. Disadvantages: impossibility to measure local potentials and presence of sampling bias. A grid of electrodes (multielectrode array) for stimulation and extracellular recording from several sites at once can be used, providing spatiotemporal information on the astrocyte signaling. Extracellular electrodes record differences in the membrane potential measured in units of microvolts (μV). The electrical activity of the astrocytes obtained from extracellular recordings is characterized by spontaneous bursts of discrete signals, initially sporadic and later synchronized in quasi-periodic patterns [8].

2.2.2 Intracellular Recording

The recording electrode is placed inside the studied cell and the reference electrode is placed outside the cell. Intracellular electrodes record differences in membrane potentials measured in units of millivolts (mV). The astrocyte-generated current is measured by two electrophysiological techniques. In the *current clamp* experiment, a known constant or time-varying current is applied and changes in membrane potential caused by the applied current are measured. In a *voltage clamp* experiment, one controls the membrane voltage (V, measured in units of volts) and measures the transmembrane current (I, measured in units of amperes) required to maintain that voltage. Data collected from voltage clamp are plotted as I/V curves [3].

2.2.3 Patch Clamp Recording

This category of recordings implies placement of a glass micropipette into direct contact with a small portion (patch) of the cell membrane. Four principal patch clamp techniques exist. Within the *cell-attached mode* a seal between the micropipette and the membrane is formed without disrupting the integrity of the membrane; this allows to study single ion channels. If retracting the pipette, a patch of membrane will remain attached to the pipette with intracellular surface exposed to medium—*inside-out mode*; this allows to study the effect of intracytoplasmic molecules on functioning of ion channels. By *whole-cell mode* the membrane is disrupted, pipette becoming continuous with the cytoplasm of the cell; this allows to study electric potentials from the whole cell. Double whole-cell patch clamp experiments allow to investigate intracellular currents for both astrocytes and neurons. If retracting the pipette while in whole-cell mode, the patch of membrane will be exposed with the extracellular surface to the bathing solution—*outside-out mode*; this allows to study the effect of extracytoplasmic molecules on ion channel activity.

In vitro recordings are performed in primary astrocyte cell cultures or brain slices, while in vivo electrophysiological recordings in awake or anesthetized animals.

2.3 Imaging Techniques

2.3.1 Structural Imaging

Magnetic resonance imaging (MRI) investigations in mouse models of MS are usually performed with narrow-bore and high-field-strength magnets (4.7–17.6 T), the 7 T platforms being most commonly employed with high resolution and signal-to-noise-ratio [9]. However, obtaining of high quality in vivo images in animal models is still hindered by inherent biological phenomena (cerebrospinal fluid motion, respiration, pulsation of blood vessels, etc.). Non-contrast as well as contrast-enhanced sequences are used to study the role of oligodendrocytes and microglia (Fig. 1).

Magnetization transfer imaging is a noninvasive technique that can be used to estimate the brain's myelin content by quantifying the magnetization transfer ratio (MTR) as a measure of in vivo demyelination severity in mouse models of MS. The MTR decreases along with increasing of demyelination in WM lesions (corpus callosum) and, to a lesser extent, in normal appearing WM (NAWM). Usually, the MRI measurements of demyelination are complemented by anti-MBP/anti-PLP immunohistochemical staining. It should be noted that in brain regions with low myelin levels like cerebral cortex, cerebellum and olfactory bulb, the MTR is not sensitive enough to detect the myelin kinetics. The MTR is influenced by axonal density, tissue edema and inflammatory cell infiltration, and should be used with caution when interpreting the results on dynamics of myelin content.

Microglia/Macrophage Imaging

In conditions of altered BBB within active MS lesions, microglial cells/CNS macrophages express proteins involved in the iron metabolism that results in intracytoplasmic iron accumulation. This property of capturing iron particles underlies noninvasive visualization techniques that allow to track microglial cells in vivo by using iron-sensitive sequences, that is, susceptibility-weighted imaging (SWI). Mapping of inflammatory infiltrates is performed by using high-field MRI scanners to evaluate both brain and spinal cord compartments. Lesions detected by SWI sequences in the WM of experimental autoimmune encephalomyelitis (EAE) correspond to areas of iron deposition and demyelination, which are validated by combined histopathology studies. Lesions detected by SWI are evidenced in the WM of spinal cord and cerebellar tracts [10]. The sophisticated post-processing pipelines necessary for this sequence limit its broader use in experimental setting.

A series of gadolinium-based contrast agents selectively binding to myelin sheaths have been introduced. For example, myelin imaging compound (MIC) is administered via intra-cerebroventricular infusion into the rodents' brain with experimental MS. MIC labels in vivo myelinated fibers in myelin-rich WM regions (corpus callosum, striatum, cerebellum) and detects areas of focal demyelination [11]. Another gadolinium-based agent with similar myelin-binding properties, DODAS, labels areas of increased/decreased myelination in vivo and in vitro conditions [12]. By using these contrast agents a higher image resolution and tissue contrast is achieved, however, the agents cannot cross an unimpaired BBB, hence, cannot bind to non-inflammatory MS lesions.

2.3.2 Molecular Imaging

Molecular neuroimaging techniques allow to detect the immune-inflammatory responses of glial cells before occurrence of neuroaxonal structural abnormalities and track the molecular patterns of disease progression in EAE models. For this, a microPET scanner, sometimes combined with MRI or computerized tomography (CT), is used in anesthetized animals. Positron emission tomography (PET) relies on the introduction of unstable positron-emitting radioisotopes, usually fluorine-18 (^{18}F) or carbon-11 (^{11}C) into the blood stream that can be readily used in animal models of MS for in vivo imaging of glial activity (astrocytes, oligodendrocytes or microglia).

Detection of astrocyte activation at molecular and cellular levels is based on their uptake of acetate by overexpressed enzyme monocarboxylate transporter. The [^{11}C] acetate PET has emerged as a tool to display the proliferation response at the borders of inflammatory demyelinating lesions and NAWM, however, its use to identify reactive astrogliosis in experimental setting of MS found limited application. Monoamine oxidase B and I2-imidazoline receptor are other potential targets for PET imaging of activated astroglia but which were not still approved for use in the setting of MS, at least in patients [13].

To detect microglial activation in murine models of MS, PET platforms make use of ^{18}F as positron emitting radioisotope and matrix metalloprotease inhibitor (MMPi) or Flutriciclamide (GE180) as tracers. The MMP or mitochondrial translocator protein (TSPO) serve as binding sites for the tracers. In humans for the same purpose, ^{18}F as a positron emitting radioisotope and vinpocetine, peripheral benzodiazepine receptor (PBR) 28, DPA-713 or (R)-PK11195 as tracers are used. In case if ^{11}C is used as a radioisotope, then the PBR111 or fluoroethyl-5-methoxybenzyl) acetamide 1106 are used as tracers. For both radioisotopes, the TSPO serves as a binding site [14]. Currently, TSPO is considered an imaging biomarker of microglia activation to detect the diffuse pathology in vivo, however, it does not provide information about the pro- or anti-inflammatory phenotype of microglial cells in MS.

Dynamics of oligodendrocyte/myelin pathology and regeneration can be visualized by PET imaging with different tracers. For example, [^{11}C]N-methyl-4,4′-diaminostilbene ([^{11}C]MeDAS) targets myelin membranes and selectively labels WM regions of the brain and spinal cord, demonstrating high sensitivity and specificity for myelin content in EAE models of MS [13]. A dual approach combining two radiotracers—[^{11}C]MeDAS targeting myelin sheaths and [^{18}F]3-F-4-AP targeting axonal potassium channels underneath the myelin—provides accurate visualization of demyelination in the spinal cord. Demyelination sites are characterized by decreased uptake of [^{11}C]MeDAS and increased [^{18}F]3-F-4-AP uptake. In recent years, PET with amyloid tracers became popular, PiB (Pittsburgh compound B) being largely used, which shows high uptake in cerebral WM but low uptake in cerebellar WM. PET imaging of myelin kinetics has the potential to monitor in a quantifiable way the remyelination processes at molecular level and repair effects of disease-modifying therapies.

3 Role of Glial cells in Multiple Sclerosis: Insights from the Bench

3.1 Role of Astrocytes

Astrocytes are actively involved in the MS pathology from the early disease stages and play multiple roles in initiation, progression and restoration of MS-induced tissue injury. On one hand, by secreting pro-inflammatory cytokines reactive astrocytes promote immune-mediated inflammation, demyelination, and neuronal damage. On the other hand, astrocytes have a protective role by supporting remyelination and glial scar formation. During acute inflammation, astrocytes regulate the passage of lymphocytes through the BBB into the brain parenchyma by modulating the expression of vascular adhesion molecule-1 (VCAM-1) and intercellular adhesion molecule-1 (ICAM-1) that bind to the very late antigen-4 (VLA4) and lymphocyte function-associated antigen-1 (LFA-1), respectively [15, 16]. Reactive astrocytes are located within the demyelinating lesions and extend into the surrounding NAWM. Acting as antigen presenting cells (APCs) and expressing MHC class II molecules, astrocytes are able to present myelin antigens and activate auto-reactive T cells. The astrocyte-secreted chemokines—CXCL10 and 12 and CCL2 recruit monocytes, T cells and microglia to the site of MS lesions. A wide range of pro-inflammatory cytokines like IL-1β, IL-6, IL-12, IL-17, and IL-23 are produced by astrocytes and are able to polarize T cells to pro-inflammatory Th1 and Th17 or to regulatory T (Treg) phenotypes. Within MS lesion IL-15 derived from reactive astrocytes activates the CD8 T lymphocytes that leads to ongoing tissue damage [17]. Astrocytes as well promote the survival and proliferation of B lymphocytes through B-cell-activating factor (BAFF) release [18]. At the site of MS injury, astrocytes release cytotoxic factors such as nitric oxide (NO), reactive oxygen and nitrogen species (ROS, NOS) and glutamate. Increased levels of extracellular glutamate is the main mechanism of excitotoxicity that results in neuronal cell damage and neurodegeneration. During restorative stages, the molecules secreted by reactive astrocytes from the glial scar—fibroblast growth factor-2, hyaluronan, prevent proper remyelination processes. Other molecules (ephrins and chondroitin sulfate proteoglycans) interfere with axonal regeneration [19]. At the same time, astrocytes secret anti-inflammatory cytokines (IL-4, IL-5, IL-10, TGF-β) and recruit microglia to sites of active demyelination that remove myelin debris before the remyelination takes place. Astrocyte-derived chemokines promote restoration of the myelin by stimulating the proliferation of NG2 glia and its differentiation into oligodendrocytes. Astrocytes display neuroprotective effects by releasing neurotrophic factors, brain-derived neurotrophic factor (BDNF) and ciliary neurotrophic factor (CNTF), which stimulate neuronal survival [20].

3.2 Role of Oligodendrocytes and NG2 Glia

Development of mature myelinating oligodendrocytes is a complex and fine-tuned process of activation, proliferation, mobilization and differentiation of OPCs. In MS, each of these steps can be affected that results into demyelination and axonal degeneration. At the sites of acute MS lesions an extensive loss and apoptosis of oligodendrocytes and demyelination occur. Oligodendrocytes secrete CXCL10, CCL2, and CCL3 that recruit microglial cells. The activated immune-inflammatory cascades orchestrated by microglia damage oligodendrocytes and create a proinflammatory microenvironment that reduces the number of OPCs, limiting their myelinating capacity and myelin repair [21]. Diffuse demyelination is characteristic of chronic MS within NAWM and GM. The remyelination process requires proliferation of NG2 glia and their differentiation into mature myelinating oligodendrocytes that are able to produce myelin and restore nerve conduction, however, remyelination is often incomplete and declines over time. To launch remyelination, OPCs require to be activated, that is, responsive to growth factors, cytokines, and chemokines that boost their proliferation, differentiation, and population of demyelinating lesions. Myelin debris inhibits differentiation of OPCs and impairs axonal repair. Myelin restoration depends on the functional state of astrocytes that create either a permissive or inhibitory milieu, promoting or suppressing remyelination, respectively. In chronic MS lesions astrocyte-secreted fibronectin stimulates OPC proliferation but alters their differentiation into myelinating oligodendrocytes

[21]. As disease progresses, persistent demyelination results from oligodendrocyte malfunction and loss from one side and impaired generation of oligodendrocytes from OPCs from the other side.

3.3 Role of Microglia   Activated microglia is characterized by a complex phenotypic diversity and immune-molecular signature under the changing conditions of emerging MS pathology. In the acute phase, after activation microglial cells acquire the M1 phenotype that produces pro-inflammatory cytokines (TNF-α, IL-1β, IL-6, IL-12, and IL-23) and chemokines (CCL4, CCL5, CCL8, CXCL2, CXCL4, CXCL9, and CXCL10) responsible for neuronal cell damage and demyelination. Microglia are one of the most abundant immune cells in active MS lesions. In MS lesions, activated microglia are an important source of ROS and NO radicals that are involved in neurodegeneration and demyelination processes [22]. Experimental studies assign microglia a prominent role in inducing neurodegeneration from the disease outbreak and maintaining it on later stages of the disease [23]. A common pathway in the neurodegenerative axis is the disruption of immune checkpoints that interfere with microglial sensing, housekeeping, and host defense functions [24]. Activated M1 microglia recruit adaptive immune cells into CNS tissue and act as APCs by expressing MHC class I and II. The costimulatory molecules expressed by microglia stimulate the proliferation, differentiation and cytokine secretion by T cells [25]. In the latter stages, microglial cells polarize into M2 state that possess protective functions: phagocytosis of cellular debris at the sites of demyelination; secretion of anti-inflammatory cytokines (IL-4, IL-10, and IL-13) and mediators promoting permissive microenvironment for subsequent regeneration; stimulation of stem cell populations required for neurogenesis and oligodendrogenesis. Microglial cells are able to clear apoptotic cellular material and myelin debris without triggering inflammation—condition essential for proper remyelination. Mediators secreted by microglia recruit phagocytic and repair-promoting cells that remove injured tissue and contribute to tissue regeneration: TNF-α induces the proliferation of oligodendrocyte-lineage precursor cells, BDNF and neurotrophin-3 (NT-3) promote remyelination and axonal regeneration [26]. In models of EAE, ablation of microglia leads to attenuation of the disease severity and demyelination [27].

4 Glial cells as Therapeutic Targets: Translation to the Bedside

Currently, pharmacological modalities available for MS treatment primarily focus on targeting immune-inflammatory pathways. Modulation of the glial cell machinery holds an enormous potential to therapeutically approach neuroinflammatory, neurodegenerative and neurorestorative processes.

4.1 Targeting Astrocytes

As astrocytes play multiple roles in MS pathology, targeting their specific functions can facilitate the development of astrocyte-oriented therapies (Fig. 2). Ideally, such treatment strategies should consider the multi-functionality of astrocytes to reduce detrimental effects on one hand and promote regeneration on the other hand. Design of such therapies is a difficult task given the dual properties of many astrocyte-secreted factors promoting inflammation and neuronal repair. As astrocyte-derived glutamate provokes neuronal injury, the neuronal and oligodendrocyte cell death can be attenuated by treatment with glutamate receptor antagonists [28]. Thus, acting on extracellular glutamate levels appears to be a prominent therapeutic strategy preventing neurodegeneration in MS, especially in progressive forms of the disease. Pharmacological antagonists of astrocyte P2X/P2Y receptors might attenuate the deleterious effects of ATP and NO on oligodendrocytes, axons,

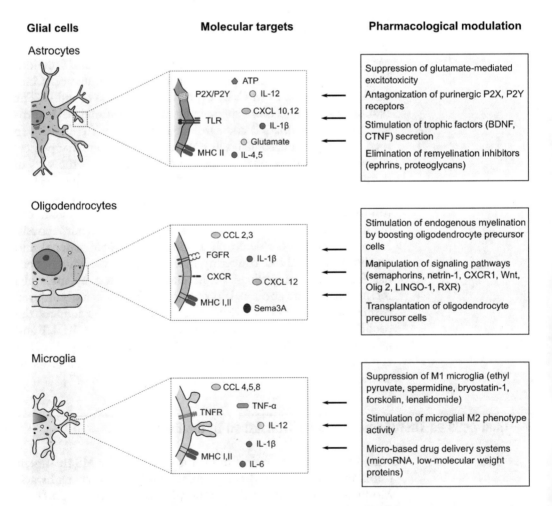

Fig. 2 Modulation of glial cell molecular machinery (factors, receptors, signaling pathways) as translatable approach for development of MS therapeutic interventions (modified from [26])

and neurons [28]. Another putative approach is to stimulate the secretion by astrocytes of trophic factors such as BDNF and CTNF, which enhance neuronal survival. Other approaches using cell-based interventions with autologous hematopoietic stem cells, mesenchymal stem cells, neural stem cells or astrocyte progenitor cells are under research. Currently, several immunomodulatory drugs used in clinical practice are able to modulate the activity of astrocytes: interferon-beta (IFN-β) inhibits the production of NO by astrocytes and promotes their survival in vitro; laquinimod downregulates the pro-inflammatory responses of astrocytes by interfering intracellular signaling pathways, thus slowing down brain atrophy and disability progression in MS patients; dimethyl fumarate upregulates the Nrf2 factor in astrocytes, which in turn upregulates the oxidative stress-induced growth inhibitor 1 (OSGIN1) that is protective against oxidative damage [20].

4.2 Targeting Oligodendrocytes

Deciphering the mechanisms of myelin restoration and the causes of remyelination failure is crucial, since efforts of drug discovery should also be directed towards promoting myelin regeneration. Within MS lesions, oligodendrocytes surviving demyelination do not proliferate and oligodendrocytes from the lesion periphery do not migrate into the lesion core, hence, do not participate in subsequent remyelination. In this context, strategies stimulating endogenous myelin repair should focus on boosting the OPCs rather than promoting oligodendrocyte survival. Identification of key molecular pathways regulating the migration and differentiation of OPCs nearby the lesion sites can offer potential targets to pharmacologically modulate remyelination [29]. Following pathways and molecules are of particular interest: semaphorins (Sema3A, 3F) and netrin-1 that inhibit OPC migration; CXCR4 and Olig 2 factor that induce the differentiation of OPCs into oligodendrocytes; LINGO-1 signaling that suppresses oligodendrocyte survival and differentiation, axonal regeneration and myelination; notch signaling that blocks OPC differentiation; Wnt proteins that modulate oligodendrocyte differentiation and myelination in a dose-dependent manner; RXR signaling pathway that upregulates oligodendrocyte differentiation (Fig. 2). A few immunomodulatory drugs routinely used for MS treatment affect also oligodendrocyte functions: alemtuzumab promotes survival of OPCs and myelination [30]; S1P modulatory agents induce differentiation, migration and survival of OPCs depending on the developmental stage [31]; laquinimod increases the number of oligodendrocyte and myelin density [32]; glatiramer acetate promotes proliferation, recruitment and maturation of OPCs [33]. Engineering of drugs that would influence specific pathways involved in remyelination and translated into efficient regenerative therapies applicable not only in relapsing-remitting but also in progressive forms of MS is of great demand. Elaboration of reliable in vivo myelin markers would facilitate the development of such treatments by noninvasive tracking of their effects. Translational research from preclinical models to MS patients is a critical concern, the efficacy of such strategies requiring further validation by clinical trials.

4.3 Targeting Microglia

Several approaches targeting microglial functions can be addressed for translational purposes (Fig. 2). A first approach to alleviate the immune-mediated damage relates to the suppression of pro-inflammatory activity of M1 microglia. A series of compounds among which ethyl pyruvate, spermidine, bryostatin-1, forskolin and lenalidomide are under preclinical investigations [26]. Ethyl pyruvate inhibits the activation of microglia and protects against the development of EAE. Spermidine inhibits the production of pro-inflammatory cytokines by microglial cells. Forskolin suppresses the expression of CD86 and attenuates the severity of EAE. Bryostatin-1 promotes the differentiation of lymphocytes into Th2 cells by acting on CNS microglia [26]. The second approach is based on the application of micro-based drug delivery systems. Such systems are designed for targeted delivery of different modulatory drugs to CNS microglia. As microglia are able to migrate, they can be used as cellular vehicles to disseminate low-molecular weight factors, RNAs and proteins within the CNS [26]. A third approach is the promotion of microglial polarization to an anti-inflammatory M2 phenotype. Particular disease-modifying drugs are thought to act in this mode: glatiramer acetate increases the expression of anti-inflammatory IL-10 and reduces the expression of pro-inflammatory TNF-α in microglia [34]; mitoxantrone exerts a broad cytotoxic activity, including immunosuppression of microglial cells [35]; fingolimod diminishes the production of pro-inflammatory cytokines and free radicals by microglia [36]; dimethyl fumarate decreases the microglial secretion of TNF-α, IL-1 β, IL-6, and NO [37]. As M2 microglia stimulate neural repair and myelination, boosting their activity can aid remyelination of MS lesions. However, still extensive research is required before the physiology of microglial subsets could be selectively manipulated for MS therapy.

5   Conclusion

Modern laboratory techniques and manipulation tools have provided invaluable evidence on glial cell involvement in the development, progression, and restoration of MS-induced injury. Modulation of glial cell activity is a promising research landscape for bench-to-bedside translation into effective intervention therapies for MS patients. However, the translational research based on glial cell targeting faces many challenges due to methodological constraints of "human" MS replication. Studying the role of glial cells in vivo and in vitro experimental setting may help bridge the translational gap between currently used preclinical models of MS and patients.

Acknowledgments

We would like to thank Lilian Şaptefraţi, Vitalie Mazuru, Tatiana Globa, and Veaceslav Fulga from the Department of Histology, Cytology and Embryology, Nicolae Testemitanu State University of Medicine and Pharmacy, Chişinău, Republic of Moldova, for helpful advices on Sects. 1.1 and 2.1.

References

1. Pawlina W, Ross MH (2018) Histology: a text and atlas: with correlated cell and molecular biology. Lippincott Williams & Wilkins, Baltimore, MD

2. Haim LB, Rowitch DH (2017) Functional diversity of astrocytes in neural circuit regulation. Nat Rev Neurosci 18(1):31

3. Brette R, Destexhe A (2012) Intracellular recording. In: Handbook of neural activity measurement, pp 44–91

4. Dimou L, Gallo V (2015) NG 2-glia and their functions in the central nervous system. Glia 63 (8):1429–1451

5. Lloyd AF, Miron VE (2019) The pro-remyelination properties of microglia in the central nervous system. Nat Rev Neurol 15(8):447–458

6. Reemst K, Noctor SC, Lucassen PJ, Hol EM (2016) The indispensable roles of microglia and astrocytes during brain development. Front Hum Neurosci 10:566

7. Kettenmann H, Hanisch U-K, Noda M, Verkhratsky A (2011) Physiology of microglia. Physiol Rev 91(2):461–553

8. Mestre AL, Inácio P, Elamine Y, Asgarifar S, Lourenço AS, Cristiano ML, Aguiar P, Medeiros MC, Araújo IM, Ventura J (2017) Extracellular electrophysiological measurements of cooperative signals in astrocytes populations. Front Neural Circuits 11:80

9. Krämer J, Brück W, Zipp F, Cerina M, Groppa S, Meuth SG (2019) Imaging in mice and men: pathophysiological insights into multiple sclerosis from conventional and advanced MRI techniques. Prog Neurobiol 182:101663

10. Nathoo N, Agrawal S, Wu Y, Haylock-Jacobs S, Yong VW, Foniok T, Barnes S, Obenaus A, Dunn JF (2013) Susceptibility-weighted imaging in the experimental autoimmune encephalomyelitis model of multiple sclerosis indicates elevated deoxyhemoglobin, iron deposition and demyelination. Mult Scler J 19 (6):721–731

11. Frullano L, Zhu J, Wang C, Wu C, Miller RH, Wang Y (2011) Myelin imaging compound (MIC) enhanced magnetic resonance imaging of myelination. J Med Chem 55(1):94–105

12. Frullano L, Zhu J, Miller RH, Wang Y (2013) Synthesis and characterization of a novel gadolinium-based contrast agent for magnetic resonance imaging of myelination. J Med Chem 56(4):1629–1640

13. Bauckneht M, Capitanio S, Raffa S, Roccatagliata L, Pardini M, Lapucci C, Marini C, Sambuceti G, Inglese M, Gallo P (2019) Molecular imaging of multiple sclerosis: from the clinical demand to novel radiotracers. EJNMMI Radiopharm Chem 4 (1):1–25

14. Bischof A, Caverzasi E, Cordano C, Hauser SL, Henry RG (2017) Advances in imaging multiple sclerosis. Semin Neurol 37:538–545

15. Gimenez MAT, Sim JE, Russell JH (2004) TNFR1-dependent VCAM-1 expression by astrocytes exposes the CNS to destructive inflammation. J Neuroimmunol 151 (1–2):116–125

16. McFarland HF, Martin R (2007) Multiple sclerosis: a complicated picture of autoimmunity. Nat Immunol 8(9):913–919

17. Saikali P, Antel JP, Pittet CL, Newcombe J, Arbour N (2010) Contribution of astrocyte-derived IL-15 to CD8 T cell effector functions in multiple sclerosis. J Immunol 185 (10):5693–5703

18. Michel L, Touil H, Pikor NB, Gommerman JL, Prat A, Bar-Or A (2015) B cells in the multiple sclerosis central nervous system: trafficking and contribution to CNS-compartmentalized inflammation. Front Immunol 6:636

19. Silver J, Miller JH (2004) Regeneration beyond the glial scar. Nat Rev Neurosci 5 (2):146

20. Ponath G, Park C, Pitt D (2018) The role of astrocytes in multiple sclerosis. Front Immunol 9:217

21. Domingues HS, Portugal CC, Socodato R, Relvas JB (2016) Oligodendrocyte, astrocyte, and microglia crosstalk in myelin development, damage, and repair. Front Cell Dev Biol 4:71

22. Dendrou CA, Fugger L, Friese MA (2015) Immunopathology of multiple sclerosis. Nat Rev Immunol 15(9):545

23. Friese MA, Schattling B, Fugger L (2014) Mechanisms of neurodegeneration and axonal dysfunction in multiple sclerosis. Nat Rev Neurol 10(4):225

24. Hickman S, Izzy S, Sen P, Morsett L, El Khoury J (2018) Microglia in neurodegeneration. Nat Neurosci 21(10):1359–1369

25. Hemmer B, Kerschensteiner M, Korn T (2015) Role of the innate and adaptive immune responses in the course of multiple sclerosis. Lancet Neurol 14(4):406–419

26. Weng Q, He Q, Yang B, Wang J, Wang J, Wang J (2019) Targeting microglia and macrophages, a potential treatment strategy for multiple sclerosis. Front Pharmacol 10:286

27. Goldmann T, Wieghofer P, Müller PF, Wolf Y, Varol D, Yona S, Brendecke SM, Kierdorf K, Staszewski O, Datta M (2013) A new type of microglia gene targeting shows TAK1 to be pivotal in CNS autoimmune inflammation. Nat Neurosci 16(11):1618

28. Correale J, Farez MF (2015) The role of astrocytes in multiple sclerosis progression. Front Neurol 6:180

29. Franklin RJ, Gallo V (2014) The translational biology of remyelination: past, present, and future. Glia 62(11):1905–1915

30. Münzel EJ, Williams A (2013) Promoting remyelination in multiple sclerosis—recent advances. Drugs 73(18):2017–2029

31. Jung C-G, Kim H, Miron V, Cook S, Kennedy T, Foster C, Antel J, Soliven B (2007) Functional consequences of S1P receptor modulation in rat oligodendroglial lineage cells. Glia 55(16):1656–1667

32. Moore S, Khalaj AJ, Yoon J, Patel R, Hannsun G, Yoo T, Sasidhar M, Martinez-Torres L, Hayardeny L, Tiwari-Woodruff SK (2013) Therapeutic laquinimod treatment decreases inflammation, initiates axon remyelination, and improves motor deficit in a mouse model of multiple sclerosis. Brain Behav 3 (6):664–682

33. Siri MR, Badaracco M, Pasquini JM (2013) Glatiramer promotes oligodendroglial cell maturation in a cuprizone-induced demyelination model. Neurochem Int 63(1):10–24

34. Weber MS, Prod'homme T, Youssef S, Dunn SE, Rundle CD, Lee L, Patarroyo JC, Stüve O, Sobel RA, Steinman L (2007) Type II monocytes modulate T cell–mediated central nervous system autoimmune disease. Nat Med 13 (8):935

35. Vollmer T, Stewart T, Baxter N (2010) Mitoxantrone and cytotoxic drugs' mechanisms of action. Neurology 74(1 Supplement 1):S41–S46

36. Hughes JE, Srinivasan S, Lynch KR, Proia RL, Ferdek P, Hedrick CC (2008) Sphingosine-1-phosphate induces an antiinflammatory phenotype in macrophages. Circ Res 102 (8):950–958

37. Wierinckx A, Brevé J, Mercier D, Schultzberg M, Drukarch B, Van Dam A-M (2005) Detoxication enzyme inducers modify cytokine production in rat mixed glial cells. J Neuroimmunol 166(1–2):132–143

Chapter 6

Translational Value of CSF and Blood Markers of Autoimmunity and Neurodegeneration

Timo Uphaus and Stefan Bittner

Abstract

Multiple sclerosis is an autoimmune disease with underlying inflammatory and neurodegenerative processes. Future biomarker research is needed to increase our understanding of the pathophysiologic substrates of the disease and thereby enable prediction of disease course and response to immunomodulatory and neuroprotective therapy strategies. Recently developed ultra-sensitive analytic techniques (single molecule array (Simoa)) enable the reliable detection of low levels of protein that are released from the central nervous system (CNS) into the peripheral blood. However, peripheral sources of markers measured with the Simoa technique have to be taken into account, when interpreting CNS-disorders. In this chapter, we provide a detailed overview of the Simoa technique and present emerging and established markers of inflammatory and neurodegenerative processes of multiple sclerosis. We particularly highlight Neurofilament light chain (NfL), a biomarker for neurodegenerative processes in various neurological disorders.

Key words Multiple sclerosis, Biomarker, Neurofilament light chain, Glial fibrillary acid protein (GFAP)

1 Introduction

Multiple sclerosis is an inflammatory and neurodegenerative neurologic disorder of the central nervous system (CNS), and is further characterized by substantial heterogeneity between patients in disease severity, therapy response, and long-term prognosis. There is currently a need to identify biomarkers that distinguish these features and help to stratify patients to the best therapy. Current research efforts focus on biomarkers and their potential to predict the future disease course and response to immunomodulatory therapy. However, preanalytical variability, insufficient standardization, and a need for precisely characterized multicenter patient cohorts are just a few of the obstacles that need to be overcome prior to translation into clinical practice. Within this context, an ideal biomarker should be measurable based on easily accessible samples, such as plasma or serum, with the possibility of

Sergiu Groppa and Sven G. Meuth (eds.), *Translational Methods for Multiple Sclerosis Research*, Neuromethods, vol. 166, https://doi.org/10.1007/978-1-0716-1213-2_6, © Springer Science+Business Media, LLC, part of Springer Nature 2021

standardized sample storage procedures that can be implemented in daily clinical practice. Moreover, an ideal biomarker should be CNS-specific and mirror a special pathophysiologic aspect of the disease.

With regard to samples accessibility, an important step towards a personalized medicine approach is the development of cutting edge technologies such as Simoa (single molecule array). Simoa technology is sensitive enough to enable the detection of low-level biomarkers in the blood, with the potential to replace the use of cerebrospinal fluid (CSF) in future biomarker studies. Simoa is 126-fold and 25-fold more sensitive than ELISA and electrochemiluminescence-based assays, respectively, for quantification of NfL [1]. This ultra-sensitive method has improved the detection of biomarkers in blood samples of the periphery in various neurologic disorders (e.g., multiple sclerosis [2] and cerebral ischemia [3]) and also from healthy controls with a strong correlation between CSF and blood protein levels [4].

2 Current Knowledge and Pitfalls in Rating Disease Activity in Patients with Multiple Sclerosis

Clinical relapses are clear signs of inflammatory disease activity and are an important parameter for clinical decision making. High rates of relapses especially after disease onset, are an independent risk factor for progressive disease stages [5]. Early initiation of disease modifying therapies (DMTs) that reduce relapse rates delays disease progression [6]. It is still a matter of debate whether the total relapse rate during the relapsing–remitting phase of the disease is related to the development of secondary progressive multiple sclerosis, or the latency of transition into progressive disease forms [7]. Moreover, a second important diagnostic measure for rating of inflammatory disease activity includes identifying new or enlarging T2-lesions on magnetic resonance imaging (MRI). More subtle MRI alterations such as grey matter lesions, chronically active lesions or diffuse white matter pathology seem to be even more important for the development of progressive disease stages [8]. However, grey matter lesions for example, are rarely identified by conventional imaging techniques, due to the small size of the cortex compared to conventional imaging voxel dimensions, and a low contrast between the lesion and the normal appearing cortical grey matter [9]. Current clinical decision making is mainly based on clinically rated disease activity (relapse rate, expanded disability status scale (EDSS)—change over time) and conventional MRI measurements, such as lesion count or breakdown of the blood–brain barrier (BBB), as reflected by gadolinium enhancement. These methods are not able to discriminate chronic diffuse inflammatory processes, known as a major factor of the neuroaxonal injury component of the disease [10–12]. Therefore, development

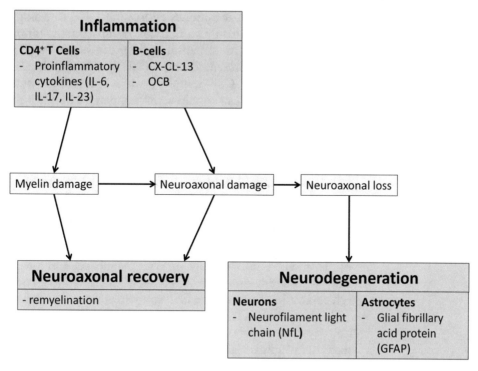

Fig. 1 Key pathophysiologic components and related biomarkers of Multiple sclerosis. MS is a demyelinating disease with inflammatory and neurodegenerative components. The inflammatory component of the disease is mainly driven by B- and T-cells, contributing directly or indirectly (through myelin damage) to neuroaxonal loss and subsequent neuroaxonal damage. Neuroaxonal damage is either results in neuroaxonal loss or neuroaxonal recovery via remyelination. Neuroaxonal loss is the underlying pathophysiologic substrate of neurodegeneration, which is most promisingly reflected by the neuronal marker NfL and the astrocytic marker GFAP. *IL-6* Interleukin 6, *IL-17* Interleukin 17, *IL-23* Interleukin 23, *CX-CL13* C-X-C motif chemokine 13, *OCB* oligoclonal Bands, *NfL* Neurofilament light chain, *GFAP* Glial fibrillary acid protein

of novel biomarker approaches that fully reflect the spectrum of disease pathology will help to foster patient-individualized treatment approaches (Fig. 1).

3 Methodological Approaches

Different techniques have been used to investigate the various underlying pathophysiologic causes of different MS disease courses. Significant progress has been made in this regard, especially in understanding the progressive component of the disease. To get further insights into the progressive phase of the disease and pathophysiologic substrate of the acute and chronic inflammatory components of multiple sclerosis, a standardized biosampling protocol and clinical follow-up evaluation of these patients is mandatory. We here provide details on the recently introduced digital ELISA-based method, the single molecule array technique (Simoa).

3.1 Single Molecule Array Technique (Simoa)

Conventional ELISA-techniques addressing CNS pathology have so far been limited mainly to the CSF compartment. Biomarkers representing the status of pathological processes in the CNS are more likely to be represented in the CSF than in the blood. However, acquisition of CSF samples requires a lumbar puncture, making it impractical to obtain serial samples and limiting its use in clinical practice. As a digital immunoassay, the Simoa technique is significantly more sensitive than a conventional ELISA, meaning that it can also be used with blood samples, allowing valid quantification and differentiation of low peripheral biomarker levels [13]. Whether all CNS biomarkers are released to the blood has to be taken into account and there is still a need for markers that reflect CNS-specific processes. Nevertheless, the advantages of this analytical platform have been underlined in neurological disorders such as multiple sclerosis [2], ischemic stroke [3], and others [4]. Numerous publications on the diagnostic and prognostic value of different candidate biomarkers, such as NfL, confirm a high correlation between the CSF compartment and periphery [4].

In the Simoa method, single protein molecules are captured on antibody-coated, paramagnetic beads. These capture proteins are linked to an enzyme label. In a next step, single beads are extracted and sealed in arrays of femtoliter (fl) wells in the presence of the enzyme substrate. This sealing step reduces the fluorescence generated by a single enzyme to a volume of around 40 fl, which can be detected by an uncooled CCD camera, by taking advantage of white light excitation. Moreover, the generation of paramagnetic beads with fluorescent dyes of various wavelengths and concentrations in order to create optically distinct subpopulations of beads, enables the development of multiplex digital ELISAs on the Simoa platform [14]. The concept of multiplex Simoa technique uses antibodies against specific proteins, which are linked to color-coded beads. Subsequently, a mixture of these beads is incubated with the sample, so that multiple specific proteins are captured by their specific bead partner. As described before, these proteins are then labeled with enzymes, suspended in substrate and placed into arrays on the Simoa microfluidic disc. Finally, they are sealed with oil and undergo fluorescent imaging to detect fluorescence intensity. The fluorescence scanner identifies the subpopulation through the fluorescent wavelength, the location of individual beads in the femtoliter well and the presence of a single enzyme associated with each bead. For each bead subpopulation an average enzyme per bead ratio (AEB) is calculated and used to calculate the concentration of each protein [13]. This technique offers the opportunity to understand biochemically complex physiological states and various pathophysiologic substrate of a specific disease using a single sample and takes into account the limited ability, and often limited quantities of biosamples. With regard to multiple sclerosis, one sample might be used to get insight into different cell types (e.g., astrocytes [glial fibrillary acid protein (GFAP)], neurons [NfL], and T cells [interleukin (IL)-17]).

4   Distinct Inflammatory Components and Biomarkers

Multiple sclerosis is a chronic, inflammatory, demyelinating disease with accompanied neuronal injury [8]. Traditionally the neuropathological substrate of MS was defined as neuroinflammation, combined with demyelination under exclusion of neuroaxonal structures [15]. These characteristic lesions are located within the white matter and are composed out of invading leucocytes, reactive astrocytes, microglia and myelin-containing macrophages [16]. Involvement of the grey matter is nowadays accepted as the neurodegenerative part of the disease. This diffuse neuroaxonal injury already starts in early disease stages [17, 18] and is even detectable in normal-appearing white matter, independent of demyelination processes and not associated with white matter plaques [10, 19, 20]. These findings suggest that neuroaxonal injury happens at least in part independently of focal demyelinating processes, representing a slow process initiated by acute lymphocytic inflammation and subsequently sustained by chronic diffuse parenchymal inflammatory processes of the innate immune system. In this section, we summarize biomarkers that either reflect focal inflammatory disease activity or chronic diffuse inflammatory and neurodegenerative processes.

4.1   Markers of Acute Inflammatory Disease Activity

CD4+ T cells and especially Th17 cells are a unique subset of T cells that are mandatory for the clearance of extracellular pathogens and, under pathological conditions, are involved in orchestrating autoimmunity [21]. Th17-related cytokines, such as IL-6, IL-17, and IL-23, were reported to be elevated in the CSF of patients with multiple sclerosis [22]. However, this observation was not supported by other groups, who demonstrated undetectable levels of the aforementioned cytokines in patients with active MS [23, 24]. Nonetheless, CSF IL-6 is reported as a useful biomarker in the differential diagnosis of MS, especially with regard to neuromyelitis optica [25]. Osteopontin (OPN) is also a pro-inflammatory cytokine, involved in physiological and pathological immunity and inflammation. Notably, the protein is abundantly expressed in tissues and body fluids. Contradictory results for CSF levels have been reported. Whereas elevated levels were found in patients with active relapsing–remitting multiple sclerosis [26, 27], they were also reported in patients with primary progressive multiple sclerosis [28] and were even associated with subsequent development of disability [28]. Higher levels of OPN were detected in patients with clinical relapse activity; however, therapy with methylprednisolone did not alter the concentration of OPN in the CSF [29]. Conversely, treatment with natalizumab, over a therapy interval of 60 weeks, reduced OPN levels significantly in patients with progressive multiple sclerosis [30]. To summarize, publications on the value of cytokines as indicators of T cell inflammation in multiple sclerosis show a high heterogeneity.

B-cell related biomarkers such as C-X-C motif chemokine 13 (**CXCL13**) represents the ligand of the B-cell receptor CXCR5, one of the most potent B-cell chemoattractants. Increased concentrations of CXCL13 were found in CSF of patients with multiple sclerosis compared to controls [31, 32]. Moreover, CXCL13 level in the CSF correlated with relapse rate [31], disability scores (EDSS) and new lesions on MRI [32]. Moreover, levels of protein might have a predictive role, as increased CXCL13 concentrations together with detection of oligoclonal bands were found in CSF of patients with clinically isolated syndrome converting to multiple sclerosis [32]. Under disease modifying therapy with natalizumab [30, 31, 33], mitoxantrone [34], and fingolimod [33], a reduction of CXCL13 levels was observed.

Chitinase 3-like protein 1 (CHI3L) expression is generally increased in inflamed tissue and increased expression has been found in multiple sclerosis and various other neurological disorders. Within the CNS, CHI3L1 is mostly expressed in astrocytes and, to a smaller extent, in activated macrophages and microglia [35]. The role of CHI3L within this context is mainly unknown, but it might be involved in tissue regeneration during neuroinflammation. CHI3L1 exerts prognostic properties, as increasing levels were found in patients with clinical isolated syndrome and patients with optic neuritis prone to convert into multiple sclerosis [36, 37]. Increased CHI3L1 levels were found to be associated with relapse rate and number of gadolinium-enhancing lesions [33, 38, 39] and with progression of disability [36] in both RRMS [33, 35, 38, 40] and SPMS [38, 39] patients. Reduction after application of disease-modifying therapy was observed upon treatment with natalizumab [38, 41], fingolimod [42], daclizumab [43], and mitoxantrone [38].

4.2 Markers of Chronic Neurodegenerative Processes

Accumulating evidence suggests that neurodegeneration occurs in parallel with inflammatory-driven neuronal injury in relapsing MS. Neurofilaments and GFAP are currently the most promising candidates for neurodegenerative processes in MS. Their major advantages are that their expression is restricted to CNS tissue and that they reflect two different pathophysiologic circumstances. Neurofilament is released into the CSF due to neuroaxonal damage and GFAP indicates astrocyte activation and astrogliosis. Exploration of other degenerative biomarkers in multiple sclerosis have been found to be either contradictory or to have no convincing link between application of therapeutic interventions and reduction of biomarker level (e.g., Tau-protein, S100A, and brain-derived neurotrophic factor). While this may change in future, we will therefore focus within this overview on neurofilaments, especially neurofilament light chain (NfL), and GFAP.

NfL levels are increased in the CSF of all disease stages of multiple sclerosis [44, 45], and during relapse activity the concentration of NfL in the CSF is increased 3- to 10-fold [45–47] compared to the remission phase. After clinical relapse activity, a peak in NfL levels in CSF is observed after 2 weeks. Within 2–3 months after the relapse, NfL values are reduced [44, 47]. Regarding association with MRI variables, it was demonstrated, that NfL in CSF is increased in patients with gadolinium-enhancing lesions [44, 48, 49] and in case of development of new T2-lesions [33, 50]. In patients with clinical isolated syndrome, NfL levels in the CSF were linked to detection of oligoclonal bands and a higher risk of conversion from clinically isolated syndrome to multiple sclerosis [45, 50]. A study of two different cohorts—one longitudinal ($n = 246$), one cross-sectional ($n = 142$)—showed higher NfL values in patients with multiple sclerosis compared to controls and further correlated NfL-values with relapse rate, EDSS worsening and development of new T2 and gadolinium-enhancing lesions [51]. Moreover, increased peripheral NfL predicted future brain [2, 52] and spinal cord atrophy [52] in two independent cohorts [2, 52]. This underlines the ability of NfL to measure acute inflammatory activity (gadolinium enhancement) and neuroaxonal damage (EDSS-worsening, brain atrophy) over time. However, concerning disease progression the correlation between NfL and EDSS is heterogenous [44, 53, 54], raising questions about the use of NfL for the progressive components of multiple sclerosis. There is accumulating evidence that NfL is reduced upon treatment with disease-modifying therapies in patients with RRMS [33, 42, 55] and even more so in progressive MS patients [48] and therefore might be used to monitor treatment effects and answer questions on treatment escalation. Specifically, NfL levels were found to be reduced in patients with RRMS after treatment with natalizumab [49], fingolimod [42, 55, 56], rituximab [57] and also in patients with progressive subforms of MS after treatment with mitoxantrone or rituximab [48].

GFAP is essential for the structure of astrocytes, as it is part of the cytoskeleton of this cell type, and is therefore widely accepted as a CNS-specific protein [58]. In patients with multiple sclerosis, GFAP-levels were not associated with relapse rate but were found in patients with increased disability, suggesting that GFAP might be a marker for chronic rather than acute inflammatory processes [59]. This is underlined by the fact that a long-term follow-up study over 9 years, detected the highest GFAP levels in patients with secondary progressive multiple sclerosis (SPMS) [44]. No change of GFAP levels was detected in response to treatment in patients with RRMS after treatment with natalizumab [49], nor in patients with progressive MS after treatment with mitoxantrone or rituximab [48].

5 Conclusion

Technical progress has recently been made in different fields of medicine by enabling measurements of low abundant proteins that could act as biomarkers within bio-fluids and tissue. Examples are the so-called liquid biopsy in case of neoplasm of any origin and measurements within peripheral blood instead of CSF in disorders of the central nervous system. However, great efforts still have to be made with regard to validation and standardization in order to pave the way for these biomarkers to be used routinely in clinical practice. In the case of multiple sclerosis, a combination of different biomarkers will be required to give insights into disease stage depending on the underlying pathophysiologic biological processes. The next steps towards developing this approach include working out the correlations between serum-based proteins and cellular-based biomarkers (determined via fluorescence-activated cell scanning (FACS)) and advanced MRI techniques.

Acknowledgments

We thank Rosie Gilchrist for proofreading and editing. The work of the authors was supported by SFB CRC TR-128 (to S.B.) and Else Kröner Memorial Stipendium (2018_EKMS.21 to T.U.)

References

1. Kuhle J, Barro C, Andreasson U et al (2016) Comparison of three analytical platforms for quantification of the neurofilament light chain in blood samples: ELISA, electrochemiluminescence immunoassay and Simoa. Clin Chem Lab Med 54:1655–1661

2. Siller N, Kuhle J, Muthuraman M et al (2019) Serum neurofilament light chain is a biomarker of acute and chronic neuronal damage in early multiple sclerosis. Mult Scler 25:678–686

3. Uphaus TB, Gröschel S, Steffen F, Wasser K, Weber-Krüger M, Zipp F, Wachter F, Gröschel K (2019) Neurofilament light chain levels as predictive marker for long-term outcome after ischemic stroke. Stroke 50(11):3077–3084

4. Khalil M, Teunissen CE, Otto M et al (2018) Neurofilaments as biomarkers in neurological disorders. Nat Rev Neurol 14:577–589

5. Scalfari A, Neuhaus A, Daumer M, Deluca GC, Muraro PA, Ebers GC (2013) Early relapses, onset of progression, and late outcome in multiple sclerosis. JAMA Neurol 70:214–222

6. Palace J, Duddy M, Bregenzer T et al (2015) Effectiveness and cost-effectiveness of interferon beta and glatiramer acetate in the UK multiple sclerosis risk sharing scheme at 6 years: a clinical cohort study with natural history comparator. Lancet Neurol 14:497–505

7. Scalfari A, Neuhaus A, Degenhardt A et al (2010) The natural history of multiple sclerosis: a geographically based study 10: relapses and long-term disability. Brain 133:1914–1929

8. Larochelle C, Uphaus T, Prat A, Zipp F (2016) Secondary progression in multiple sclerosis:

neuronal exhaustion or distinct pathology?
Trends Neurosci 39:325–339

9. Geurts JJ, Barkhof F (2008) Grey matter
pathology in multiple sclerosis. Lancet Neurol
7:841–851

10. Kutzelnigg A, Lucchinetti CF, Stadelmann C
et al (2005) Cortical demyelination and diffuse
white matter injury in multiple sclerosis. Brain
128:2705–2712

11. Magliozzi R, Howell O, Vora A et al (2007)
Meningeal B-cell follicles in secondary progressive multiple sclerosis associate with early onset
of disease and severe cortical pathology. Brain
130:1089–1104

12. Magliozzi R, Howell OW, Reeves C et al
(2010) A gradient of neuronal loss and meningeal inflammation in multiple sclerosis. Ann
Neurol 68:477–493

13. Rissin DM, Kan CW, Campbell TG et al (2010)
Single-molecule enzyme-linked immunosor-
bent assay detects serum proteins at subfemto-
molar concentrations. Nat Biotechnol
28:595–599

14. Rivnak AJ, Rissin DM, Kan CW et al (2015) A
fully-automated, six-plex single molecule
immunoassay for measuring cytokines in
blood. J Immunol Methods 424:20–27

15. Lassmann H, van Horssen J, Mahad D (2012)
Progressive multiple sclerosis: pathology and
pathogenesis. Nat Rev Neurol 8:647–656

16. Stadelmann C (2011) Multiple sclerosis as a
neurodegenerative disease: pathology, mechanisms and therapeutic implications. Curr Opin
Neurol 24:224–229

17. Trapp BD, Ransohoff R, Rudick R (1999)
Axonal pathology in multiple sclerosis: rela-
tionship to neurologic disability. Curr Opin
Neurol 12:295–302

18. De Stefano N, Matthews PM, Filippi M et al
(2003) Evidence of early cortical atrophy in
MS: relevance to white matter changes and
disability. Neurology 60:1157–1162

19. Frischer JM, Bramow S, Dal-Bianco A et al
(2009) The relation between inflammation
and neurodegeneration in multiple sclerosis
brains. Brain 132:1175–1189

20. DeLuca GC, Williams K, Evangelou N, Ebers
GC, Esiri MM (2006) The contribution of
demyelination to axonal loss in multiple sclerosis. Brain 129:1507–1516

21. Buehler U, Schulenburg K, Yurugi H et al
(2018) Targeting prohibitins at the cell surface
prevents Th17-mediated autoimmunity.
EMBO J 37:e99429

22. Wen SR, Liu GJ, Feng RN et al (2012)
Increased levels of IL-23 and osteopontin in
serum and cerebrospinal fluid of multiple sclerosis patients. J Neuroimmunol 244:94–96

23. Matsushita T, Tateishi T, Isobe N et al (2013)
Characteristic cerebrospinal fluid cytokine/
chemokine profiles in neuromyelitis optica,
relapsing remitting or primary progressive multiple sclerosis. PLoS One 8:e61835

24. Bielekova B, Komori M, Xu Q, Reich DS, Wu
T (2012) Cerebrospinal fluid IL-12p40,
CXCL13 and IL-8 as a combinatorial bio-
marker of active intrathecal inflammation.
PLoS One 7:e48370

25. Uzawa A, Mori M, Ito M et al (2009) Markedly increased CSF interleukin-6 levels in neuromyelitis optica, but not in multiple sclerosis.
J Neurol 256:2082–2084

26. Braitch M, Nunan R, Niepel G, Edwards LJ,
Constantinescu CS (2008) Increased osteopontin levels in the cerebrospinal fluid of
patients with multiple sclerosis. Arch Neurol
65:633–635

27. Chowdhury SA, Lin J, Sadiq SA (2008) Specificity and correlation with disease activity of
cerebrospinal fluid osteopontin levels in
patients with multiple sclerosis. Arch Neurol
65:232–235

28. Bornsen L, Khademi M, Olsson T, Sorensen
PS, Sellebjerg F (2011) Osteopontin concentrations are increased in cerebrospinal fluid
during attacks of multiple sclerosis. Mult Scler
17:32–42

29. Ratzer R, Iversen P, Bornsen L et al (2016)
Monthly oral methylprednisolone pulse treat-
ment in progressive multiple sclerosis. Mult
Scler 22:926–934

30. Romme Christensen J, Ratzer R, Bornsen L
et al (2014) Natalizumab in progressive MS:
results of an open-label, phase 2A, proof-of-
concept trial. Neurology 82:1499–1507

31. Sellebjerg F, Bornsen L, Khademi M et al
(2009) Increased cerebrospinal fluid concentrations of the chemokine CXCL13 in active
MS. Neurology 73:2003–2010

32. Khademi M, Kockum I, Andersson ML et al
(2011) Cerebrospinal fluid CXCL13 in multiple sclerosis: a suggestive prognostic marker for
the disease course. Mult Scler 17:335–343

33. Novakova L, Axelsson M, Khademi M et al
(2017) Cerebrospinal fluid biomarkers as a
measure of disease activity and treatment efficacy in relapsing-remitting multiple sclerosis. J
Neurochem 141:296–304

34. Axelsson M, Mattsson N, Malmestrom C,
Zetterberg H, Lycke J (2013) The influence
of disease duration, clinical course, and
immunosuppressive therapy on the synthesis
of intrathecal oligoclonal IgG bands in multiple
sclerosis. J Neuroimmunol 264:100–105

35. Bonneh-Barkay D, Wang G, Starkey A, Hamilton RL, Wiley CA (2010) In vivo CHI3L1

(YKL-40) expression in astrocytes in acute and chronic neurological diseases. J Neuroinflammation 7:34

36. Modvig S, Degn M, Roed H et al (2015) Cerebrospinal fluid levels of chitinase 3-like 1 and neurofilament light chain predict multiple sclerosis development and disability after optic neuritis. Mult Scler 21:1761–1770

37. Comabella M, Fernandez M, Martin R et al (2010) Cerebrospinal fluid chitinase 3-like 1 levels are associated with conversion to multiple sclerosis. Brain 133:1082–1093

38. Malmestrom C, Axelsson M, Lycke J, Zetterberg H, Blennow K, Olsson B (2014) CSF levels of YKL-40 are increased in MS and replaces with immunosuppressive treatment. J Neuroimmunol 269:87–89

39. Burman J, Raininko R, Blennow K, Zetterberg H, Axelsson M, Malmestrom C (2016) YKL-40 is a CSF biomarker of intrathecal inflammation in secondary progressive multiple sclerosis. J Neuroimmunol 292:52–57

40. Correale J, Fiol M (2011) Chitinase effects on immune cell response in neuromyelitis optica and multiple sclerosis. Mult Scler 17:521–531

41. Stoop MP, Singh V, Stingl C et al (2013) Effects of natalizumab treatment on the cerebrospinal fluid proteome of multiple sclerosis patients. J Proteome Res 12:1101–1107

42. Novakova L, Axelsson M, Khademi M et al (2017) Cerebrospinal fluid biomarkers of inflammation and degeneration as measures of fingolimod efficacy in multiple sclerosis. Mult Scler 23:62–71

43. Komori M, Kosa P, Stein J et al (2017) Pharmacodynamic effects of daclizumab in the intrathecal compartment. Ann Clin Transl Neurol 4:478–490

44. Malmestrom C, Haghighi S, Rosengren L, Andersen O, Lycke J (2003) Neurofilament light protein and glial fibrillary acidic protein as biological markers in MS. Neurology 61:1720–1725

45. Norgren N, Sundstrom P, Svenningsson A, Rosengren L, Stigbrand T, Gunnarsson M (2004) Neurofilament and glial fibrillary acidic protein in multiple sclerosis. Neurology 63:1586–1590

46. Alping P, Frisell T, Novakova L et al (2016) Rituximab versus fingolimod after natalizumab in multiple sclerosis patients. Ann Neurol 79:950–958

47. Lycke JN, Karlsson JE, Andersen O, Rosengren LE (1998) Neurofilament protein in cerebrospinal fluid: a potential marker of activity in multiple sclerosis. J Neurol Neurosurg Psychiatry 64:402–404

48. Axelsson M, Malmestrom C, Gunnarsson M et al (2014) Immunosuppressive therapy reduces axonal damage in progressive multiple sclerosis. Mult Scler 20:43–50

49. Gunnarsson M, Malmestrom C, Axelsson M et al (2011) Axonal damage in relapsing multiple sclerosis is markedly reduced by natalizumab. Ann Neurol 69:83–89

50. Teunissen CE, Iacobaeus E, Khademi M et al (2009) Combination of CSF N-acetylaspartate and neurofilaments in multiple sclerosis. Neurology 72:1322–1329

51. Disanto G, Barro C, Benkert P et al (2017) Serum Neurofilament light: a biomarker of neuronal damage in multiple sclerosis. Ann Neurol 81:857–870

52. Barro C, Benkert P, Disanto G et al (2018) Serum neurofilament as a predictor of disease worsening and brain and spinal cord atrophy in multiple sclerosis. Brain 141:2382–2391

53. Varhaug KN, Barro C, Bjornevik K et al (2018) Neurofilament light chain predicts disease activity in relapsing-remitting MS. Neurol Neuroimmunol Neuroinflamm 5:e422

54. Kuhle J, Malmestrom C, Axelsson M et al (2013) Neurofilament light and heavy subunits compared as therapeutic biomarkers in multiple sclerosis. Acta Neurol Scand 128:e33–e36

55. Kuhle J, Disanto G, Lorscheider J et al (2015) Fingolimod and CSF neurofilament light chain levels in relapsing-remitting multiple sclerosis. Neurology 84:1639–1643

56. Kappos L, Radue EW, O'Connor P et al (2010) A placebo-controlled trial of oral fingolimod in relapsing multiple sclerosis. N Engl J Med 362:387–401

57. Alvarez E, Piccio L, Mikesell RJ et al (2015) Predicting optimal response to B-cell depletion with rituximab in multiple sclerosis using CXCL13 index, magnetic resonance imaging and clinical measures. Mult Scler J Exp Transl Clin 1:2055217315623800

58. Kepes JJ, Perentes E (1988) Glial fibrillary acidic protein in chondrocytes of elastic cartilage in the human epiglottis: an immunohistochemical study with polyvalent and monoclonal antibodies. Anat Rec 220:296–299

59. Axelsson M, Malmestrom C, Nilsson S, Haghighi S, Rosengren L, Lycke J (2011) Glial fibrillary acidic protein: a potential biomarker for progression in multiple sclerosis. J Neurol 258:882–888

Part II

Animal Models of Neuroinflammation

Chapter 7

Spontaneous Mouse Models of Neuroinflammation

Shin-Young Na and Gurumoorthy Krishnamoorthy

Abstract

Animal models featuring various components of neuroinflammation are widely used for investigating the pathogenic processes that contribute to the development and regulation of inflammatory lesions in multiple sclerosis (MS), a central nervous system (CNS)-specific autoimmunity disease. The most commonly used model of neuroinflammation is the experimental autoimmune encephalomyelitis (EAE) model, which is induced either by immunization with myelin autoantigen or by the adoptive transfer of preactivated myelin-specific T cells. While these models are extremely useful to understand the inflammatory mechanisms of tissue damage and regulation in the CNS, these models are not ideal to identify immunological mechanisms during the initiation phase of CNS autoimmunity. To overcome these limitations, T cell receptor (TCR) transgenic mice specific to various myelin antigens have been generated. Several reports have shown that myelin-specific TCR transgenic mice develop varying degrees of spontaneous neurological symptoms and clinical patterns mimicking some aspects of MS. In this chapter, we describe a detailed methodology to generate and characterize TCR transgenic mouse strains recognizing a myelin autoantigen, myelin oligodendrocyte glycoprotein (MOG). This approach can be applied to generate any TCR transgenic mouse strain of interest to enable a deeper understanding of the mechanisms of disease initiation in inflammatory diseases.

Key words Experimental autoimmune encephalomyelitis, Spontaneous EAE, TCR transgenic mice, Myelin oligodendrocyte glycoprotein

1 Introduction

Multiple sclerosis (MS) is a complex neuroinflammatory disease resulting from the demyelination induced by the infiltration of various immune cells into the central nervous system (CNS). The CNS lesions in MS are dominated by CD8$^+$ T cells although other immune cells such as CD4$^+$ T cells, B cells, and other innate immune cells are also frequently observed [1]. Owing to this complexity of immunopathology and pathogenic mechanisms, no single animal model is sufficient to reflect all aspects of clinical and pathological phenotypes. Instead, a number of complementary models have been developed that exhibit different clinical features and pathology in the CNS [2]. The most common model that has

Sergiu Groppa and Sven G. Meuth (eds.), *Translational Methods for Multiple Sclerosis Research*, Neuromethods, vol. 166, https://doi.org/10.1007/978-1-0716-1213-2_7, © Springer Science+Business Media, LLC, part of Springer Nature 2021

been frequently used for the study of MS pathogenesis is called experimental autoimmune encephalomyelitis (EAE) which is induced by either immunization with myelin autoantigen or by transfer of activated T cells [3]. Although these EAE models have been instrumental in understanding the autoimmune pathogenic processes, these are not ideal models for understanding of the triggers of the disease.

An ideal model should develop autoimmune disease spontaneously mimicking features of MS observed in humans. However, unfortunately, there are no naturally occurring spontaneous models representing autoimmunity against CNS tissue. To circumvent this problem, a number of transgenic models have been developed that present spontaneous neurological disease symptoms [4]. These models are predominantly engineered mouse strains that express a T cell receptor (TCR) specific to myelin or neuronal antigens. The first reported spontaneous mouse model of CNS autoimmunity is a transgenic B10.PL mouse strain that expresses a rearranged TCS for myelin basic protein (MBP) Ac1-11 isolated from a $CD4^+$ T cell clone [5]. These transgenic mice developed spontaneous EAE when housed in conventional housing conditions at different ages and the incidence of spontaneous EAE was below 50% [5]. Another research group generated RAG-1-deficient MBP-specific $CD4^+$ TCR transgenic mice and observed 100% EAE incidence [6]. Similarly, $CD4^+$ TCR transgenic mice against proteolipid protein (PLP) 139-151 on an SJL/J genetic background were also shown to develop spontaneous EAE [7]. Following these earlier TCR transgenic models, a transgenic mouse model recognizing myelin oligodendrocyte glycoprotein (MOG) 35-55 (2D2 mice) in the commonly used C57BL/6 genetic background has been developed. 35% of 2D2 mice developed spontaneous optic neuritis and 4% showed paralytic EAE symptoms [8]. The spontaneous EAE was further enhanced when 2D2 mice were crossed with MOG-specific immunoglobulin heavy chain knockin mice. About 50% of these double transgenic mice develop spontaneous opticospinal EAE [9]. A similar high incidence of spontaneous EAE was noted in MOG92-106-specific $CD4^+$ TCR transgenic mice in the SJL/J background. These mice with or without the combination of MOG-specific immunoglobulin heavy chain knockin mice show more than 80% spontaneous disease incidence and interestingly also a relapsing–remitting disease resembling the western type of MS [10].

Unlike $CD4^+$ TCR transgenic models that develop spontaneous EAE, $CD8^+$ TCR transgenic models that develop a spontaneous neuroinflammatory disease are scarce. MBP-specific $CD8^+$ TCR transgenic mice did not result in spontaneous EAE [11] unless activated by a viral infection which expresses for both viral antigens and MBP [12]. A humanized mouse model that carries a human MBP 84-102-specific $CD8^+$ TCR and the corresponding human leukocyte antigen (HLA) DR2 was shown to develop spontaneous EAE in 4% of mice [13]. Later, human PLP45-53 CD8[+] TCR transgenic mice crossed with human HLA-A3 transgenic mice were shown to develop spontaneous MS-like disease [14]. Due to the difficulty of the generation of human CD8[+] TCR and their matching HLA transgenic mice, alternative models expressing neo-self-antigen transgenic mice were developed which express ovalbumin (OVA) under the MBP promoter in oligodendrocytes (ODC-OVA) [15]. When ODC-OVA mice were crossed with the well-characterized ovalbumin-specific CD8[+] TCR transgenic (OT-I) mice, the spontaneous neuroinflammatory disease was observed in 90% of the mice between 12 and 19 days of life [16]. However, a similar strategy using an influenza Hemagglutinin (HA) transgenic mouse crossed with HA peptide-specific CD8[+] TCR transgenic mice did not develop spontaneous EAE symptoms [17]. Recently, CD8[+] TCR transgenic mice reactive to glial fibrillary acidic protein (GFAP), a protein that is expressed in astrocytes, were reported to develop spontaneous relapsing–remitting EAE [18].

Overall, several reports suggest that spontaneous neuroinflammatory symptoms are commonly observed in TCR transgenic mice that recognize a myelin antigen. In the present chapter, we describe a detailed protocol that has been used to generate a myelin antigen-specific TCR transgenic mouse model [10]. This protocol can be used to generate new animal models of neuroinflammation.

2 Materials

2.1 Animals

The choice of the inbred mouse strain depends on their response to chosen myelin antigen and susceptibility to EAE. Commonly used strains for EAE studies are C57BL/6 and SJL/J which recognize MOG 35-55/I-A^b and MOG92-106/I-A^s, respectively. The animals should be housed according to the institutional and governmental rules and regulations concerning animal health and welfare. The stress and suffering should be minimized as per the approved protocols. If animals were purchased from commercial vendors, the animals should be kept at the local colony for a minimum of 1–2 weeks in the new environment before experimentation. 8–12-week-old SJL/J mice were used in the experiments described in this chapter, but the procedure can be applied to other strains and older mice.

2.2 Antigen

Purified N-terminal extracellular domain of rMOG (1-125), which was produced as a HIS-tagged recombinant protein in *E. coli* [19], is used. Alternatively, a synthetic MOG 92-106 (DEGGYT CFFRDHSYQ) peptide could be used (core facility of the Max Planck Institute of Biochemistry or Biotrend, Germany). All antigens should have >95% purity.

2.3 Growth Medium

Mouse primary cell culture: Roswell Park Memorial Institute (RPMI) medium containing 10% fetal bovine serum (FBS) (Sigma), 1× nonessential amino acids (Gibco), 1× sodium pyruvate (Gibco), 1× penicillin–streptomycin (Gibco), and 2 mM L-glutamine (Gibco).

Hybridoma and 58α⁻β⁻ cell line: Dulbecco's Modified Eagle Medium (DMEM)-GlutaMAX containing 10% FBS, 1× nonessential amino acids, 1× sodium pyruvate, 1× penicillin–streptomycin.

2.4 Primers

All primers are purchased as desalted from either Sigma or Metabion.

2.5 Other Reagents

1. Incomplete Freund's adjuvant (Difco).

2. *Mycobacterium tuberculosis* (strain H37Ra) (Difco).

3. NycoPrep 1.077 (Alere Technologies).

4. Polyethylene glycol solution Hybri-Max (Sigma).

5. Cell proliferation dye eFluor 450 (eBioscience).

6. Antibodies: CD3ε PE-Cy7 (145-2C11), CD4 PerCP-Cy5.5 (RM4-5), mouse Vβ TCR screening Panel, and anti-mouse TCR antibodies (BD Biosciences or BioLegend).

7. Recombinant mouse IL-2 carrier-free (BioLegend).

8. Flow cytometry staining (FACS) buffer: 1% BSA, 0.1% sodium azide in PBS.

9. 2× BES: 50 mM N,N-bis(2-hydroxyethyl)-2-aminoethanesulfonic acid (BES), 280 mM NaCl, 1.5 mM Na_2HPO_4 pH 6.95.

10. Hypoxanthine–aminopterin–thymidine (HAT) 50× media supplement (Sigma).

11. Hypoxanthine–thymidine (HT) 50× media supplement (Sigma).

12. Verso cDNA Kit (Thermo Scientific).

13. Q5 high fidelity DNA polymerase (New England Biolabs).

14. Restriction enzymes (New England BioLabs): *MfeI*, *EcoRI*, and *BamHI*.

15. Hexadimethrine bromide (Polybrene) (Sigma).

16. Chloroquine diphosphate salt (Sigma).

3 Method

Here we shall provide protocols for the commonly used method to generate TCR transgenic mice which involves multiple steps that include isolation of an antigen-specific T cell clone, identification of TCR, and pronuclear injection. A schematic illustration of the sequential steps of the following protocol is depicted in Fig. 1.

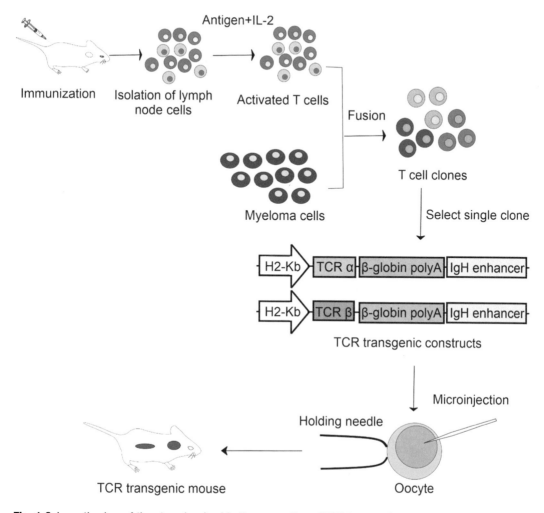

Fig. 1 Schematic view of the steps involved in the generation of TCR transgenic mice

3.1 Preparation of Emulsion and Immunization

Naïve T cells specific for certain antigens are often present at very low frequencies. Some estimates suggest that in many cases these are around one cell per million T cells or lower, which makes it hard to isolate directly from naïve animals. The first preferred step is to expand the antigen-specific T cells by immunization with the antigen of interest together with immune adjuvants such as complete Freund's adjuvant (CFA). This allows for the expansion of T cells that react with specific antigen considerably and facilitates the isolation of a specific clone for further characterization.

1. Dilute 100 µg antigen (rMOG protein) in 100 µl PBS.

2. Prepare an equal volume of Freund's adjuvant supplemented with 5 mg/ml *Mycobacterium tuberculosis* (strain H37Ra).

3. Emulsify both solutions together using 1 ml Luer Lock syringes (BD Bioscience) and micro-emulsifying needles with reinforcement.

4. Leave emulsion for at least for 30 min at 4 °C. The properly formed emulsion will not separate into two phases.

5. Inject the emulsion subcutaneously using a 20G syringe (BRAUN), at left and right lower back, 100 μl/site.

3.2 Isolation and Culture of T Cells

A critical step to the successful generation of TCR transgenic mice is the selection of a T cell clone that is specific to the antigen of interest. These T cells must be evaluated for their specificity and functional capacity prior to making transgenic mice. The availability of the population of T cells that react to the specific antigen from the same progenitor clone with a unique phenotype and function in sufficient numbers is important to perform these assays. The strategy to isolate autoantigen-specific T cell clones has been established earlier [20]. This involves repeated cycles of stimulation with the specific antigen presented by antigen-presenting cells (APCs) followed by the expansion with T cell growth factor, IL-2. The following steps describe the typical culture of antigen-specific T cells.

1. 10 days post-immunization, harvest the draining lymph nodes in 5 ml RPMI, mesh through 40 μm cell strainer (corning) using the backside of a 1 ml syringe and centrifuge at $300 \times g$ for 10 min at 4 °C.

2. Resuspend lymph node cells in RPMI complete medium and count using a hemocytometer. Adjust the cell density to 2×10^6/ml and add rMOG (20 μg/ml) and rIL-2 (20 ng/ml). Seed the cell suspension in a suitable cell culture dish (*see* **Note 1**) and incubate at 37 °C/5% CO_2 incubator.

3. 3 days later add rIL-2 (20 ng/ml).

4. After 5–7 days of stimulation, collect cells by centrifugation at $300 \times g$ for 10 min at 4 °C. To isolate live cells, resuspend cells in 5 ml RPMI, overlay on 3 ml NycoPrep 1.077, and centrifuge $800 \times g$ for 30 min at room temperature without break.

5. Collect the interphase live cells and restimulate with irradiated (5Gy) splenocytes as APCs (1:5 ratio of T cells and APCs (*see* **Note 2**)) together with 20 μg/ml antigen and rIL-2 (20 ng/ml).

6. Repeat the cycles of restimulation as above until a specific T cell line is established. Typically, it takes at least 3–4 restimulation cycles until a stable antigen-specific T cell line can be obtained.

7. To check for antigen-specificity, add 4×10^4 live T cells and 2×10^5 irradiated splenocytes in a 96-well round bottom plate with or without 20 μg/ml antigen for 72 h. Collect the supernatants and measure IFNγ levels by ELISA. Alternatively, thymidine incorporation or dye-based methods can be used to asses antigen-specific proliferation.

8. Once the antigen specificity has been established, freeze multiple aliquots of 5×10^6 T cells in freezing medium (10% FCS + 90% FBS). The specific T cell line may contain multiple T cell clones that can be used for the isolation of a clone containing single TCR (*see* **Note 3**).

3.3 T Cell Hybridoma Generation

While antigen-specific T cells can be maintained by repeated restimulation cycles, they often cease to expand after several restimulation cycles. There is also considerable variability of the capacity of T cells from different mouse strains that can be expanded in vitro. In our experience, T cells from the C57BL/6 strain are often difficult to maintain for a longer time compared to T cells from SJL/J mice. An alternative way to keep the selected antigen-specific T cells is to immortalize them by fusing them with a T cell lymphoma cells. The T cell hybridoma generation described below is very similar to the protocols used for making B cell hybridoma.

1. Split fusion partner BW5147 cells (ATCC, TIB-48) 1 day before fusion (*see* **Note 4**).

2. Place 1 ml of PEG, 5 ml DMEM, and DMEM complete at 37 °C.

3. Collect BW5147 cells, count and mix with preactivated Nyco-Prep purified T cells (day 3) at 1:1 ratio.

4. Centrifuge at $300 \times g$ for 10 min at 4 °C and aspirate the supernatant carefully to completely remove all the medium and leave only the cell pellet.

5. Add 400 μl warm PEG slowly into the cell pellet and mix gently with the tip of the pipette.

6. Add 1 ml of warm DMEM slowly dropwise over 1 min and incubate at 37 °C for 1 min and repeat this step twice.

7. Add warm DMEM complete and centrifuge at $300 \times g$ for 10 min at 4 °C.

8. Resuspend cells in DMEM complete medium and plate 100 μl/well containing 1×10^5 cells in a 96-well flat bottom plate.

9. The next day add 100 μl of 2× HAT solution in DMEM complete medium.

10. Starting at 5 days, monitor each well and transfer growing hybridoma clones into a 24-well plate and add 1 ml of 1× HT in DMEM complete medium.

3.4 Selection of Antigen-Specific T Cell Hybridoma Clones

T cells are a heterogeneous pool of cells with different characteristics. Every T cell harbors unique TCR generated through the rearrangement of its TCR genes. Thus, there is an extreme heterogeneity of T cells in the naïve immune repertoire that recognize different epitopes. A clone will consist of cells which share the same phenotypic characteristics and function. Hence, to generate antigen-specific TCR transgenic mice, it is imperative to obtain an antigen-specific T cell clone.

1. Transfer an aliquot of actively growing T cell hybridoma clones into a 96-well plate and incubate with 20 μg/ml antigen together with irradiated splenocytes (5×10^5 cells/well) as APCs.

2. After 24 h, collect 100 μl of the supernatant and perform IL-2 ELISA (BioLegend) to determine specificity.

3. Subclone selected hybridoma clones by limiting dilution to 0.3 cells/well into 96-well plates and screen once more for the antigen specificity.

3.5 TCR Repertoire Analysis

TCRs are highly diverse heterodimers consisting of α and β chains (αβ TCR) expressed by the majority of T cells. The TCR chains are composed of a variable region that is important for antigen recognition and a constant region. The variable region of TCRα chain is derived by a combination of a number of variable (V) and joining (J) genes, while TCRβ chain is composed of variable (V) and joining (J) and diversity (D) genes. There are different techniques that were developed that enable the analysis of the TCR. For instance, monoclonal antibodies allow the analysis of specific V gene subgroups by flow cytometry. However, there is only a limited number of antibodies that are available and they do not provide a resolution at the level of D and J genes. The other methods that are commonly used to obtain an overview of the repertoire are quantitative polymerase chain reaction (PCR), next-generation sequencing (NGS), and spectratyping techniques. We typically use the combination of flow cytometry and Sanger sequencing of cloned TCR chains.

1. *By flow cytometry:*

    (a) Transfer 1×10^5 hybridoma cells to a 96-well V-bottom plate.

    (b) Centrifuge at $300 \times g$ for 5 min at 4 °C.

    (c) Remove the supernatant and resuspend cell pellets in 50 μl FACS buffer containing 1:200 diluted fluorochrome-labeled antibodies (CD3-PE-Cy7, CD4 PerCP-Cy5.5, anti-TCRβ-APC, and each FITC-labeled anti-TCR Vα or anti-TCR Vβ antibody (*see* **Note 5**)).

    (d) Mix well and incubate for 20 min on ice and in the dark.

    (e) Add 150 μl of FACS buffer and centrifuge plate at $300 \times g$ for 5 min at 4 °C.

    (f) Repeat the washing once more and resuspend cell pellets in 200 μl of FACS buffer.

(g) Acquire the data using FACSCanto or other suitable flow cytometer and analyze data using FlowJo software.

2. *By PCR*: Since there are only a limited number of anti-TCR Vβ and Vα antibodies commercially available, PCR-based screens can be performed using a set of primers designed to target specific TCR V regions (*see* Subheading 2 for TCR V region-specific primers; Tables 1 and 2). Primers are designed based on the sequences available from Immunogenetics (IMGT) website [21] (http://www.imgt.org/IMGTrepertoire/index.php?section=LocusGenes&repertoire=genetable&species=Mus_musculus&group=TRAV and http://www.imgt.org/IMGTrepertoire/index.php?section=LocusGenes&repertoire=genetable&species=Mus_musculus&group=TRBV).

(a) Isolate total RNA from the selected hybridoma clone by RNeasy mini kit (QIAGEN) and convert it into cDNA using the Verso cDNA kit (Thermo Scientific).

5× cDNA synthesis buffer	4 µl
dNTP Mix	2 µl
RT Enhancer	1 µl
Oligo dT	1 µl
Template (RNA)	X µl (100 ng to 1 µg)
Verso enzyme mix	1 µl
Water	*Variable*
Total	**20 µl**

Incubate at 42 °C for 30 min and 95 °C for 2 min.

(b) Set up the PCR reaction with the cDNA using TCR V region-specific primers together with corresponding TCR α- or β-constant region-specific reverse primer as follows (*see* **Note 6**).

Template cDNA	1 µl
2× Q5® High-Fidelity master mix	25 µl
Forward primer (10 µM)	2.5 µl
Reverse primer (10 µM)	2.5 µl
dH$_2$O	19 µl

Table 1
Primers for TCR Vα analysis

Primer name	TCR Gene	IMGT Nomenclature	Oligo sequence (5′ → 3′)
TRAV1 sense		TRAV1	AGGTGTCGACTGGTGGTGTCATGC TGCAG
mTCR Vα2 sense	Vα2	TRAV2	AGGTGTCGACGAGAATATGAAGCAGG TGGCAAAAGTGACTGT
mTCR Vα2 sense #2	Vα2	TRAV2	AGGTGTCGACTGCTTACAAAGAGAATA TGAAGCAGGTGGCAAAAG
TRAV3 sense	Vα5	TRAV3	AGGTGTCGACACAGAGGCATCTTGTC TGGCT
mTCR Vα11 sense		TRAV4	ATTGTCGACCCAGCGATTGGACAGG
TRAV5 sense		TRAV5	AGGTGTCGACGGAAGCACAA TGAAGACA
TRAV6-1-4 sense		TRAV6-1,6-2,6-3,6-4, 6D-3, 6D-4	AGGTGTCGACTCATTTCTTTATG TGAAGAGTTG
TRAV6-6-7 sense		TRAV6-6, 6-7, 6D-5, 6D-6	AGGTGTCGACATGCTAAGCA TCAAGACCACT
TRAV6-5 sense		TRAV6-5, 6D-5	AGGTGTCGACATATTTGTA TTCACACACTACAG
TRAV7-3 sense	Vα1	TRAV7-2, 7-3, 7D-2, 7D-3	AGGTGTCGACCTTAGAGTTGAGGATC TCAGT
TRAV7-4 sense	Vα1.1	TRAV7-4, & 7D-4, 7-1	AGGTGTCGACTCCACAG TGAAGAGGGAAGAG
TRAV7-5 sense	Vα1.2	TRAV7-5, 7D-5	AGGTGTCGACAGAATCCCCAG TGGAGAGAGA
TRAV7-6 sense		TRAV7-6, 7D-6	AGGTGTCGACGCTGGCTTGAAGTG TGAATCT
TRAV8-1 sense		TRAV8-1, 8D-1	AGGTGTCGACTGGAGCTGTATCTC TTGCGA
TRAV8-2 sense		TRAV8-2, 8D-2	AGGTGTCGACGCCTTTCCTGTGACA TCAAT
mTCR Vα 3.2 sense	Va3.2	TRAV9D-2, 9D-3	AGGTGTCGACCTCAGCCATGCTCC TGGCGCTC
TRAV9-1 sense	Vα3.2	TRAV9-1, 9D-1	AGGTGTCGACAGGCACCAGAGCTG TTTCCAG
TRAV9-4 sense	Vα3.1, 3.3, 3.4	TRAV9-4, 9-2, 9D-2, 9D-4	AGGTGTCGACAAGGCTCAGCCATGC TCCTGG

(continued)

Table 1
(continued)

Primer name	TCR Gene	IMGT Nomenclature	Oligo sequence (5′ → 3′)
mTCR Vα15 sense	Vα15	TRAV10	AGGTGTCGACAGGAAGAATGA TGAAGACATCCCTTCACAC
TRAV11 sense		TRAV11, 11D	AGGTGTCGACGCAGA TTTGAGGCAGGCTTCT
mTCR Vα8.3 sense	Vα8.3	TRAV12	AGGTGTCGACCTTCCATGAACATGCG TCCTGACACC
mTCR Vα 10-U07879 sense	Vα10	TRAV13D-3	AGGTGTCGACAGCACCATGAAGAGGC TGCTG
mTCR Vα 10-M34216 sense	Vα10	TRAV13-1,13-33,13D-1,13-D3	AGGTGTCGACAGCACCATGCAGTGG TTTTATCAACGTCCTGG
TRAV14 sense	Vα2	TRAV14	AGGTGTCGACAGTCTAGGAGGAA TGGACAAG
TRAV16 sense		TRAV16, 16D, 16N	AGGTGTCGACAGAGCTAACAGTATGC TGATTC
TRAV17 sense		TRAV17	AGGTGTCGACACTGCCTAGCCATG TTCCTA
TRAV18 sense		TRAV18	AGGTGTCGACAGGGAGAAAAGATGC TCCTGA
TRAV19 sense	Vα20	TRAV19	AGGTGTCGACTCAGAGAAGACATGAC TGGCT
mTCR Cα anti-sense	TCR α constant region	TRAC	ATAGGATCCTCAACTGGACCACAGCC TCAGC

Cycle number	Denature	Anneal	Extend
1	98 °C, 3 min		
2–34	98 °C, 20 s	60 °C, 30 s	72 °C, 1 min
35			72 °C, 5 min

(c) Analyze the amplified products in an agarose gel (approximate size 850–950 bp).

(d) Clone the amplified TCR chains into pGEM-T or another suitable vector (Promega) using standard molecular biology techniques.

Table 2
Primers for TCR Vβ analysis

Primer name	TCR Gene	IMGT nomenclature	Oligo sequence (5′ → 3′)
mTCR Vβ4 sense	Vβ4	TRBV2	AGGTGTCGACTGACACTGCTATGGGC TCCATTTTCCTC
mTCR Vβ2 sense	Vβ2	TRBV1	AGGTGTCGACCCTGAGGTCTCAGAGA TGTGGCA
mTCR Vβ5.1/ 5.2 sense	Vβ5.1/5.2	TRBV12-2; TRBV12-1	AGGTGTCGACCTGAGAGGAAGCATGTC TAACAC
mTCR Vβ6 sense	Vβ6	TRBV19	AGGTGTCGACCCAAACTATGAACAAG TGGGT
mTCR Vβ7 sense	Vβ7	TRBV29	AGTGTCGACACCACCATGAGAG TTAGGCTCATC
mTCR Vβ8.2 sense	Vβ8.2	TRBV13-2	AGTGTCGACGTCCCAAGATGGGC TCCAGGCT
mTCR Vβ8.1 sense	Vβ8.1	TRBV13-3	AGTGTCGACTAGTTCTGAGATGGGC TCCAGACT
mTCR Vβ8.3 sense	Vβ8.3	TRBV13-1	AGTGTCGACTCGCGAGATGGGC TCCAGGCT
mTCR Vβ9 sense	Vβ9	TRBV17	AGTGTCGACCTGGACCATCCA TGGACCCTAGAC
mTCR Vβ10 sense	Vβ10	TRBV4	AGTGTCGACTGACCCAACTATGGGCTG TAGGCT
mTCR Vβ11 sense	Vβ11	TRBV16	AGTGTCGACTCATCTTGCCA TGGCCCCCAGGCT
mTCR Vβ12 sense	Vβ12	TRBV15	AGTGTCGACATCCCACTATGGGCA TCCAGACC
mTCR Vβ13 sense	Vβ13	TRBV14	AGTGTCGACAATCTGCCA TGGGCACCAGGCT
mTCR Vβ14 sense	Vβ14	TRBV31	AGTGTCGACCTTGAACTATGCTGTAC TCTCTC
mTCR Vβ1 sense	Vβ1	TRBV5	AGTGTCGACCTGATTCCACCATGAGC TGCAG
mTCR Vβ17 sense	Vβ17	TRBV24	AGGTGTCGACTGATCATGGG TGCAAGACTGCTC
mTCR Vβ3 sense	Vβ3	TRBV26	AGTGTCGACCCCAGGAAGTGTAA TGAGGAAGATTGAGAAC
mTCRβ C2 anti-sense	TCR β C2 constant region	TRBC	ATAGGATCCGGGTGAAGAACGGC TCAGGATGC
mTCRβ C1 anti-sense	TCR β C1 constant region	TRBC	ATAGGATCCCAGCTCTTGTATTCATC TTCACATCTGGCTTC

(e) Sequence the cloned TCR chains and analyze the V regions using IMGT/V-Quest (https://www.imgt.org/IMGT_vquest/vquest).

3.6 Verification of Expression and Response in 58α-β-Cells

One approach to studying TCR properties is to clone TCR genes into retroviral vectors and transfer antigen specificity into other cells. In this way, the specificity of TCR that is cloned can be reassessed and if necessary TCR sequences can be modified by standard molecular biological techniques to generate alternative TCR constructs that have altered affinity for the specific antigen. The strategy involves the fusion of TCR α and β chains into a retroviral vector and expresses them in a TCR, deficient hybridoma which has all T cell components except the TCR.

3.6.1 Cloning of Fused TCR α and β Chains

To verify if the identified TCR V α and β are matching pairs, it is necessary to re-express them in a TCR-deficient cell line. This is achieved by fusing specific TCR α and β chains with a T2A self-cleaving linker peptide. As a template, sequence-verified TCRα and β clones can be used.

1. Set up PCR reactions for α and β chains with the specific primers separately as described above. The common TCH Cα rev. and TCH β C2 rev. primers are given in the Table 3. Forward primers for TCRα should be designed to include necessary restriction sites and the TCR β with part of the T2A gene for fusion. An example forward primer is given in Table 3.

Cycle number	Denature	Anneal	Extend
1	98 °C, 3 min		
2–34	98 °C, 30 s	72 °C, 1 min	
35			72 °C, 2 min

2. After verification of the PCR products in a gel, add 10 ng each of the PCR products and perform fusion PCR with TCH Vα FOR and TCH beta C2 rev primers.

3. Purify the PCR products (~2 kb fragment) and digest with *MfeI* and *BamHI* enzyme and clone into the *EcoRI*- and *BamHI*-digested pMSCV-IRES2-GFP vector. Alternatively, any other retroviral vector may also be used.

3.6.2 Retroviral Expression of TCR in 58α-β- Cells

The retroviral packaging cell line, phoenix cells (ATCC, CRL-3213), and TCR deficient $58\alpha^-\beta^-$ cells [22] are used for the reconstitution of TCR.

1. One day prior to transfection, seed 2×10^6 phoenix cells in 10 cm cell culture dish in DMEM complete medium.

Table 3
Primers for the fusion of TCR α and β chains

Primer	Oligo Sequence (5′–3′)	Restriction enzyme
TCH Vα FOR	ATTTCAATTGCCACCATGAAGACAGTGACTGGACCTTTGT	MfeI
TCH Cα rev	TCACCGCATGTTAGAAGACTTCCTCTGCCCTCAGCTATGCA TCCAACG	–
TCH T2A Vβ	TTCTAACATGCGGTGACGTGGAGGAGAATCCCGGCCCTATGGGC TCCAGGCTCTTCTT	–
TCH β C2 rev	ATAGGATCCGGGTGAAGAACGGCTCAGGATGC	BamHI

2. Replace medium with 10 ml DMEM complete containing 25 μM chloroquine.

3. For each plate, prepare a mixture of 12 μg pMSCV-TCRα-β-IRES-GFP vector, 3.5 μg pCL-Eco retrovirus packaging vector in 438 μl H_2O and add 62 μl of 2 M $CaCl_2$. Add 500 μl 2× BES dropwise while vortexing, incubate for 20 min at 37 °C, and add dropwise to phoenix cells.

4. Next day, change the medium with an 8 ml fresh DMEM complete medium.

5. 2 days after transfection, collect the supernatant containing virus particles, add 4 ml fresh DMEM medium.

6. Filter the supernatant through 45 μm filter and add polybrene at a final concentration of 8 μg/ml.

7. In a 6-well plate transfer 1×10^6 58α⁻β⁻ cells and add 2–4 ml virus supernatant.

8. Centrifuge the plate for 90 min at 600 × g at RT and return the plate to the incubator.

9. Repeat the spin infection procedure on day 3 using the fresh virus supernatant.

10. Fourty eight hours posttransfection, collect cells and stain for the TCRβ and α expression by flow cytometry with anti-TCRβ antibodies.

3.7 Pronuclear Injection and Characterization of TCR Transgenic Mice

The final step in the generation of TCR transgenic mice is to stably integrate the cloned antigen-specific TCR into the mouse genome. A variety of plasmid vectors containing either endogenous or foreign regulatory elements has been used to drive the expression of TCR genes. This includes endogenous TCR promoter, CD2 promoter, and ubiquitous promoters such as MHC class I promoter. In our model, we have routinely used the rearranged TCRα and β chains subcloned into the pHSE3' vector, which drives the expression of TCR under the control of the transgenic MHC class I H-2Kb promoter [23]. Linearized TCR-containing plasmids are coinjected into the pronuclei of fertilized FVB oocytes and transgenic founders are identified by PCR using specific primers for TCR Vα and Vβ chain. Individual founder transgenic mice were backcrossed for at least 10 generations into the SJL/J background (*see* **Note 7**).

The expression of specific TCR Vα and Vβ chain and antigen specificity by the transgenic T cells were assessed by flow cytometry and proliferation assay.

1. Take thymus and spleen from the transgenic mice; mesh through a cell strainer; stain with CD3, CD4, and TCR Vβ/Vα antibodies; and analyze by flow cytometry. A typical CD4$^+$ TCR transgenic mice will show higher frequencies of CD4$^+$ T cells expressing specific TCR chains in both the thymus and spleen.

2. To confirm antigen specificity, label splenocytes (2×10^7/ml) with 5 μM eFluor 450 proliferation dye (Invitrogen) in PBS for 10 min at 37 °C, wash with RPMI complete medium twice and plate 2×10^5 cells into 96 round-well plate with the specific antigen. Three days later, measure the proliferation (dye dilution) by flow cytometry.

3.8 Evaluation of Spontaneous EAE

Transgenic mice should be monitored regularly and evaluation of clinical neurological disease is performed according to the classic EAE disease determination. The infiltration of leukocytes can be analyzed by flow cytometry or by histopathology.

3.8.1 Classic Paralytic EAE

Score 0, healthy animal; 1, limp tail; 1.5, limp tail and impaired righting reflex; 2, limp tail, impaired righting reflex, and weakness of hind legs; 2.5, one hind leg paralysis; 3, both hind leg paralyzed with residual mobility in both legs; 3.5, complete paralysis of both hind legs; 4, both hind legs paralyzed and beginning front limb paralysis; 5, moribund animal or death of the animal after preceding clinical disease.

3.8.2 Ataxic EAE

Score 0, healthy; 1, mouse partly tilted, feet fall into cage fence; 2, tilted and tumbles; 3, mouse heavily tilted and moves in circles; 4, inability to walk, is only rolling; 5, moribund.

4 Notes

1. We typically use 60 mm cell culture dishes or multi-well plates to culture lymph node cells. We avoid using 6-well plates since cells tend to accumulate in the middle of the well in these

plates. It is also very important to keep the right density of lymph node cells to ensure proper presentation of antigens to T cells by the APC's.

2. It is critical to add sufficient irradiated splenocytes as APCs. At least a 1:5 ratio of T cells and APCs is recommended. For a single 60 mm dish, we use 2–4×10^6 T cells and 10–20×10^6 irradiated splenocytes.

3. The T cell lines are often a mixture of several individual T cell clones that respond to the same antigen. Depending upon the growth capacity of a particular antigen-specific T cell line, there are a variety of ways one can isolate a single T cell clone for TCR cloning. We typically attempt to generate individual T cell clones by limiting dilution or single-cell sorting and subsequent restimulation with APCs for multiple rounds. Alternatively, we generate T cell hybridoma by fusing T cell lines with BW5147 cells and select the specific hybridoma for antigen-specific TCR isolation.

4. It is very important to keep the BW5147 cells in exponential growth conditions in order to achieve an efficient fusion. Overgrown cells and cells that were not in exponential growth conditions will lead to drastically reduced fusion efficiency.

5. To determine specific TCR receptor genes expressed by T cell hybridoma, cells are stained with commercially available anti-mouse TCR antibodies and analyzed by flow cytometry. For TCR Vβ chain, mouse Vβ TCR screening panel antibodies available from BD Biosciences can be used. For TCR Vα chain there are only a limited number of antibodies available. It is ideal to select a hybridoma clone that can be stained with the anti-TCR antibodies as that would facilitate the identification by flow cytometry. If none of the available antibodies stain the TCR, the expression of the TCR chains can be determined by primers targeting individual TCR α or β chains.

6. The use of a high-fidelity polymerase is highly recommended to avoid mutations due to PCR amplification. For the initial screening of an unknown TCR chain, pools of up to four different forward primers can be used.

7. Instead of pHSE3' vector, cassette vectors that have natural regulatory elements can be used [24]. Although we typically express α and β chains in different plasmids, TCRα and β chains fused together with the T2A peptide can be used. With the emerging CRISPR/Cas9 technology, it is also possible to knockin TCR chains into specific locus, thus avoiding random integration due to pronuclear injection.

5   Limitations

Although myelin antigen-specific TCR transgenic models offer a possibility to study the role of antigen-specific T cells in the neuroinflammatory disease process, there are potential limits to this approach which should be kept in mind for interpretation. First, transgenic T cell precursor frequencies in the recipients are usually much higher than normal. This could cause T cell competition for APCs resulting in abnormal responses. Second, the proportion of other T cell subsets are often altered due to the forced expression of rearranged TCR chains. The skewed proportion of $CD4^+/CD8^+$ T cells, as well as regulatory T cells, has been commonly observed in TCR transgenic mice. This could potentially alter the balance between regulatory and proinflammatory immune responses. Third, the model may represent the behavior of one T cell clone which may not be representative of the different clones of T cells in the repertoire.

6   Concluding Remarks

Until now, EAE is the most frequently used model that simulates at least some aspects of MS. While the commonly used actively induced EAE model using C57BL/6 has revealed many interesting mechanisms of neuroinflammation, there is a need to develop alternative models that closely resemble clinical and pathological aspects of MS. In particular, spontaneous EAE models are needed to address the mechanisms of disease triggers. Furthermore, it is possible to use these models to evaluate the effect of various specific genes in the regulation of neuroinflammation. The protocol described here serves as a guide to generate new TCR transgenic mouse models in different genetic backgrounds, which may develop spontaneous EAE. Using such an approach we have developed a spontaneous relapsing–remitting EAE model that closely resembles a western type of MS [10].

Acknowledgments

The experimental work in the author's laboratory is supported by European Research Council starting grant (GAMES; 635617), German research foundation (DFG) SFB TR-128 (Project A1), and the Max Planck Society.

References

1. Lassmann H (2019) Pathogenic mechanisms associated with different clinical courses of multiple sclerosis. Front Immunol 9:3116

2. Fuller KG, Olson JK, Howard LM, Croxford JL, Miller SD (2004) Mouse models of multiple sclerosis: experimental autoimmune encephalomyelitis and Theiler's virus-induced demyelinating disease. Methods Mol Med 102:339–361

3. Krishnamoorthy G, Wekerle H (2009) EAE: an immunologist's magic eye. Eur J Immunol 39:2031–2035

4. Krishnamoorthy G, Holz A, Wekerle H (2007) Experimental models of spontaneous autoimmune disease in the central nervous system. J Mol Med 85:1161–1173

5. Goverman J, Woods A, Larson L, Weiner LP, Hood L, Zaller DM (1993) Transgenic mice that express a myelin basic protein-specific T cell receptor develop spontaneous autoimmunity. Cell 72:551–560

6. Lafaille JJ, Nagashima K, Katsuki M, Tonegawa S (1994) High incidence of spontaneous autoimmune encephalomyelitis in immunodeficient anti-myelin basic protein T cell receptor transgenic mice. Cell 78:399–408

7. Waldner H, Whitters MJ, Sobel RA, Collins M, Kuchroo VK (2000) Fulminant spontaneous autoimmunity of the central nervous system in mice transgenic for the myelin proteolipid protein-specific T cell receptor. Proc Natl Acad Sci U S A 97:3412–3417

8. Bettelli E, Pagany M, Weiner HL, Linington C, Sobel RA, Kuchroo VK (2003) Myelin oligodendrocyte glycoprotein-specific T cell receptor transgenic mice develop spontaneous autoimmune optic neuritis. J Exp Med 197:1073–1081

9. Krishnamoorthy G, Lassmann H, Wekerle H, Holz A (2006) Spontaneous opticospinal encephalomyelitis in a double-transgenic mouse model of autoimmune T cell/B cell cooperation. J Clin Invest 116:2385–2392

10. Pöllinger B, Krishnamoorthy G, Berer K, Lassmann H, Bösl M, Dunn R, Domingues HS, Holz A, Kurschus FC, Wekerle H (2009) Spontaneous relapsing-remitting EAE in the SJL/J mouse: MOG-reactive transgenic T cells recruit endogenous MOG-specific B cells. J Exp Med 206:1303–1316

11. Perchellet A, Stromnes I, Pang JM, Goverman J (2004) CD8 + T cells maintain tolerance to myelin basic protein by "epitope theft". Nat Immunol 5:606–614

12. Ji QY, Perchellet A, Goverman JM (2010) Viral infection triggers central nervous system autoimmunity via activation of CD8 + T cells expressing dual TCRs. Nat Immunol 11:628–634

13. Madsen LS, Andersson EC, Jansson L, Krogsgaard M, Andersen CB, Engberg J, Strominger JL, Svejgaard A, Hjorth JP, Holmdahl R et al (1999) A humanized model for multiple sclerosis using HLA-DR2 and a human T-cell receptor. Nat Genet 23:343–347

14. Friese MA, Jakobsen KB, Friis L, Etzensperger R, Craner MJ, McMahon RM, Jensen LT, Huygelen V, Jones EY, Bell JI et al (2008) Opposing effects of HLA class I molecules in tuning autoreactive CD8 + T cells in multiple sclerosis. Nat Med 14:1227–1235

15. Cao Y, Toben C, Na SY, Stark K, Nitschke L, Peterson A, Gold R, Schimpl A, Hünig T (2006) Induction of experimental autoimmune encephalomyelitis in transgenic mice expressing ovalbumin in oligodendrocytes. Eur J Immunol 36:207–215

16. Na SY, Cao Y, Toben T, Nitschke L, Stadelmann C, Gold R, Schimpl A, Hünig T (2008) Naïve CD8 T-cells initiate spontaneous autoimmunity to a sequestered model antigen of the central nervous system. Brain 131:2353–2365

17. Saxena A, Bauer J, Scheikl T, Zappulla J, Audebert M, Desbois S, Waisman A, Lassmann H, Liblau RS, Mars LT (2008) Cutting edge: multiple sclerosis-like lesions induced by effector CD8 T cells recognizing a sequestered antigen on oligodendrocytes. J Immunol 181:1617–1621

18. Sasaki K, Bean A, Shah S, Schutten E, Huseby PG, Peters B, Shen ZT, Vanguri V, Liggitt D, Huseby ES (2014) Relapsing–remitting central nervous system autoimmunity mediated by GFAP-specific CD8 T cells. J Immunol 192:3029–3042

19. Adelmann M, Wood J, Benzel I, Fiori P, Lassmann H, Matthieu JM, Gardinier MV, Dornmair K, Linington C (1995) The N-terminal domain of the myelin oligodendrocyte glycoprotein (MOG) induces acute demyelinating experimental autoimmune encephalomyelitis in the Lewis rat. J Neuroimmunol 63:17–27

20. Ben-Nun A, Wekerle H, Cohen IR (1981) Vaccination against autoimmune encephalomyelitis using attenuated cells of a T lymphocyte line reactive against myelin basic protein. Nature 292:60–61

21. Giudicelli V, Chaume D, Lefranc MP (2005) IMGT/GENE-DB: a comprehensive database for human and mouse immunoglobulin and T cell receptor genes. Nucleic Acids Res 33: D256–D261

22. Letourneur F, Malissen B (1989) Derivation of a T cell hybridoma variant deprived of functional T cell receptor α and ß chain transcripts reveals a nonfunctional α-mRNA of BW5147 origin. Eur J Immunol 19:2269–2274

23. Pircher H, Mak TW, Ballhausen W, Rüedi E, Hengartner H, Zinkernagel RM, Bürki K (1989) T cell tolerance to Mls a encoded antigens in T cell receptor Vß8.1 chain transgenic mice. EMBO J 8:719–727

24. Kouskoff V, Signorelli K, Benoist C, Mathis D (1995) Cassette vectors directing expression of T cell receptor genes in transgenic mice. J Immunol Methods 180:273–280

Chapter 8

Chronic Progressive Models

Maren Lindner, Ann-Katrin Fleck, and Luisa Klotz

Abstract

Even though various animal models for multiple sclerosis (MS) exist, studies with long-term chronic progressive models, reflecting key aspects of progressive forms of MS are still quite sparse.

However, to understand the mechanisms underlying chronic progressive CNS inflammation, where neurodegenerative processes become more prominent and disease is not directly fueled by a targeted immune response from the periphery, the only way to go forward is using these chronic progressive models to improve our understanding of the pathophysiology of progressive MS and to study treatment approaches, which might exert neuroprotective and/or remyelinating effects.

Here, we will discuss several chronic progressive animal models and focus thereafter on two different animal models in more detail, the genetically driven, primary progressive, OSE model and the secondary progressive NOD-EAE model.

Key words MS, Chronic animal models, OSE, NOD-EAE

1 Introduction

A wide range of animal models of multiple sclerosis exists using various species (e.g., zebrafish, mice, rats, or nonhuman primates) and/or different disease induction schemes (reviewed in [1, 2]). The most widely used and best studied animal model for MS is the "classical" murine experimental autoimmune encephalomyelitis (EAE), such as the active-induced myelin oligodendrocyte glycoprotein peptide (MOG_{35-55}) EAE and the adoptive-transfer EAE. Here, disease course and resulting CNS pathology largely depend on the animal strain and the antigen used [3]. Most of the studies were performed with C57BL/6 mice, due to the fact that these mice can be easily genetically modified. However, these "classical" EAE models only mimic an acute, monophasic severe disease course (clinically resembling occurrence of one severe relapse with sequelae) with lesions in the spinal cord, but only limited immune cell infiltration in the brain.

However, the majority of patients with MS show a relapsing–remitting disease course, which manifests in distinct relapses, which

Sergiu Groppa and Sven G. Meuth (eds.), *Translational Methods for Multiple Sclerosis Research*, Neuromethods, vol. 166, https://doi.org/10.1007/978-1-0716-1213-2_8, © Springer Science+Business Media, LLC, part of Springer Nature 2021

are followed by remission phases. In addition, over time, the relapsing–remitting MS (RRMS) converts into secondary progressive MS (SPMS), where neurological deficits develop gradually over time, often without occurrence of acute and focal "inflammatory" disease activity. Hence, there is an obvious need to develop more suitable animal models, which reflect to a greater extent the different disease courses of MS, and which allow to particularly focus on the progressive phases of disease.

Overcoming the limitations of a monophasic model, the proteolipid protein (PLP)-induced EAE in SJL mice can be used to focus on varying diseases stages before, during and after relapses, therefore more closely resembling the relapsing–remitting course of human MS. Upon immunization with $PLP_{139-151}$ peptide, SJL mice develop a series of relapses with remission phases in between [4]. After around 8 weeks, these mice histologically show an increasing axonal and neurofilament pathology [5].

Another open question that is not yet addressable by the above mentioned EAE models is the impact of de- and remyelination on the process of axonal degeneration, which is the major cause of long-term disability in MS patients. To overcome this limitation, several in vivo studies are performed with toxic demyelination models such as cuprizone [6, 7], where the remyelination phase is fast and efficient, at least in mice [8]. However, due to the toxic nature of these models, an influence of immune cells is completely absent. As remyelination is known to be impaired in the context of inflammation, this represents a major drawback of these models. To overcome this limitation, a combination of cuprizone and MOG immunization, the so-called Cup-EAE model, was developed [9]. In this model, the mice are initially fed for 2 weeks with a normal cuprizone diet, then switched to a low dose cuprizone diet for 3 weeks, while immunized with MOG. Here, the mice show T cell infiltration and axonal damage on top of the severe demyelination. Another approach, instead of using an active immunization with MOG, is the transfer of myelin-reactive T cells into cuprizone-treated mice [10]. Interestingly, in this study the authors could demonstrate that the presence of T helper 17 cells (Th 17 cells) in the brain is associated with an impaired remyelination process at the corpus callosum, making this approach a valuable tool, particularly in light of the fact that in the human situation remyelination often fails and is incomplete [11, 12].

As a consequence of failed or impaired remyelination in human MS, most lesions in MS patients show large areas of demyelination, which in consequence result in chronic neuronal damage and/or axonal degeneration. This neurodegenerative process, characterized by an increase in disability score, plays a key role in the chronic progressive phase of MS. Progressive MS can be divided in two distinct disease entities based on the clinical disease course, either characterized by complete absence of any relapses (PPMS) or an initially relapsing–remitting phase followed by secondary progression (SPMS). Both forms are of high interest in MS research due to the fact that for these disease courses there are only limited treatment options available so far [13]. In the following section, we introduce animal models addressing the chronic progressive disease course of MS.

2   Animal Models

2.1   Active EAE Induced by a Fusion Protein (MBP-PLP)

An easy way to modify a "classically" active EAE is to adjust the peptide or protein used for immunization. The usage of a fusion protein, consisting of two different proteins of the myelin, the myelin basic protein (MBP) and the phospholipid protein (PLP) is known as MBP-PLP or MP4 [14, 15], turned out to be beneficial. In this model, immunized mice develop first signs of paralysis between day 5 and day 10 after disease induction with a peak of disease between day 15 and day 20. Thereafter the mice develop a chronic disease course without clear remission and relapses. By day 40, regions of demyelination can be found in the proximity of immune cell infiltration in the brain and spinal cord [14]. Here, the frequencies and sizes of CNS lesions show a correlation with clinical disease severity [15] and after long-term disease course (6 months after onset) profound demyelination and axonal pathology can be observed [16]. Even though this model shows more neuropathological features of the human disease, for example, also involvement of $CD8^+$ T cells, B cells, and myelin-specific antibodies [17, 18], in contrast to the classical MOG peptide-EAE, where the disease is mainly $CD4^+$ T cell driven, one major drawback of this model is its monophasic disease course.

2.2   Theiler's Murine Encephalomyelitis Virus Infection of SJL Mice

Another not so commonly used model is the Theiler's murine encephalomyelitis virus (TMEV) infection of SJL mice. TMEV is a naturally occurring enteric pathogen of the mouse. Infection of SJL mice with the Daniels strain of TMEV results in persistent viral infections of the CNS where the mice develop clinical and histopathological features similar to the human disease. Initially the virus predominantly infects neurons in the grey matter, such as the hippocampus and the cerebral cortex [19]. Increased infiltration of $CD4^+$ and $CD8^+$ T cells, monocytes, and a few B cells and plasma cells can be observed in the grey matter of the brain [20]. After the acute phase, the viral load decreases but the immune systems fails to clear the virus completely, which consequently leads to a chronic phase of the disease [21]. During the chronic phase of the disease, mice (1 month after TMEV inoculation) start to show signs of disease presented by weakness of the hind limbs, which develops to a severe paralysis with no recovery. In contrast to the acute phase, the virus is not found in neurons anymore, but persists in oligodendrocytes, astrocytes, and microglia/macrophages during the chronic phase [22–24]. The infiltrating immune cells, predominantly CD4$^+$ and CD8$^+$ T cells and to a lesser extent, macrophages, B cells, and plasma cells, persist in the white matter while demyelination in the spinal cord and axonal damage can be observed.

In this model the TMEV persistence is important to induce demyelination; however, the trigger mechanism for demyelination is inflammation and induction of autoimmunity. On account of the discussion, if a viral infection can trigger neuroinflammation, this model is certainly suitable.

In our working group two other chronic models are well established and will be discussed in the following two paragraphs in detail; the double-transgenic mouse model OSE and the NOD-EAE model, a model for SPMS.

2.3   OSE Model

The opticospinal encephalomyelitis/EAE model (OSE model, also referred to as Devic model) is a chronic progressive mouse model for CNS inflammation without relapses or a remission phase; described for the first time in 2006 [25, 26]. Due to their genetically modified background, OSE mice possess T and B cells with a specificity against the autoantigen MOG, a protein which is expressed in myelin layers of the CNS. Based on this predestined immune responses, inflammatory demyelinating lesions almost exclusively occur in the optic nerve and the spinal cord of the mice and result in the development of neurologic symptoms that are comparable to those of human CNS autoimmunity [25, 26].

2.3.1   Material and Methods

Mice and Housing

OSE mice are generated by intercrossing hemizygous transgene 2D2 mice with homozygous Th mice (Fig. 1a). The majority of the CD4$^+$ T cells of transgene 2D2 mice (also referred to as TCR^{MOG} mice; C57BL/6 background) have a specific transgene T cell receptor (TCR) against the MOG$_{35-55}$ peptide, resulting in around 40% of the 2D2 mice in spontaneous, isolated optical neuritis and the ranking as burdened mouse line [27]. For more detailed background information of the exact generation and hemizygous breeding recommendations concerning the 2D2 mice, visit the link [28]. In contrast, approximately 30% of the B cells in a Th mouse (also known as IgH^{MOG} mice; C57BL/6 background) are producing MOG-specific immunoglobulin H [29]. In sum, the combination of MOG-specific T and B cell responses results in an immunologically driven CNS inflammation in OSE mice and entail the categorization as burdened mouse line. In conventional and specific pathogen free housing, the OSE mice spontaneously develop EAE-like symptoms with incidence rates that range between 50% and 90% (Fig. 1b), but can differ for germ-free mice, as described for other spontaneous EAE models [25, 26, 30–32].

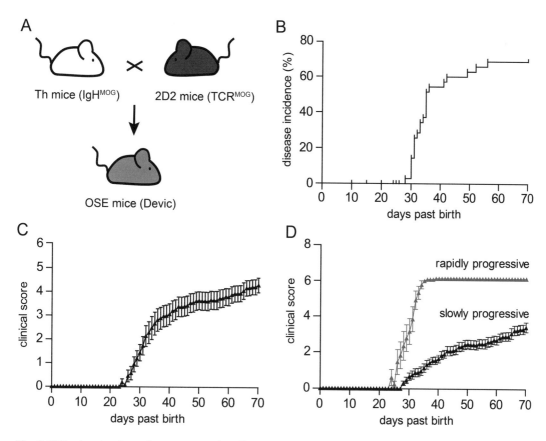

Fig. 1 OSE mice develop primary progressive disease without an exogenous trigger. Mating scheme of OSE mice, which have T cells and B cells with a specificity against MOG (**a**). Data from the cohort of ref. 23 ($n = 45$; mean ± SEM) demonstrate the phenotype of OSE mice (**b–d**). In this cohort, the disease incidence was almost 80% (**b**). Around an age of 4 weeks, the mice start to develop a primary progressive disease course. In (**c**), the disease course of all animals in the cohort is pooled, and in (**d**), the cohort is split according to the speed of development (slowly or rapidly progressive) of neurodegenerative symptoms

Phenotyping of Blood Specimen Via Flow Cytometry

Due to the nature of intercrossing two different mouse lines to breed OSE mice, only 50% of the OSE offspring are carriers of the transgene 2D2 TCR (Vα3.2 and Vß11). Therefore, each individual pup of 2D2 and OSE breeding needs to be phenotyped. Around 2–3 weeks after birth, some drops (<50 μl) of blood are withdrawn from the vena facialis. The blood is collected in Heparin-enriched phosphate buffered saline (PBS) and red blood cell lysis is performed. Afterward, residual immune cells are stained with anti-Vß11 and/or anti-Vα3.2 antibodies as well as anti-CD4 antibody. Via flow cytometric analysis the percentage of CD4+ T cells with the specific TCR Vß11 and/or Vα3.2 is assessed. Mice with a percentage of more than 90% of the specific 2D2 TCR are defined as 2D2 TCR transgene and can develop the characteristic phenotypes. The other littermates are defined as "wild type" and do not develop any symptoms. The "wild type" 2D2 mice are used for hemizygous 2D2 mating, whereas "wildtype" OSE mice cannot be used for further breeding.

Phenotype and Classifying Criteria

Due to the fact that OSE mice have a spontaneous disease onset and belong to the category of burdened mouse lines, all mice have to be monitored strictly. For detailed and standardized characterization of disease symptoms, we are using an eight-point scoring system as depicted in Table 1. In contrast to a "classical" EAE, the symptoms start spontaneously at around 4–8 weeks after birth with a moderate ataxia (Score 1) characterized by slightly atonic hind limbs, while the tail tone is initially unaffected (Fig. 1c). The neurologic symptoms worsen gradually with increasing age without any relapses or a remission phase; however, in the first 2 weeks after disease onset the severity of disease symptoms increases more rapidly as compared to later stages (Fig. 1c; complete cohort). Finally, the mice may develop a quadriplegia (Score 7) or in the worst case die (Score 8) [25, 26, 31, 32]. Depending on country-specific, current legal regulations of the animal welfare, it has to be considered that a complete paralysis of the hindquarters (Score 6) over several days is declared as severe burden for the mouse and the animal should be sacrificed. Moreover, with regard to the kinetics of disease development, OSE littermates can be divided into two distinct disease courses. The (1) slowly progressive disease course and the (2) rapidly progressive phenotype (Fig. 1d). OSE mice with a fast progression (2) develop at an age of around 30–40 days after birth severe symptoms up to a Score 6 in 1–3 days.

Table 1
Scoring criteria (eight-point-scoring system)

Score	Description
0	Healthy
1	Slight waddle; moderate ataxia
2	Slight unilateral paresis of hind limbs
3	Moderate unilateral or slight bilateral paralysis of hind limbs
4	Moderate bilateral paralysis of hind limbs; lowered hindquarters
5	Severe paralysis of hind limbs, lowered and broad hindquarters
6	Complete paralysis of hind limbs
7	Complete quadriplegia
8	Moribund animal

2.3.2 Pathological Manifestations

Histopathological analysis confirmed that in OSE mice inflammatory lesions are mainly focal and localized in the optic nerve and the leptomeninges of the spinal cord compared to a few diffuse lesions in the brain parenchyma (cerebrum, midbrain, brainstem, or cerebellum). The lesions of OSE mice are characterized by demyelination and axonal loss associated with massive infiltration of immune cells comprising mononuclear cells (not granulocytes, eosinophils), inflammatory CD4$^+$ T cells (expressing CD69, CD25, and CD44), B cells as well as a few CD8$^+$ T cells [25, 26, 31]. Moreover, the majority of the infiltrating immune cells are organized in distinct follicle-like structures in the leptomeninges. Notably, these lymphoid follicle-like structures are also found in progressive MS [26, 33].

2.3.3 Advantages and Limitations of the OSE Model

Advantages

1. The OSE model demonstrates that besides the undisputed role of inflammatory T cells at least in animal models, also B cells display an important function in the induction of CNS autoimmunity in the OSE model, probably via MHC II-dependent antigen presentation [25, 26, 33].

2. The OSE model is a genetically driven model, thus no exogenous trigger, such as injection of Freund's Complete Adjuvant (CFA) or pertussis toxin (PTX) is required to unspecifically boost the responsiveness and break tolerance toward autoantigens as used in active EAE models.

3. The genetic susceptibility of OSE mice is precisely defined by the genetically modified backgrounds of 2D2 and Th mice. This allows for the investigation of influences (e.g., environmental factors) potentially involved in the development of autoimmunity [34].

Limitations of the Model

1. It is a monophasic model with a primary progressive phase and the mice exhibit a strong phenotype of a paralytic disease. Hence, therapeutic approaches with a rather mild impact or a longer onset of efficacy are likely to be "overwritten" by the strong immune response.

2. Due to the early onset of disease (4–8 weeks after birth), it might not be feasible to assess the influence of distinct external factors, such as in the study of Enzmann et al. investigating the influence of smoking in different EAE models [32]. In this case, an investigation of smoke exposure in OSE mice was not feasible, as the mice were already sick before smoke exposure was started. An exposure already in utero had to be rejected due to insufficient immune system maturation or inadequate comparison to the other EAE model setups [32].

3. The individual disease onset, incidence, and course varies between the different previous publications depending on the cohort (breeding pairs), animal facility conditions, individual

clinical scoring systems, and analysis methods of the incidence rate. A highly plausible rationale behind this phenomenon are divergent microbial exposures along with differences in the gut microbiome composition, which have been described to strongly influence autoimmune processes in OSE mice [34].

4. As a consequence of these variations, a higher number of animals has to be calculated for an individual experiment to account for the variability of the model.

2.4 NOD-EAE Model

The NOD/ShiLtJ strain was originally established as a model for type I diabetes mellitus [35] and was derived from female mice spontaneously exhibiting diabetic symptoms without obesity. This strain is susceptible to inducible autoimmune diseases, such as EAE, experimental thyroiditis, or systemic lupus erythematosus like diseases due to its unique MHC haplotype [36]. Classical immunization with the immunogenic myelin peptide MOG_{35-55} with CFA is required for disease induction. In addition, the use of purified PTX is needed as auxiliary adjuvant for targeted breakdown the blood–brain barrier and enhancement of inflammatory infiltration [37]. Interestingly, in this mouse strain the susceptibility for disease and disease course is similar in female and male mice [38].

The NOD-EAE model is considered as a model for secondary progression due to the fact that after a first relapse with partial remission, the mice develop a progressive clinical disease course accompanied by lesions in the spinal cord as well as in the brain [39, 40].

2.4.1 Phenotypic Characteristics and Pathological Manifestations

The MOG_{35-55} peptide NOD-EAE is quite unique because inflammatory CNS lesions occur in the corpus callosum, external capsule, fimbria, internal capsule, and thalamus, which is similar to what can be found in MS [41]. Notably, the lesion characteristics mirror much more closely features seen in progressive MS patients, where extensive demyelination and axonal loss can be observed. In the animal model already from day 12 onward postimmunization, there is a coincidence of axonal injury and myelin damage with the first parenchymal infiltration of T cells. Surprisingly, at a time when parenchymal T cells are barely detectable, a high degree of astrocytic response, extensive axonal injury, and myelin damage can be observed, hence suggesting that only a small number of infiltrated T cells are required to elicit substantial injury [42]. During the progressive phase of NOD-EAE, activated microglia, CNS-infiltrating macrophages, and astrocytes produce tumor necrosis factor alpha TNFα and nitric oxide, which have direct neurotoxic effects [43, 44]. In addition, the production of the chemokine CCL2 contributes to disease pathogenesis due to recruitment of peripheral macrophages into the CNS.

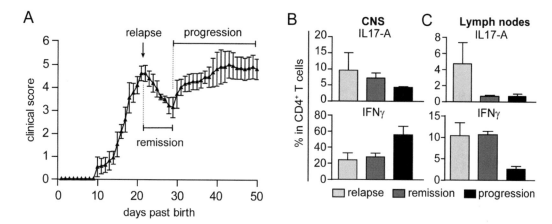

Fig. 2 NOD EAE mice develop chronic disease without clearance of pathogenic T cells from the CNS EAE in NOD mice shows three stages: first relapse, remission, and chronic phase (**a**). Mice develop stable disease score from day 30 onwards, $n = 15$. Pro-inflammatory Th1 (producing IFNγ) and Th17 cells (producing IL-17A) are cleared from the periphery, $n = 3$ (**b**). In the CNS, Th17 cells are diminishing during disease course, but Th1 cells are resistant in the CNS during chronic disease stage, $n = 3$ (**c**)

Due to the discussed neuropathological features and the two-phasic disease course with an initial relapse with partial remission followed by a chronic progressive phase, the model is clearly coming more and more into the focus of MS research. In our hands, however, all mice develop symptoms (disease incidence = 100%) to different extents, but with similar disease course (Fig. 2a, disease score shown with SD) with a peak around day 20, which is followed by a remission phase. From day 30 onward, the mice develop a stable chronic disease course (Fig. 2a). In line with the OSE model, limpness of tail is not the first manifesting clinical symptom and depending on laboratory or animal facility conditions and disease courses, onset as well as incidence can vary [45, 46].

2.4.2 Material and Methods

Mice

The NOD/ShiLtJ strain (commonly called NOD, [47]) is a polygenic model for autoimmune type 1 diabetes. Marked decreases in pancreatic insulin content occur in females at about 12 weeks of age and several weeks later in males, however if immunized with CFA these mice are resistant to diabetes [48].

Procedure of Disease Induction

At day zero, 8–10 week old female mice are immunized s.c. with 100 μg MOG per 4 mg CFA in 100 μl PBS in one flank under isoflurane anesthetic. In addition, mice also receive 200 ng PTX in 100 μl PBS i.p. after MOG injection and 2 days later. According to animal facility-inherent individual differences, the concentrations of MOG or PTX have to be adjusted to induce a proper EAE phenotype; as it is the case also for the "classical" EAE.

Handling and Classifying Criteria

Since NOD mice are extremely lively and fast, proper monitoring is recommended in order to avoid overlooking subtle clinical signs. Scoring criteria are similar to the OSE mice, as described above in Table 1.

2.4.3 Advantages and Limitations of the Model

Advantages

1. The NOD-EAE model is the only model with one relapse followed by a remission phase, which then transforms into a chronic progressive disease course.

2. The disease course is homogeneous between mice.

3. The neuropathology reflects common features of the human disease.

Limitations of the Model

1. The NOD mice are very lively and fast moving. If you consider doing behavior testing, you have to take into account that some tests might not be feasible with this strain. In addition, NOD mice get deaf with age, so behavior tests involving sounds are not applicable.

2. These mice tend to show more obvious injection wounds than C57BL/6 mice due to the white fur color. Hence, topic treatment of the wounds is recommended to reduce the risk of severe infection of the wound, which can be an exclusion criterion.

3. The NOD background has an immune phenotype which consist of defects in antigen presentation, T lymphocyte repertoire, natural killer cell function, macrophage cytokine production, wound healing, and complement 5. Depending on your immunological research questions, you might need to consider a full screen of the immune cell subsets before and during disease.

4. Limited use of the NOD mouse strain as host for genetic manipulation.

3   Concluding Remarks

In this chapter, we summarize a variety of chronic murine EAE models and highlight two particular chronic progressive models. These models allow us to address new scientific questions that deal with the identification of new targets in the context of progression and to test novel treatment strategies to prevent or delay disease progression. In the end, it should be emphasized that the choice of the optimal animal model primarily depends on the scientific question to be addressed. Before choosing a model, you should carefully study the literature about advantages and limitations of the individual models. However, always keep in mind that even the best animal model is still a model and not the human disease.

Acknowledgments

We would like Annika Engbers and Andrea Pabst for technical assistance with the genotyping. A special thanks is also given to Kathrin Koch for the support by monitoring the mice. This work is supported by the SFB CRC TR128 provided by the DFG and the Innovative Medical Research of the University Münster.

References

1. Lassmann H, Bradl M (2017) Multiple sclerosis: experimental models and reality. Acta Neuropathol 133:223–244

2. Burrows DJ, McGown A, Jain SA et al (2019) Animal models of multiple sclerosis: from rodents to zebrafish. Mult Scler 25:306–324

3. Kuerten S, Angelov DN (2008) Comparing the CNS morphology and immunobiology of different EAE models in C57BL/6 mice - a step towards understanding the complexity of multiple sclerosis. Ann Anat 190:1–15

4. Kuerten S, Wunsch M, Lehmann PV (2013) Longitudinal T cell-derived IFN-gamma/IL-17 balances do not correlate with the disease course in two mouse models of experimental autoimmune encephalomyelitis. J Immunol Methods 398–399:68–75

5. Soellner IA, Rabe J, Mauri V et al (2013) Differential aspects of immune cell infiltration and neurodegeneration in acute and relapse experimental autoimmune encephalomyelitis. Clin Immunol 149:519–529

6. Skripuletz T, Bussmann JH, Gudi V et al (2010) Cerebellar cortical demyelination in the murine cuprizone model. Brain Pathol 20:301–312

7. Lindner M, Fokuhl J, Linsmeier F et al (2009) Chronic toxic demyelination in the central nervous system leads to axonal damage despite remyelination. Neurosci Lett 453:120–125

8. Lindner M, Heine S, Haastert K et al (2008) Sequential myelin protein expression during remyelination reveals fast and efficient repair after central nervous system demyelination. Neuropathol Appl Neurobiol 34:105–114

9. Boretius S, Escher A, Dallenga T et al (2012) Assessment of lesion pathology in a new animal model of MS by multiparametric MRI and DTI. NeuroImage 59:2678–2688

10. Baxi EG, DeBruin J, Tosi DM et al (2015) Transfer of myelin-reactive Th17 cells impairs endogenous remyelination in the central nervous system of cuprizone-fed mice. J Neurosci 35:8626–8639

11. Patrikios P, Stadelmann C, Kutzelnigg A et al (2006) Remyelination is extensive in a subset of multiple sclerosis patients. Brain 129:3165–3172

12. Kuhlmann T, Miron V, Cuo Q et al (2008) Differentiation block of oligodendroglial progenitor cells as a cause for remyelination failure in chronic multiple sclerosis. Brain 131:1749–1758

13. Dendrou CA, Fugger L, Friese MA (2015) Immunopathology of multiple sclerosis. Nat Rev Immunol 15(9):545–558

14. Kuerten S, Lichtenegger FS, Faas S et al (2006) MBP-PLP fusion protein-induced EAE in C57BL/6 mice. J Neuroimmunol 177:99–111

15. Kuerten S, Kostova-Bales DA, Frenzel LP et al (2007) MP4- and MOG:35-55-induced EAE in C57BL/6 mice differentially targets brain, spinal cord and cerebellum. J Neuroimmunol 189:31–40

16. Prinz J, Karacivi A, Stormanns ER et al (2015) Time-dependent progression of demyelination and axonal pathology in MP4-induced experimental autoimmune encephalomyelitis. PLoS One 10:e0144847

17. Kuerten S, Pauly R, Rottlaender A et al (2011) Myelin-reactive antibodies mediate the pathology of MBP-PLP fusion protein MP4-induced EAE. Clin Immunol 140:54–62

18. Kuerten S, Javeri S, Tary-Lehmann M et al (2008) Fundamental differences in the dynamics of CNS lesion development and composition in MP4- and MOG peptide 35-55-induced experimental autoimmune encephalomyelitis. Clin Immunol 129:256–267

19. Tsunoda I, Kurtz CI, Fujinami RS (1997) Apoptosis in acute and chronic central nervous system disease induced by Theiler's murine encephalomyelitis virus. Virology 228:388–393

20. Tsunoda I, Fujinami RS (2010) Neuropathogenesis of Theiler's murine encephalomyelitis virus infection, an animal model for multiple sclerosis. J Neuroimmune Pharmacol 5:355–369

21. DePaula-Silva AB, Hanak TJ, Libbey JE et al (2017) Theiler's murine encephalomyelitis virus infection of SJL/J and C57BL/6J mice:

models for multiple sclerosis and epilepsy. J Neuroimmunol 308:30–42

22. Clatch RJ, Miller SD, Metzner R et al (1990) Monocytes/macrophages isolated from the mouse central nervous system contain infectious Theiler's murine encephalomyelitis virus (TMEV). Virology 176:244–254

23. Oleszak EL, Chang JR, Friedman H et al (2004) Theiler's virus infection: a model for multiple sclerosis. Clin Microbiol Rev 17:174–207

24. Rodriguez M, Leibowitz JL, Lampert PW (1983) Persistent infection of oligodendrocytes in Theiler's virus-induced encephalomyelitis. Ann Neurol 13:426–433

25. Krishnamoorthy G, Lassmann H, Wekerle H et al (2006) Spontaneous opticospinal encephalomyelitis in a double-transgenic mouse model of autoimmune T cell/B cell cooperation. J Clin Invest 116:2385–2392

26. Bettelli E, Baeten D, Jäger A et al (2006) Myelin oligodendrocyte glycoprotein-specific T and B cells cooperate to induce a Devic-like disease in mice. J Clin Invest 116:2393–2402

27. Bettelli E, Pagany M, Weiner HL et al (2003) Myelin oligodendrocyte glycoprotein-specific T cell receptor transgenic mice develop spontaneous autoimmune optic neuritis. J Exp Med 197:1073–1081

28. 006912—C57BL/6-Tg(Tcra2D2,Tcrb2D2) 1Kuch/J

29. Litzenburger T, Fässler R, Bauer J et al (1998) B lymphocytes producing demyelinating autoantibodies: development and function in gene-targeted transgenic mice. J Exp Med 188:169–180

30. Berer K, Mues M, Koutrolos M et al (2011) Commensal microbiota and myelin autoantigen cooperate to trigger autoimmune demyelination - with comments. Nature 479:538–541

31. Klotz L, Kuzmanov I, Hucke S et al (2016) B7-H1 shapes T-cell-mediated brain endothelial cell dysfunction and regional encephalitogenicity in spontaneous CNS autoimmunity. Proc Natl Acad Sci U S A 113:E6182–E6191

32. Enzmann G, Adelfio R, Godel A et al (2019) The genetic background of mice influences the effects of cigarette smoke on onset and severity of experimental autoimmune encephalomyelitis. Int J Mol Sci 20:1433

33. Molnarfi N, Schulze-Topphoff U, Weber MS et al (2013) MHC class II–dependent B cell APC function is required for induction of CNS autoimmunity independent of myelin-specific antibodies. J Exp Med 210:2921–2937

34. Berer K, Martínez I, Walker A et al (2018) Dietary non-fermentable fiber prevents autoimmune neurological disease by changing gut metabolic and immune status. Sci Rep 8:1–12

35. Makino S, Kunimoto K, Muraoka Y et al (1980) Breeding of a non-obese, diabetic strain of mice. Jikken Dobutsu 29:1–13

36. Serreze DV, Leiter EH (1994) Genetic and pathogenic basis of autoimmune diabetes in NOD mice. Curr Opin Immunol 6:900–906

37. Munoz JJ, Sewell WA (1984) Effect of pertussigen on inflammation caused by Freund adjuvant. Infect Immun 44:637–641

38. Papenfuss TL, Rogers CJ, Gienapp I et al (2004) Sex differences in experimental autoimmune encephalomyelitis in multiple murine strains. J Neuroimmunol 150:59–69

39. Levy H, Assaf Y, Frenkel D (2010) Characterization of brain lesions in a mouse model of progressive multiple sclerosis. Exp Neurol 226:148–158

40. Dang PT, Bui Q, D'Souza CS et al (2015) Modelling MS: chronic-relapsing EAE in the NOD/Lt mouse strain. Curr Top Behav Neurosci 26:143–177

41. Levy Barazany H, Barazany D, Puckett L et al (2014) Brain MRI of nasal MOG therapeutic effect in relapsing-progressive EAE. Exp Neurol 255:63–70

42. Wang D, Ayers MM, Catmull DV et al (2005) Astrocyte-associated axonal damage in pre-onset stages of experimental autoimmune encephalomyelitis. Glia 51:235–240

43. Farez MF, Quintana FJ, Gandhi R et al (2009) Toll-like receptor 2 and poly(ADP-ribose) polymerase 1 promote central nervous system neuroinflammation in progressive EAE. Nat Immunol 10:958–964

44. Basso AS, Frenkel D, Quintana FJ et al (2008) Reversal of axonal loss and disability in a mouse model of progressive multiple sclerosis. J Clin Invest 118:1532–1543

45. Colpitts SL, Kasper EJ, Keever A et al (2017) A bidirectional association between the gut microbiota and CNS disease in a biphasic murine model of multiple sclerosis. Gut Microbes 8:561–573

46. Baker D, Nutma E, O'Shea H et al (2019) Autoimmune encephalomyelitis in NOD mice is not initially a progressive multiple sclerosis model. Ann Clin Transl Neurol 6:1362–1372

47. 001976—NOD/ShiLtJ

48. McInerney MF, Pek SB, Thomas DW (1991) Prevention of insulitis and diabetes onset by treatment with complete Freund's adjuvant in NOD mice. Diabetes 40:715–725

Chapter 9

Electrophysiological Measurements for Brain Network Characterization in Rodents

Manuela Cerina, Luca Fazio, and Sven G. Meuth

Abstract

Applying electrophysiological measurements to MS diagnosis and follow-up is very efficacious for developing hypotheses and identifying potential pathophysiological mechanisms of the disease that subsequently can be targeted in animal models in an attempt to achieve "reverse translation." Accordingly, canonical translational research can be performed with animal models and electrophysiological readouts to identify biomarkers or screen new compounds. An increasing number of electrophysiological results obtained from a variety of MS animal models shows altered neuronal activity at single-cell and network levels that can be associated with cytomorphological and behavioral changes. Therefore, analyzing neuronal activity in patients and animal models is the key to understanding the mechanisms underlying the disease. In this chapter, we review some electrophysiological methods that are applied to ex vivo and in vivo studies that can be employed to achieve this scope. Moreover, some exemplary results produced by taking advantage of MS animal models are presented, and limitations and pitfalls of the methods are briefly discussed.

Key words Multiple Sclerosis, MS animal models, Cuprizone, Lysolecithin, Demyelination, Remyelination, In vivo electrophysiology, Single-unit activity, Voltage-sensitive dye, Single-cell recordings, Patch-clamp, Network activity, Altered excitability, Thalamus, Auditory cortex, Thalamocortical system

1 Introduction

Investigating and understanding the pathophysiology behind autoimmune diseases like multiple sclerosis (MS) is very complex. Pathophysiological hallmarks including activation of the immune system, damage to the blood–brain barrier, and consequent infiltrations into the central nervous system (CNS) coincide to induce inflammation and neurodegeneration associated with myelin loss and synthesis [1]. Additional variables that render the disease even more complex are the time of lesion occurrence and lesion location, extension, and number, which normally define the degree of locomotor and cognitive symptoms observed in patients [2, 3]. Experimental studies in human patients are possible but challenging due to the high number of limitations [4]. However, valuable

Sergiu Groppa and Sven G. Meuth (eds.), *Translational Methods for Multiple Sclerosis Research*, Neuromethods, vol. 166, https://doi.org/10.1007/978-1-0716-1213-2_9, © Springer Science+Business Media, LLC, part of Springer Nature 2021

information can be obtained in longitudinal studies by collecting imaging data [2, 5–7], by analyzing the cerebrospinal fluid and blood, and by performing neuropsychological assessments [8] associated with the analysis of neuro-axonal functionality, for example, measuring visual evoked potentials [9]. In this respect, applying electrophysiological measurements to MS diagnosis and follow-up is very efficacious for developing relevant hypotheses and identifying potential pathophysiological mechanisms of the disease that can be targeted in animal models in an attempt to achieve "reverse translation" [10]. Accordingly, canonical translational research can be performed with animal models and electrophysiological readouts to identify biomarkers or screen new compounds [10, 11]. There is increasing evidence for changes in neuronal network activity in the CNS of MS patients which often occur earlier than locomotor deficits [2, 12–16]. Indeed, while locomotor deficits are mainly associated with lesions in the spinal cord, early grey matter lesions might be responsible for altered cognition [17–20]. Along with these findings, an increasing number of electrophysiological measures obtained from a variety of MS animal models reveal altered neuronal activity at single-cell and network level, either as dampened excitability due to myelin loss [3, 21–23] or increased excitability due to demyelinating and/or neuroinflammatory insults [21–24]. Consequently, analyzing neuronal activity in patients and animal models is the key to understanding the mechanisms introduced above.

In this chapter, we review some electrophysiological methods that are applied to ex vivo and in vivo animal studies that can be employed to achieve this scope. Some of the exemplary results presented here are obtained from well-established animal models of MS. Specifically, we will briefly refer to the cuprizone model of general demyelination and remyelination [25, 26] as well as to the lysolecithin model which mimics focal demyelination and remyelination [27, 28]. The compounds that are used to induce both experimental models target oligodendrocytes and thereby facilitate demyelination [25, 27, 28]. However, once the compounds are finally metabolized or their administration is terminated, spontaneous remyelination occurs, with the advantage that it is largely independent of adaptive immune system activation [25, 27–30].

2   Electrophysiology In Vitro (Ex Vivo)

2.1   Materials and Methods

This section will provide information on equipment and appropriate solutions required to perform electrophysiological recordings in ex vivo models of MS, preferentially using brain slices.

Table 1
The composition in mM of the different solutions used for the electrophysiological experiments described in this chapter

Composition (mM)	Cutting solution	Incubation solution-ACSF	Composition (mM)	K-gluconate-based intracellular solution	High Cl⁻ intracellular solution
NaCl	–	120	NaCl	10	10
KCl	2.5	2.5	K-gluconate	88	–
NaH$_2$PO$_4$	1.25	1.25	K$_3$-citrate	20	–
NaHCO$_3$	–	22	KCl	–	110
C$_6$H$_{12}$O$_6$	10	25	EDTA	–	11
CaCl$_2$	0.5	2	HEPES	10	10
MgSO$_4$	10	2	BAPTA	3	–
PIPES	20	–	Phosphocreatine	15	15
Saccharose	200	–	MgCl$_2$	1	1
			CaCl$_2$	0.5	0.5
			Mg-ATP	3	3
			Na-GTP	0.5	0.5

2.1.1 Brain Slice Preparation

Certain conditions have to be met to prepare acute living brain slices. Briefly, animals are decapitated and the brain is quickly removed and transferred to a cold "cutting solution" (*see* Table 1 for composition details) set to a pH of 7.35 with carbogen (95% O$_2$ + 5% CO$_2$) and maintained at 4 °C [31]. Cutting is performed with the vibratome Leica VT 1200 S. Depending on the scientific question to be answered, orientation and thickness of the slices can vary between 250 and 300 μm (*see* Fig. 1a and Subheading 3 for details). In our laboratory, we prepare parasagittal slices containing a preserved auditory thalamocortical system by cutting the brain at an angle of 25° with the help of an agar ramp, as described by Broicher and colleagues [3, 32]. The auditory thalamocortical system consists of two very well organized grey matter structures (the cortex and the thalamus) reciprocally connected by one white-matter fiber tract and provides an ideal basis to investigate the consequences of demyelination and remyelination as well as cognitive impairment [33, 34]. Next, the slices are quickly transferred to a custom-made incubation chamber filled with artificial cerebrospinal fluid (ACSF, *see* Table 1 for composition details) set to a pH of 7.35 with carbogen and continuously carbonated to keep the slices alive. For further measurements, slices are transferred into the recording chamber underneath the objective of an upright microscope (Axioexaminer, Zeiss, Germany). The region of interest is identified visually, using the Paxinos and Franklin Mouse Brain Atlas where necessary [35].

2.1.2 Single-Cell Electrophysiological Recordings

Electrophysiological recordings are carried out with glass-pipette electrodes pulled from borosilicate glass (GC150TF-10; Clark Electromedical Instruments, Pangbourne, UK) that are connected to an EPC-10 double amplifier (HEKA Elektronik, Lambrecht, Germany), as described previously [31, 36]. The pipette is filled with a solution composed to resemble the intracellular milieu. Mostly, we use K-gluconate-based intracellular solutions (*see* Table 1 for composition details) set to a pH of 7.25 with KOH and an osmolarity of 295 mOsmol/kg. However, depending on the kind of recording to be carried out, other components can be added or substituted to answer a determined scientific question. In this context, it may be useful to mention one example: substituting K-gluconate with a high concentration of KCl (*see* Table 1 for composition details) would alter the intracellular environment by modifying the Cl^- gradient and inverting the reversal potential of GABAergic receptors, thereby allowing the recording of GABAergic activity at potentials close to the resting membrane potential of the cell [37]. Similar considerations occur for electrode resistance, which depends on the size of the tip of the borosilicate pipette. Typically, neurons are measured using a resistance of 3–4 MΩ, with a series resistance in the range of 5–15 MΩ (compensation $\geq 25\%$).

2.1.3 Patterns of Single-Cell Recordings

An initial characterization of neuronal functionality is based on analyzing intrinsic neuron excitability, which can be affected by neuroinflammatory and neurodegenerative events [24, 38–40]. In the CNS, we can identify and investigate two main mechanisms responsible for excitability, namely, the generation of action potentials (APs) in response to given stimuli of a certain intensity and the balance between the glutamatergic and GABAergic system [41, 42]. Here, we focus on analyzing AP generation. In acute brain slices, AP recordings are performed in current clamp mode mainly by performing an input/output protocol as shown in Fig. 1b: series of increased currents are applied to the cell via the patch pipette (input) and the number of APs produced in a period of 2.5 s was recorded (output). Ideally, a physiological response would include an increase in the number of APs proportional to the increased stimulus. This protocol provides information on basic neuronal excitability (e.g., membrane resistance and capacitance) and the functionality of the ion channels involved in AP generation or resting membrane potential maintenance such as voltage-activated Na^+, Ca^{2+}, and K^+ channels [21, 43, 44].

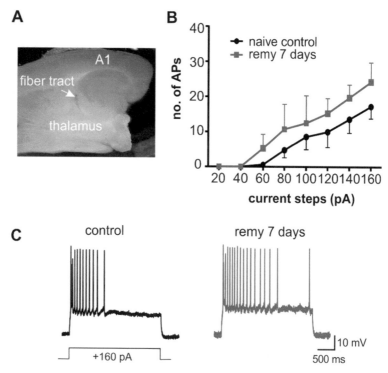

Fig. 1 Patterns of single-cell electrophysiological recordings. (**a**) Microphotograph of a parasagittal murine brain slice containing an intact thalamocortical system consisting of the primary auditory cortex (**A1**), the thalamus, and the white matter fiber tract connecting both grey matter regions. Electrophysiological recordings were performed in **A1**. (**b**) Input/output graph showing the number of action potentials (APs) generated in response to depolarizing current steps of increasing intensity (from +20 pA to +160 pA). (**c**) Exemplary traces showing the APs generated in response to +160 pA for control and remyelinated mice (black and red traces, respectively) in the cuprizone model of general demyelination and remyelination

2.1.4 Performing Neuronal Network Electrophysiological Recordings

Results and data collected from single neuron analysis in different animal models pave the ground for understanding neuronal network level events since these result from synchronized activities involving various neuronal populations. Such activities can also be analyzed in ex vivo preparations, obtained as described above, by combining local field potential (LFP) [45] and voltage-sensitive dye imaging (VSDi) [32, 46] experiments. For human slices, these techniques have been extensively described by Broicher and Speckmann in a previous issue of Neuromethods [47]. Murine brain slices containing an intact auditory thalamocortical system are produced as described above (thickness of 500 μm, Fig. 2a) and incubated with the voltage-sensitive dye RH-795 (Invitrogen, Karlsruhe, Germany; 12 μg/ml) at 31 °C for 60 min with continuous oxygenation. Then, slices are transferred to a holding chamber (oxygenated ACSF, 31 °C) and allowed to rest before the

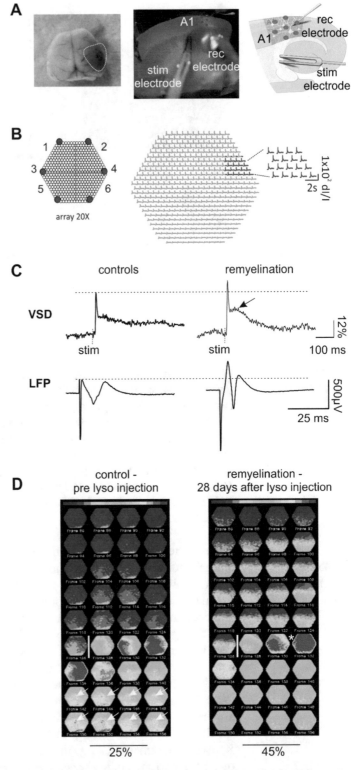

Fig. 2 Patterns of neuronal network electrophysiological recordings. (**a**, left) Photograph of an intact brain, showing the extension of the lysolecithin injections in the cortex reflected by the blue area and delimited by the dashed area

experiments started. Recordings were performed in a submerged chamber in an inverted microscope (Zeiss, Göttingen, Germany). Optical recordings are performed in A1 as described previously [32] and governed by the software Neuroplex (newest version: TurboSM, Redshirt Imaging, Decatur, GA, USA). Fluorescence changes are viewed with a hexagonal photodiode array composed of 464 elements through a $20\times$ objective detecting an area of 0.416 mm^2 that anatomically corresponds to the auditory cortex in each given slice (Fig. 2a, b). The parameter sampling interval, the length of the recordings, and the fluorescence specifications of the dye are extensively described elsewhere and were maintained in an identical manner [22, 32, 47]. Optical signals are recorded in parallel with electrical signals induced by a single electrical stimulation conveyed via a custom-made bipolar electrode placed within the thalamocortical projections onto the cortex, in an attempt to mimic incoming subcortical stimuli (Fig. 2). Stimulation is triggered by the software Axoscope (Molecular Devices, Sunnyvale, CA, USA) and responses are detected with two recording electrodes filled with ACSF and positioned in the auditory cortex. Pulse amplitude for individual stimulation ($\sim$500 µA) and stimulation length (150 µs) are set to obtain optimal responses in our experimental settings.

Fig. 2 (continued) (lysolecithin is coinjected together with Evans blue dye to identify the injection site during slice preparation. (**a**, middle) Microphotograph of a parasagittal murine brain slice containing an intact thalamocortical system consisting of the primary auditory cortex (**A1**), the thalamus, and the white matter fiber tract connecting both grey matter regions. To mimic incoming cortical stimuli, the fiber tract is stimulated with a bipolar electrode (stim electrode) while local field potential (LFP) responses are recorded with glass pipettes (rec electrodes); and voltage-sensitive dye imaging (VSDi) is depicted by the hexagonal structure represented by the six red dots in **A1** (scheme in **a**, right). (**b**) Schematic representation of the organization of 464 diodes forming the hexagonal array (left panel) and an exemplary array response to electrical stimulation in the cortex; single diodes highlighted in red on the left part of the panel. (**c**) Exemplary VSDi and LFP traces (upper and lower row, respectively) depicting a cortical response to electrical stimulation in control animals (left) and animals subjected to lysolecithin injections and allowed to remyelinate for 28 days. Of note, both VSDi and LFP responses are increased in lysolecithin treated animals in comparison to control. (**d**) Pseudocolored fluorescence maps showing the spatiotemporal pattern of activation and propagation of the cortical response in control animals (map on the left) and animals subjected to lysolecithin injections and allowed to remyelinate for 28 days (map on the right). Lysolecithin treated animals show an abnormal response upon stimulus (time point of occurrence indicated by the vertical white line), reflected by the preponderance of red colors (indicated by the white asterisk) and absence of propagation which is instead observed in the control map (indicated by the white arrows). Directionality of maps: left to right, interval between maps: 2.5 ms.

Choosing the thalamocortical system as a model for our investigations had the advantage that we could focus on structured regions like the cortex which consists of functionally distinct layers. Each layer contains different neuronal populations and/or receives certain connections and is therefore activated following a specific hierarchy upon stimulus arrival [3, 22, 48].

Optical signals were expressed as fractional changes of the fluorescence from the resting light intensity ($I_{rest} - I_{recording}/I_{rest}$; dI/I in the text). Data analysis was performed manually using Neuroplex (Redshirt Imaging, USA) and custom-made MATLAB scripts. Amplitude values were analyzed by taking single diodes to calculate the difference between baseline and peak and by using information from all diodes to build spatiotemporal fluorescence pseudocolored maps scaled to the maximum signal of the recording session. The field potential amplitude simultaneously recorded with the optical signals was analyzed as distance in μV from the most negative to the most positive peak of the recordings ("peak-to-peak" analysis).

2.2 Exemplary Results

Increasing evidence associates MS pathophysiology with altered neuronal excitability and network activity both in humans and animal models [22–24, 49, 50]. Numerous studies have demonstrated neuronal damage via direct cytotoxic attacks operated by $CD8^+$ T cells, which dampen intrinsic excitability and lead to neuronal apoptosis [51–53]. More recently, new evidence has been gathered—for MS models—on novel mechanisms as the cause for neuronal functional impairment. These mechanisms depend on (1) an altered Ca^{2+} concentration resulting from altered glutamatergic synaptic activity [24, 54, 55], or (2) an altered expression of ion channels on the neuronal soma or in the axon, which becomes exposed after the loss of myelin sheaths [21, 40, 56]. The ion channel contribution was tested several times using electrophysiological approaches in isolated neurons and slices [43, 44]. Accordingly, and by taking advantage of the cuprizone model, we detected general demyelination and dampened cellular excitability [3, 57], while we also found a transitory period of hyperexcitability of the thalamocortical network [3, 22, 23] characterized by an increased number of APs produced upon application of the input/output protocol, as described above (see Fig. 1b, c). Remarkably, altered excitability occurred at early time points after immunization as well as at later time points.

We applied such techniques to murine slices obtained from animals treated with cuprizone and measured abnormally increased LFP amplitudes in the auditory cortex of mice in the early phases of demyelination [3, 22]. This was corroborated by simultaneously recorded VSDi and seemed to depend on neurodegenerative rather than immune-mediated mechanisms [22, 23, 58]. Moreover, in our attempt to render the animal model experiment more translational, we opted for inducing focal inflammatory [59] or demyelinating lesions [27, 28] in white and grey matter regions to better mimic the occurrence of such events in humans. Interestingly, we observed increased excitability in the cortex of focally demyelinated mice during remyelination both upon VSDi and LFP analysis (Fig. 2c). After lysolecithin is metabolized no more harm is exerted on oligodendrocytes and spontaneous remyelination can occur within 28 days following the injection, similar to the cuprizone model. Here we show that even in thalamocortical brain slices obtained from lysolecithin-treated mice, a period of hyperexcitability can be observed upon remyelination (Fig. 2c, d).

3   Drawbacks and Limitations

The preparation of slices can be hindered by several factors; these include the following:

1. The temperature of the solutions and the time interval between brain removal and cutting procedure: in this initial step the brain needs to be cooled down to diminish metabolic rates and thereby apoptotic events.

2. The choice of slice thickness and cutting orientation: if the target is an intact system consisting of white and grey matter regions, it is important to consider the orientation of slice preparation (which nuclei will be present in the same preparation) and the thickness (to preserve a number of fibers sufficient to connect and convey information between two regions). However, a thickness of 500 μm should not be exceeded as maintaining slice oxygenation becomes difficult.

3. The usage of carbogen to maintain living slices: a composition of 95% O_2 and 5% CO_2 needs to be maintained and constantly provided to the slices; when recordings involving pH-sensitive targets are planned, choosing a HEPES based solution that requires only O_2 might be a better option.

4. Depending on the model of choice for investigating MS, reproducibility rates might be low: models requiring intraparenchymal injections either of cytokines or demyelinating compounds can damage the brain, making it difficult to record healthy living neurons in the proximity of the injection site; while under a fluorescence microscope, autofluorescence can be produced.

5. The careful and unbiased interpretation of data might be difficult, despite a very well-controlled environment, as narrowing-down mechanisms or specific events to certain pathways or cell types can be misleading.

4   Electrophysiology In Vivo

Some of the limitations described above can be overcome by performing the electrophysiological approaches in vivo. Implanting electrodes in targeted areas and recording in freely behaving mice allows us to combine several readouts and associate neuronal activity with specific stimuli and/or behavioral responses. For investigating other diseases, it is a very well-established technique [31, 60, 61].

4.1 Material and Methods

4.2 Implantation of Electrodes for Single-Unit Activity Recordings

Microwire arrays (one array, eight electrodes, and one reference/array per brain region; Stablohm 650; California Fine Wire, USA) are implanted under stereotaxic control (David Kopf Instruments, USA) in deeply anesthetized mice (2% isoflurane in O_2). Implants can be placed unilaterally or bilaterally depending on the targeted regions. Here we implant them in the auditory thalamocortical system (according to Paxinos coordinates [35]). The tip of each wire is gold plated by passing a cathodal current of 1 μA while wires are submerged in a gold solution. This procedure reduces the impedance to a range of 150–300 kΩ to record single-unit activity and increases the signal-to-noise ratio. Electrodes are fixed with dental cement (Pulpdent-GlassLute, Corporation Watertown, MA; USA). An additional grounding electrode is positioned close to the midline over the cerebellar region (5.8/0.5 mm from bregma) in the right hemisphere.

Animals are allowed to recover from surgery for 1 week before starting any model inducing processes (MOG_{35-55} immunization or cuprizone diet) and before starting the first recordings. Given our focus on the thalamocortical system, we take advantage of the tonotopic organization of the auditory thalamus and cortex [23, 62]. For this purpose, we implant electrodes in tonotopic regions known to be responsive to high frequencies and presented the mice with alternating low- and high-frequency tones (2.5 kHz and 10 kHz, respectively, both at 85 dB) to observe changes in response to the high-frequency tones in frequency-related neurons.

4.3 Neuronal Activity Recording and Analysis

Neuronal activity is recorded with a Multichannel Amplifier System (Alpha Omega, Israel) and stored on a personal computer (IBM). Unit activities are bandpass-filtered at 100 Hz to 20 kHz, at a sampling rate of 40 kHz. Spikes of individual neurons are sorted by time-amplitude window discrimination and principal component analysis (Offline Sorter, Plexon Inc., Dallas, TX, USA) and verified through quantification of cluster separation, as described

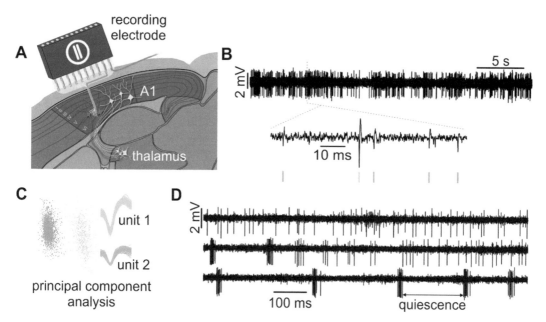

Fig. 3 Electrophysiology in vivo. (**a**) Representative scheme showing a parasagittal view of the auditory thalamocortical system with an implanted electrode for single unit activity recording in the primary auditory cortex. The same experimental approach is used to record thalamic neurons. (**b**) Exemplary traces of extracellularly recorded APs recorded in the thalamus show typical single units waveforms when on an expanded time scale. This allows spike sorting (green and yellow vertical lines). (**c**) After spike identification, waveforms were clustered and sorted into units based on principal component analysis. (**d**) Exemplary extracellular recordings representing the two distinct thalamocortical firing patterns: tonic firing (upper trace) characterized by single APs rhythmically occurring and burst firing (bottom trace) characterized by grouped APs and high frequency spiking separated by a quiet period of 50 ms duration. The two firing patterns can also occur at the same time (middle trace). (The figures presented in panels (**b–d**) were modified from Fig. 1 published in Cerina et al., BJP, 2015)

previously [3, 60, 63, 64]. For exemplary results, refer to Fig. 3b, c. Additional analyses of single-unit activity, which can be performed on sorted neurons, include basal- and stimulus-evoked activity, z-scores, and firing-latencies. Taken together, the information gathered provide useful insights to classify neuronal populations and delivered a basis to distinguish between responsive and nonresponsive neurons [23, 31]. In more detail, by identifying neurons that change their z-score $\geq$ 1.96-fold from baseline upon stimulus presentation, we could assess frequency-specific neurons and longitudinally follow the effects of inflammation and/or demyelination. We perform laboratory analysis with custom-made MATLAB (MathWorks) scripts, but it is also possible to use built-in routines from the NeuroExplorer software provided by Plexon instead. The burst-spike analysis is an additional analysis that can be performed in the thalamus but not in the cortex. It takes advantage of the ability of thalamic nuclei to switch between two different firing patterns, namely, tonic and bursting [65, 66]. While the tonic pattern is characterized by the rhythmic production of action potentials and associated with the process of relaying information to interconnected cortical regions [66, 67], the bursting pattern is characterized by periods of high-frequency action potential production alternating with periods of quiescence (Fig. 3d) [31, 68, 69]. Bursting is associated with sleep oscillations but has also been described as a mechanism that interrupts the faithful relay of information in the thalamocortical system [70–72]. Investigating pathophysiological-dependent switches between the two patterns gave us important insights for understanding burst-like neuronal activity outside sleep and upon demyelination. This could be one of the mechanisms underlying the cognitive impairment we observed in the mouse model of cuprizone [23], as has been demonstrated for other conditions such as anxiety [60, 61] and hyperexcitability-related diseases including epilepsy [73]. Single-unit (identified by principal component analysis) spike patterns are analyzed using NeuroExplorer (Plexon Inc., Dallas, TX, USA). Burst spikes are defined by the presence of a preceding silent period (>50 ms) followed by at least four high-frequency spikes (125–500 Hz) with a maximum first interspike interval (ISI) of 4 ms and a maximum final ISI of 8 ms Fig. 3d [31].

5 Conclusions

Taken together, investigating neuronal activity at the single-cell or network level is a straightforward approach to gain a deeper understanding of physiological and pathophysiological mechanisms of CNS diseases such as MS. However, employing just one technique, as well as investigating only one animal model, cannot provide sufficient information when trying to unravel complex mechanisms. Therefore, a gradual proceeding is necessary to start dissecting the underlying mechanisms. Moreover, several readouts need to be combined to extensively characterize the model and, thus, the disease. In this respect, combining in vivo electrophysiology and behavioral measures with ex vivo experiments, including histology and immune assays in addition to single-cell and/or network electrophysiology, is an important practice to build a solid basis for potential translational approaches.

Acknowledgments

We would like to thank Ms. Monika Wart and Mr. Frank Kurth for the excellent technical assistance, Ms. Sophie Gothan for taking some of the pictures used in this chapter, and Dr. Venu Narayanan for performing the lysolecithin injections. Moreover, we thank Dr. Lisa Epping, Dr. Stjepana Kovac, and Prof. Erwin-Josef Speckmann for their precious scientific feedback.

References

1. Compston A, Coles A (2008) Multiple sclerosis. Lancet 372:1502–1517

2. Gamboa OL, Tagliazucchi E, von Wegner F et al (2014) Working memory performance of early MS patients correlates inversely with modularity increases in resting state functional connectivity networks. NeuroImage 94:385–395

3. Cerina M, Narayanan V, Göbel K et al (2017) The quality of cortical network function recovery depends on localization and degree of axonal demyelination. Brain Behav Immun 59:103–117

4. Krämer J, Brück W, Zipp F et al (2019) Imaging in mice and men: pathophysiological insights into multiple sclerosis from conventional and advanced MRI techniques. Prog Neurobiol 182:101663

5. Muthuraman M, Fleischer V, Kolber P et al (2016) Structural brain network characteristics can differentiate CIS from early RRMS. Front Neurosci 10:14

6. Droby A, Yuen KSL, Muthuraman M et al (2015) Changes in brain functional connectivity patterns are driven by an individual lesion in MS: a resting-state fMRI study. Brain Imaging Behav 10(4):1117–1126

7. Deppe M, Müller D, Kugel H et al (2013) DTI detects water diffusion abnormalities in the thalamus that correlate with an extremity pain episode in a patient with multiple sclerosis. NeuroImage Clin 2:258–262

8. Giffroy X, Maes N, Albert A et al (2016) Multimodal evoked potentials for functional quantification and prognosis in multiple sclerosis. BMC Neurol 16:83

9. Silbermann E, Wooliscroft L, Bourdette D (2017) Using the anterior visual system to assess neuroprotection and Remyelination in multiple sclerosis trials. Curr Neurol Neurosci Rep 18:49

10. 't Hart BA (2015) Reverse translation of failed treatments can help improving the validity of preclinical animal models. Eur J Pharmacol 759:14–18

11. Lightfoot J, Bamman M, Booth F (2017) Translation goes both ways: the power of reverse translation from human trials into animal models. Transl J Am Coll Sport Med 2:29

12. Deppe M, Krämer J, Tenberge J-G et al (2016) Early silent microstructural degeneration and atrophy of the thalamocortical network in multiple sclerosis. Hum Brain Mapp 37(5):1866–1879

13. Calabrese M, Agosta F, Rinaldi F et al (2009) Cortical lesions and atrophy associated with cognitive impairment in relapsing-remitting multiple sclerosis. Arch Neurol 66:1144–1150

14. Yu HJ, Christodoulou C, Bhise V et al (2012) Multiple white matter tract abnormalities underlie cognitive impairment in RRMS. NeuroImage 59:3713–3722

15. Shu N, Duan Y, Xia M et al (2016) Disrupted topological organization of structural and functional brain connectomes in clinically isolated syndrome and multiple sclerosis. Sci Rep 6:29383

16. Sumowski JF, Benedict R, Enzinger C et al (2018) Cognition in multiple sclerosis: state of the field and priorities for the future. Neurology 90:278–288

17. Schlaeger R, Papinutto N, Panara V et al (2014) Spinal cord gray matter atrophy correlates with multiple sclerosis disability. Ann Neurol 76:568–580

18. Sampaio-Baptista C, Khrapitchev AA, Foxley S et al (2013) Motor skill learning induces changes in white matter microstructure and myelination. J Neurosci 33:19499–19503

19. Chiaravalloti ND, DeLuca J (2008) Cognitive impairment in multiple sclerosis. Lancet Neurol 7:1139–1151

20. Calabrese M, Filippi M, Gallo P (2010) Cortical lesions in multiple sclerosis. Nat Rev Neurol 6:438–444

21. Hamada MS, Kole MHP (2015) Myelin loss and axonal ion channel adaptations associated with gray matter neuronal hyperexcitability. J Neurosci 35:7272–7286

22. Cerina M, Narayanan V, Delank A et al (2018) Protective potential of dimethyl fumarate in a mouse model of thalamocortical demyelination. Brain Struct Funct 223(7):3091–3106

23. Narayanan V, Cerina M, Göbel K et al (2018) Impairment of frequency-specific responses associated with altered electrical activity patterns in auditory thalamus following focal and general demyelination. Exp Neurol 309:54–66

24. Ellwardt E, Pramanik G, Luchtman D et al (2018) Maladaptive cortical hyperactivity upon recovery from experimental autoimmune encephalomyelitis. Nat Neurosci 21:1392–1403

25. Matsushima GK, Morell P (2001) The Neurotoxicant , Cuprizone , as a model to study demyelination and Remyelination in the central. Brain Pathol 11:107–116

26. Skripuletz T, Gudi V, Hackstette D et al (2011) De- and remyelination in the CNS white and grey matter induced by cuprizone: the old, the new, and the unexpected. Histol Histopathol 26:1585–1597

27. Hall SM (1983) The response of the (myelinating) Schwann cell population to multiple episodes of demyelination. J Neurocytol 12:1–12

28. Blakemore WF, Franklin RJM (2008) Remyelination in experimental models of toxin-induced demyelination. Curr Top Microbiol Immunol 318:193–212

29. Döring A, Sloka S, Lau L et al (2015) Stimulation of monocytes, macrophages, and microglia by amphotericin B and macrophage colony-stimulating factor promotes remyelination. J Neurosci 35:1136–1148

30. Woodruff RH, Franklin RJ (1999) The expression of myelin protein mRNAs during remyelination of lysolecithin-induced demyelination. Neuropathol Appl Neurobiol 25:226–235

31. Cerina M, Szkudlarek HJ, Coulon P et al (2015) Thalamic Kv7 channels: pharmacological properties and activity control during noxious signal processing. Br J Pharmacol 172:3126–3140

32. Broicher T, Bidmon H-J, Kamuf B et al (2010) Thalamic afferent activation of supragranular layers in auditory cortex in vitro: a voltage sensitive dye study. Neuroscience 165:371–385

33. Groenewegen HJ, Witter MP (2004) Thalamus: The Rat Nervous System.- Oxford: Academic Press - Chapter 17, Pp. 407–453

34. Sherman SM (2012) Thalamocortical interactions. Curr Opin Neurobiol 22:575–579

35. Paxinos G, and Franklin, KB (2019) Paxinos and Franklin's the mouse brain in stereotaxic coordinates. Academic Press

36. Bista P, Meuth SG, Kanyshkova T et al (2012) Identification of the muscarinic pathway underlying cessation of sleep-related burst activity in rat thalamocortical relay neurons. Pflugers Arch 463:89–102

37. Blaesse P, Goedecke L, Bazelot M et al (2015) μ-opioid receptor-mediated inhibition of intercalated neurons and effect on synaptic transmission to the central amygdala. J Neurosci 35:7317–7325

38. Zobeiri M, Chaudhary R, Datunashvili M et al (2018) Modulation of thalamocortical oscillations by TRIP8b, an auxiliary subunit for HCN channels. Brain Struct Funct 223:1537–1564

39. Ayache SS, Créange A, Farhat WH et al (2015) Cortical excitability changes over time in progressive multiple sclerosis. Funct Neurol 30:257–263

40. Hundehege P, Fernandez-Orth J, Römer P et al (2018) Targeting voltage-dependent calcium channels with pregabalin exerts a direct neuroprotective effect in an animal model of multiple sclerosis. Neurosignals 26:77–93

41. McCormick DA, Wang Z, Huguenard J (1991) Neurotransmitter control of neocortical neuronal activity and excitability. Cereb Cortex 3:387–398

42. Turrigiano G (2011) Too many cooks? Intrinsic and synaptic homeostatic mechanisms in cortical circuit refinement. Annu Rev Neurosci 34:89–103

43. Freeman SA, Desmazières A, Fricker D et al (2016) Mechanisms of sodium channel clustering and its influence on axonal impulse conduction. Cell Mol Life Sci 73:723–735

44. Coman I, Aigrot MS, Seilhean D et al (2006) Nodal, paranodal and juxtaparanodal axonal proteins during demyelination and remyelination in multiple sclerosis. Brain 129:3186–3195

45. Crawford DK, Mangiardi M, Tiwari-Woodruff SK (2009) Assaying the functional effects of demyelination and remyelination: revisiting field potential recordings. J Neurosci Methods 182:25–33

46. Chemla S, Chavane F (2010) Voltage-sensitive dye imaging: technique review and models. J Physiol Paris 104:40–50

47. Broicher T, Speckmann E-J (2012) Living human brain slices: network analysis using voltage-sensitive dyes. Neuromethods 73:285–300

48. Winkowski DE, Kanold PO (2013) Laminar transformation of frequency organization in auditory cortex. J Neurosci 33:1498–1508

49. Mori F, Nisticò R, Nicoletti CG et al (2016) RANTES correlates with inflammatory activity and synaptic excitability in multiple sclerosis. Mult Scler 22(11):1405–1412

50. Merkler D, Klinker F, Jürgens T et al (2009) Propagation of spreading depression inversely correlates with cortical myelin content. Ann Neurol 66:355–365

51. Göbel K, Bittner S, Melzer N et al (2012) CD4 (+) CD25(+) FoxP3(+) regulatory T cells suppress cytotoxicity of CD8(+) effector T cells: implications for their capacity to limit inflammatory central nervous system damage at the parenchymal level. J Neuroinflammation 9:41

52. Meuth SG, Herrmann AM, Simon OJ et al (2009) Cytotoxic CD8+ T cell-neuron

interactions: perforin-dependent electrical silencing precedes but is not causally linked to neuronal cell death. J Neurosci 29:15397–15409

53. Göbel K, Bittner S, Cerina M et al (2015) An ex vivo model of an oligodendrocyte-directed T-cell attack in acute brain slices. J Vis Exp 96:52205

54. Mandolesi G, Grasselli G, Musumeci G et al (2010) Cognitive deficits in experimental autoimmune encephalomyelitis: neuroinflammation and synaptic degeneration. Neurol Sci 31:S255–S259

55. Centonze D, Muzio L, Rossi S et al (2010) The link between inflammation, synaptic transmission and neurodegeneration in multiple sclerosis. Cell Death Differ 17:1083–1091

56. Piaton G, Gould RM, Lubetzki C (2010) Axon-oligodendrocyte interactions during developmental myelination, demyelination and repair. J Neurochem 114:1243–1260

57. Ghaffarian N, Mesgari M, Cerina M et al (2016) Thalamocortical-auditory network alterations following cuprizone-induced demyelination. J Neuroinflammation 13:160

58. Crawford DK, Mangiardi M, Xia X et al (2009) Functional recovery of callosal axons following demyelination: a critical window. Neuroscience 164:1407–1421

59. Hundehege P, Cerina M, Eichler S et al (2019) The next-generation sphingosine-1 receptor modulator BAF312 (siponimod) improves cortical network functionality in focal autoimmune encephalomyelitis. Neural Regen Res 14:1950

60. Daldrup T, Lesting J, Meuth P et al (2016) Neuronal correlates of sustained fear in the anterolateral part of the bed nucleus of stria terminalis. Neurobiol Learn Mem 131:137–146

61. Lesting J, Geiger M, Narayanan RT et al (2011) Impaired extinction of fear and maintained amygdala-hippocampal theta synchrony in a mouse model of temporal lobe epilepsy. Epilepsia 52:337–346

62. Hackett TA, Barkat TR, O'Brien BMJ et al (2011) Linking topography to tonotopy in the mouse auditory thalamocortical circuit. J Neurosci 31:2983–2995

63. Narayanan RT, Udvary D, Oberlaender M (2017) Cell type-specific structural organization of the six Layers in rat barrel cortex. Front Neuroanat 11:91

64. Lesting J, Daldrup T, Narayanan V et al (2013) Directional theta coherence in prefrontal cortical to Amygdalo-hippocampal pathways signals fear extinction. PLoS One 8:e77707

65. Weyand TG, Boudreaux M, Guido W (2001) Burst and tonic response modes in thalamic neurons during sleep and wakefulness. J Neurophysiol 85:1107–1118

66. Steriade M, McCormick DA, Sejnowski TJ (1993) Thalamocortical oscillations in the sleeping and aroused brain. Science 262:679–685

67. Huguenard JR, McCormick DA (1992) Simulation of the currents involved in rhythmic oscillations in thalamic relay neurons. J Neurophysiol 68:1373–1383

68. Huh Y, Cho J (2013) Urethane anesthesia depresses activities of thalamocortical neurons and alters its response to nociception in terms of dual firing modes. Front Behav Neurosci 7:141

69. Huh Y, Bhatt R, Jung D et al (2012) Interactive responses of a thalamic neuron to formalin induced lasting pain in behaving mice. PLoS One 7:e30699

70. Llinás RR, Steriade M (2006) Bursting of thalamic neurons and states of vigilance. J Neurophysiol 95:3297–3308

71. Fanselow EE, Sameshima K, Baccala LA et al (2001) Thalamic bursting in rats during different awake behavioral states. Proc Natl Acad Sci U S A 98:15330–15335

72. McCormick DA (1999) Are thalamocortical rhythms the rosetta stone of a subset of neurological disorders? Nat Med 5:1349–1351

73. Steriade M (2005) Sleep, epilepsy and thalamic reticular inhibitory neurons. Trends Neurosci 28:317–324

Part III

Translation of Functional Domain Abnormalities from Human to Mice

Chapter 10

Translation of Functional Domain Abnormalities from Human to Mouse Motor System

Muthuraman Muthuraman, Dumitru Ciolac, and Venkata Chaitanya Chirumamilla

Abstract

Multiple sclerosis (MS) is characterized by frequent impairment of motor skills, with the most prominent manifestations being spasticity, gait impairment, fatigue, or disabling tremor, all highly important determinants of physical disability. The optimization of the reverse translation from humans to mice of relevant readouts of abnormal motor function could definitively improve actual models of neuroinflammation to better match human functional outcomes in MS patients. The aim of this chapter is to provide a descriptive methodological framework to approach human motor function abnormalities that could ground a translational pathway from patients to mice.

Key words Motor system, fMRI, EEG, MEG, EMG, Evoked potentials, Human–mice translation

1 Introduction

1.1 Anatomical Organization, Circuitry, and Functionality of the Motor System

1.1.1 Primary Motor Cortex (M1)

The M1 cortex is a primordial brain area involved in motor function (Fig. 1), located within the precentral gyrus of the frontal lobe and organized in a somatotopic map of the body [1, 2]. Like the primary somatosensory cortex (S1), the M1 is composed of a series of modules formed by vertical columns of neural cells. The M1 cortex is composed of six layers, the most distinctive being the V layer, which contains the giant pyramidal (Betz) cells (upper motor neurons) [2]. The output projections of the Betz cells make up approximately 30% of the corticospinal tract fibers that end on the lower motor neurons of the spinal cord [3]. The M1 area is reciprocally connected with the S1 area (mainly with proprioceptive area 2) and with the superior parietal lobule, and receives inputs from the dorsal and ventral premotor cortices and supplementary motor area [4, 5]. Cerebellar projections to M1 originate in the dentate and the interposed nuclei and are relayed in the ventral lateral nucleus of the thalamus. The M1 cortex encodes the force, direction, speed,

Sergiu Groppa and Sven G. Meuth (eds.), *Translational Methods for Multiple Sclerosis Research*, Neuromethods, vol. 166,
https://doi.org/10.1007/978-1-0716-1213-2_10, © Springer Science+Business Media, LLC, part of Springer Nature 2021

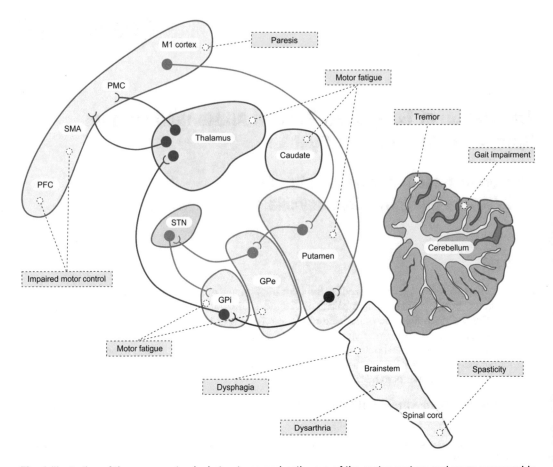

Fig. 1 Illustration of the neuroanatomical structures and pathways of the motor system and some presumable symptoms arising from their damage in MS. *GPe/i* external/internal globus pallidus, *M1* primary motor cortex, *PFC* prefrontal cortex, *PMC* premotor cortex, *SMA* supplementary motor area, *STN* subthalamic nucleus

and magnitude of voluntary movements [4]. The parameters of individual movements or simple movement sequences are encoded by distinct subpopulations of synchronized pyramidal neurons during the preparation and execution of the motor program [6]. In mice, the motor cortex is subdivided into primary (M1) and secondary (M2) motor cortices; the M1 is further subdivided into vibrissal M1 and somatic M1 [7, 8].

1.1.2 Supplementary Motor Area (SMA)

The SMA lies in front of the M1 cortical leg representation (Fig. 1), has rich connections with both cortical (M1 cortex, premotor cortex, cingulate cortex) and subcortical (basal ganglia, thalamus, cerebellum) structures, and generates fibers to the pyramidal tract [9]. Like in the premotor cortex, the body somatotopic representation in the SMA is less precisely organized than in the M1 cortex [10]. At least 10% of corticospinal tract fibers originate in the SMA. It is subdivided into SMA proper and the more rostral pre-SMA [11], which have different connectivity profiles. The SMA proper is densely connected to the M1 area and projects to the spinal cord, whereas pre-SMA has few connections with the M1 and does not project to the spinal cord. Both SMA proper and pre-SMA receive inputs from basal ganglia and cerebellum [12], while projections to these two areas originate from spatially distinct regions of globus pallidus and dentate nucleus [9]. The SMA is involved in self-initiated actions rather than movements occurring in response to external stimuli [13], regulation of body posture or bimanual coordination of the movements. It appears that the SMA primarily controls the organization and planning of sequential activation of muscle groups required to perform a movement [9, 13], whereas the M1 cortex is mainly responsible for the execution of the movement [14]. In mice, the rostral representations of the limbs within the motor cortex, the so-called rostral forelimb area (RFA), are considered the homologous structure to human SMA [7].

1.1.3 Premotor Cortex (PMC)

The PMC lies rostral to the M1 cortex (Fig. 1), is comprised of dorsal (dPM) and ventral (vPM) subdivisions [15] and contains a similar but less precise somatotopic representation of the body musculature than the M1 region. The PMC receives massive inputs from the sensory areas and has extensive reciprocal connections with the M1 cortex [11]. The PMC outputs project to the pyramidal neurons, basal ganglia, brainstem, and spinal cord [16]. Based on the information from other cortical regions, the PMC selects a specific movement or a sequence of movements that are appropriate to the actual motion context [17, 18] and like the SMA is involved in planning and initiation of voluntary motor movements [19]. The vPM mediates visually guided motor behavior and controls limb movements based on visual and somatosensory information [18]. The vPM neurons selectively fire during movement trajectory in external space, attention to visuospatial stimuli and proprioceptive stimulation [18]. In humans, the dPM is composed of F2 and F7 areas [16], which encode action instructions associated with a symbolic cue, localization of external stimuli and movement dynamics [20]. In mice, it has been speculated that the M2 cortex is a possible anatomic homolog to the human PMC, SMA or frontal eye field [7].

1.1.4 Prefrontal Cortex (PFC)

The PFC is located in front of the PMC (Fig. 1), and is a key structure in planning complex cognitive tasks, working memory, speech, logical reasoning, and social behavior and is also involved in motor activity [21]. Functional subdivisions of the PFC include the dorsolateral prefrontal cortex (DLPFC), orbitofrontal cortex, and ventromedial prefrontal cortex (vmPFC) [22]. The PFC is extensively and reciprocally interconnected with thalamus, basal ganglia, hypothalamus, amygdala, hippocampus, and temporal and parietal association cortices [11]. The PFC receives indirect inputs from the basal ganglia through mediodorsal, ventral anterior, and anterior medial thalamic nuclei [23]. Prearcuate prefrontal cortices project to neighboring premotor areas, which in turn are connected with the motor cortex [5]. The posterior orbitofrontal cortex and, especially, the anterior cingulate cortex is involved in motor control associated with emotional expression [24]. Medial prefrontal cortices underpin the motor control of emotional vocalization [25]. The PFC as well plays a role in balance and motor inhibition control [26]. In rodents (including mice), much of the controversy exists regarding the presence and nomenclature of the PFC. In rodent studies, the "prefrontal" areas are considered to be the three cytoarchitectonically defined regions of the frontal cortex, namely, the prelimbic, infralimbic, and anterior cingulate areas [27].

1.1.5 Basal Ganglia

Basal ganglia comprise a set of brain structures that include caudate nucleus, putamen, nucleus accumbens, globus pallidus (GP), subthalamic nucleus (STN), and substantia nigra (SN) (Fig. 1). Excitatory afferent inputs to the basal ganglia come from the entire cerebral cortex and from thalamic intralaminar nuclei; the M1 and the S1 cortices project mainly to the putamen, while the PMC and SMA project to the caudate [5, 28]. The main efferent outputs from basal ganglia emerge from GP and STN, which carry inhibitory GABAergic projections [29]. Several intrinsic connections are evidenced: striatopallidal pathway (GABAergic), striatonigral pathway (GABAergic), nigrostriatal pathway (dopaminergic), GP–STN (GABAergic), and STN–GP (glutamatergic) [29]. The interactions between cortex and basal ganglia are mediated through two pathways. The *direct pathway* (cortex–striatum–internal GP–thalamus–cortex [30]) facilitates the information flow through thalamus, while the *indirect pathway* (striatum–external GP–STN–thalamus–cortex [30, 31]) inhibits this flow [32]. Motor functions of basal ganglia include the regulation of voluntary motor movements, modulation of motor programs initiated by the motor cortex, realization of "primitive" motor programs, gating the execution of automated programs and the suppression of inappropriate motor acts [33].

1.1.6 Thalamus

The thalamus, the most investigated grey matter structure in MS [34], is strategically located within the projections connecting motor areas of the cerebral cortex and motor-related subcortical structures of the cerebellum and basal ganglia (Fig. 1). The motor thalamus incorporates the ventral anterior and the ventral lateral thalamic nuclei (also known as the VA-VL complex) [5]. Motor thalamus receives major inputs from three sources: (1) cerebral cortex (associative, motor, and premotor areas), (2) cerebellar nuclei, that is, the dentate and interposed nuclei, and (3) basal ganglia structures, substantia nigra (pars reticulata), and globus pallidus [5, 35]. The M1 cortical afferents to motor thalamus originate from V and VI layer neurons [36]; layer V corticothalamic neurons exert a driving activity, while layer VI neurons exert a modulating activity. Most of the output thalamic motor neurons are glutamatergic and project to the pyramidal neurons of layers I and II of the motor cortex [37]. Distinct populations of motor neurons encode initiation, selection, velocity, orientation, and intensity of the movement activity [38]. In conjunction with other motor structures, thalamic nuclei form open feedback loops that facilitate the integration of information, encoding the preparatory and executory components of the movement [35].

1.1.7 Cerebellum

The cerebellum is an integral part of the motor control system and primarily is involved in movement coordination (Fig. 1). Cerebellar neurons receive inputs from other regions via two pathways—the *mossy fibers* and the *climbing fibers* [39]. Mossy fibers carry the information from multiple sources such as cerebral cortex (via pontocerebellar pathway), spinal cord, vestibular nuclei, and reticular formation [40]. Through climbing fibers cerebellum receives afferent inputs from the inferior olive of the brainstem, which end up with excitatory synapses on Purkinje cells [5]. Purkinje cells are the output cells of the cerebellum and convey inhibitory projections to cerebellar nuclei (the *fastigial*, the *dentate*, and the *interpositus nuclei*), which in turn project to almost all motor-related structures, including spinal cord, basal ganglia, reticular and vestibular nuclei, red nucleus, and cortical areas (through thalamus) [41]. The two major loops involving cerebellum are (1) cortex–pons–dentate nucleus–thalamus–cortex and (2) inferior olive–interpositus nucleus–red nucleus–olive [40]. Neural activity in cerebellar nuclei controls body movement coordination, sustains temporal pattern generation, encodes proprioceptive information, predicts the upcoming motor behavior, and processes prediction errors in order to optimize the motor performance [42, 43]. The cerebellum is also involved in motor learning and motor adaptation [44, 45].

1.1.8 Spinal Cord

The anterior horns of the spinal cord grey matter contain lower motor neurons (Fig. 1), which form neuromuscular junctions with the skeletal muscles and are topographically organized according to the innervated muscle groups. Motor neurons receive peripheral sensory feedback from the skeletal muscles, forming the substrate for a number of spinal reflexes. The activity of motor neurons is controlled as well by descending projections from the cerebral cortex, brainstem and cerebellum within the corticospinal, rubrospinal, reticulospinal, and vestibulospinal tracts [5]. Within the anterior horn, the activity of motor neurons is regulated by excitatory and inhibitory influences of interneurons [46]. Motor neurons are classified as *alpha*, *beta*, and *gamma* [47]. Alpha motor neurons innervate the working fibers of the skeletal muscles, called the extrafusal fibers, gamma motor neurons innervate the intrafusal fibers found within the muscle spindles, and beta neurons innervate both fiber types [2]. More specifically, alpha motor neurons encode the contraction force of muscle fibers using the size principle [48], while gamma motor neurons regulate the level of tension of muscle fibers via the *gamma loop*, which consists of gamma motor neurons, intrafusal fibers, alpha motor neurons, and extrafusal fibers [48].

Structural components of the motor system and symptoms emerging from their damage in MS are presented in Fig. 1.

2   Materials and Methods for Measuring Motor System Impairment in MS

Rapid advances in modern technologies have allowed us to study human brain abnormalities noninvasively in MS patients using a variety of methods, namely functional magnetic resonance imaging (fMRI), electroencephalography (EEG), magnetoencephalography (MEG), visual evoked potential (VEP), and motor evoked potential (MEP) (Fig. 2) [49, 50]. As an indirect measure of brain activity, fMRI measures the blood oxygen level-dependent (BOLD) signal by assessing the local blood oxygenation changes, and has an excellent spatial resolution in millimeters [51]. EEG measures the brain's electrical fields, while MEG records the brain's magnetic

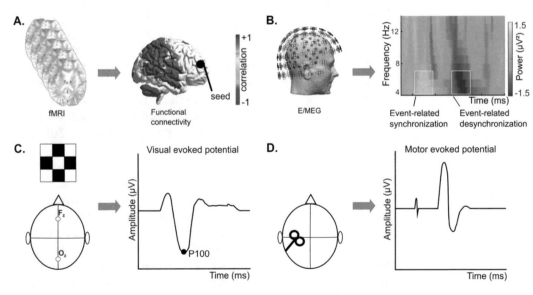

Fig. 2 The measurement of motor abnormalities in patients with MS using the measures (**a**) functional connectivity, (**b**) event-related synchronization and desynchronization, (**c**) visual evoked potential, and (**d**) motor evoked potential. fMRI: functional magnetic resonance imaging, E/MEG: electro−/ magnetoencephalography

fields [52]. Both the EEG and MEG signals are a direct measure of neuronal activity. Both methods have excellent temporal resolution on a sub-millisecond scale; however, the spatial resolution is relatively poor compared to other functional imaging techniques such as fMRI [53].

On the other hand, VEPs are useful to measure the integrity of the visual system and optic neuritis in MS patients [54]. MEP is the action potential elicited by noninvasive stimulation of the motor cortex through the scalp using transcranial magnetic stimulation (TMS), that assesses the conductivity of human motor pathways [55]. Several previous functional imaging (fMRI, EEG, and EMG) and evoked potential (VEP and MEP) studies determined the abnormal properties of the motor system in MS patients and their relationship with CNS damage and clinical measures.

2.1 Functional Abnormalities in Motor Regions: Evidence from Functional Imaging

Functional imaging methods can be used to obtain specific information about functional abnormalities in motor regions. fMRI allows for the precise spatial localization of the involved brain regions, while E/MEG provides accurate timing information of the neural activity.

2.1.1 fMRI

fMRI has been extensively used to study the abnormal connectivity patterns of motor networks in MS patients both at rest and during motor tasks. A study by Zhong and colleagues demonstrated that impaired hand function in MS patients was associated with changes in motor connectivity [56]. In clinically motor-impaired patients, the functional connectivity of the sensorimotor and somatosensory cortices was lower compared to motor-preserved MS patients and healthy controls. In another study, Jaeger and colleagues showed that fatigue in MS patients was negatively correlated with the functional connectivity of the caudate nucleus and ventral striatum with the sensorimotor cortex [57]. Taken together, these results highlight the alterations of motor network connectivity in MS patients and their relation to fatigue. Thus, functional connectivity represents a potentially important tool in the study of MS and clinically relevant alterations in motor regions.

2.1.2 E/MEG

In a seminal study, Leocani and colleagues showed cortical dysfunction in MS patients using EEG during a simple motor task involving self-paced extensions of the right thumb [58]. In MS patients exhibiting fatigue, the sensorimotor 18–22 Hz ERS (event-related synchronization) was lower relative to MS patients without fatigue and healthy controls. In another study, Barratt and colleagues showed the increased latency to peak beta rebound (increased beta power) during post-movement in the visuomotor task, which was negatively correlated with information processing speed in MS patients [59]. These results show the potential value of electrophysiological imaging studies for understanding motor system abnormalities in MS patients, hence supporting the inclusion of E/MEG-based investigational tools (ERS, and oscillatory power in specific frequency bands) in future MS studies. Further studies are required to investigate the information flow dynamics in the corticosubcorticocortical loops in MS patients [60].

2.2 Abnormalities in Evoked Potentials and Their Relation to Motor System

Evoked potentials studies have been used in the diagnosis of MS for many years [61]. However, recent research suggests that evoked potentials can also be used to gain insight into the etiology of the motor system abnormalities in MS [62]. Specifically, VEPs help to assess the abnormality in conduction in the visual pathway in MS patients. Increased latency of the cortical response is believed to be caused by demyelination in patients with MS. Previous studies have already found the prolonged latencies of the P100 responses in MS patients; these properties were related to visual and motor dysfunction [63]. On the other hand, MEPs are well established in the clinical practice to assess the changes in the motor system due to MS [64]. Several previous studies have found a significant decrease in MEP amplitude in MS patients relative to healthy controls [65, 66]. Moreover, in another study, MEP from MS patients correlated with motor impairment and disability as measured by the Expanded Disability Status Scale (EDSS) [67]. Thus, evoked potentials are sensitive tools for diagnosing MS and studying clinically relevant alterations comprising visual and motor domains in this disease.

3 Where Do We Stand in the Human-to-Mouse Translation?

The translation of scientific discoveries from MS patients to mouse models is fairly difficult and challenging. A considerable amount of data available on motor impairment in MS patients cannot be directly translated and reproduced in mouse models due to existing constraints.

The first barrier for direct translation is the complex organization of the human motor system. In humans, the caudate and putamen are separate structures, while in mice there is one analogous anatomical structure, the caudoputamen. The corticospinal tract is the predominant motor tract in humans, whereas in mice the rubrospinal tract prevails and the corticospinal tract plays a minor role in motor activity. Compared to mice, the amount of white matter in humans is considerably larger; the cerebellar nuclei of mice are smaller and poorly delimited in comparison to the human analogues [68]. Moreover, even among mouse strains differences in neuroanatomy (presence of corpus callosum) and behavior (motor and stereotyped patterns) exist.

The second hurdle is the difficulty of experimental manipulations to elicit in mice the relevant clinical and pathological aspects of patients' motor impairment. Mice with experimental autoimmune encephalomyelitis (EAE) do not exhibit the ongoing functional deterioration that MS patients experience, and hence, cannot mirror the long-term consequences of motor disturbances. The mouse EAE is predominantly a disease of the spinal cord white matter, while human MS is mainly a brain disease with prominent demyelination of cerebral and cerebellar cortex, thus making it difficult to model the human motor abnormalities caused by the damage to supraspinal structures of the motor system. Such important pathological hallmarks of human MS as demyelination and degeneration within the grey matter rarely or do not occur in mouse EAE models [69].

Third, the arsenal of technical and computational tools available to acquire and analyze functional data from humans surpasses the techniques available for studying the murine system. Several specific methodological concerns need to be emphasized:

1. Human fMRI studies are performed on awake patients, while mice usually are scanned under anesthetized conditions (to keep them restrained during the experiments) that can affect the hemodynamic responses [70, 71]. The four commonly used anesthetic drugs in mice (isoflurane, medetomidine, propofol, and urethane) determine the shape of the hemodynamic response function and influence the spatial specificity of BOLD signals across brain regions [72] and interconnected areas [73, 74]. Temporal stability of the fMRI signal is also dependent on the route and dose of anesthetic drug administration [75]. However, minimization of anesthetic effects and approximation to awake state of spontaneous neural activity in the experimental setting can be achieved by using light urethane anesthesia [76].

2. Patients benefit from noninvasive EEG recording, whereas mice usually undergo invasive electrophysiological recordings. Recordings from freely moving mice are preferred over those in a fixed setting, since the unrestrained conditions allow the study of the natural behavior, including motor, and offers valid insights into underlying neural activity [77]. Electrophysiological techniques provide targeted single- or multi-site recordings of single−/multi-unit activity and local field potentials from motor cortex, subcortical structures, and cerebellum [78, 79]. However, in freely moving mice, extracellular electrophysiological recordings can be challenging: loss of signal being a critical problem, especially in studies involving either extensive behavioral training (lasting weeks and months) or in animal

models aimed to develop chronic human diseases [80]. Technical advancements (reimplantable, feedback-controlled, lightweight microdrives; wireless systems) can help overcome these constraints [80–83] and enable reliable in vivo electrophysiological recordings to be conducted in chronically implanted experiments.

3. MEG is relatively easy applied in patients but still is not adapted for mice.

4. At the same time, some of the techniques available in mouse models but not in humans grant an access for a more in-depth exploration of the neural activity. One of these techniques is in vivo brain activity imaging in freely behaving mice that offers a cellular level resolution through minimally invasive (implantable microlenses) or more invasive (cranial window) approaches. In vivo imaging is performed with a miniature head-mounted microscope and calcium imaging to track neural activity during free locomotion and motor tasks [84]. This technique allows the calcium dynamics of neuronal populations to be visualized and recorded in a multi-layer fashion within the cortex [85] or subcortical structures as the mice are engaged in behavioral paradigms. A stable longitudinal recording from the same neuronal population enables the long-term effects of the applied treatment to be studied. By one- and two-photon microscopy, spatial dynamics of cellular activity patterns at high spatial and temporal resolution can be investigated [86–88]. Likewise, a chronic cranial window allows for long-term and high-resolution imaging of the mouse neocortex [89].

5. The analytical pipelines for processing mouse structural and functional data are less standardized and validated in comparison to those employed for human data [90].

Fourth, many of the clinical tests used to measure functional motor abnormalities in patients with MS do not have analogues implemented in mice models of MS, which usually assess the motor skills by mechanistic quantification of locomotor activity (the open field test) or balance (rotarod performance test). Such discrepancies indicate that the mice-based disease models currently used in preclinical research cannot sufficiently reproduce the essential aspects of the human MS motor symptomatology or appropriately reflect the pathophysiological mechanisms of motor abnormalities.

Currently, the translational research linking humans and mice lacks a disease model that could address a wide range of motor abnormalities experienced by MS patients. To closely approach and reproduce human MS clinical manifestations, including motor abnormalities, future experimental mouse platforms must be more "humanized," while preserving both internal and external validity.

4   Conclusions

The human motor system is complicated and multifaceted; however, emerging techniques have permitted the noninvasive study of motor abnormalities in MS patients. Given that MS is a complex and heterogeneous disease, there is currently no a unique mouse model that can capture the entire spectrum of clinical motor abnormalities observed in MS patients. Motor impairment is one of the principal determinants of disease-related disability; its successful back-translation is therefore of utmost importance for the development of representative models. Ultimately, the progress in translational continuum from human to mouse could be advanced by the gradual identification of critical methodological issues and dedicated efforts directed toward the development of more "humanized" animal models.

References

1. Squire L, Berg D, Bloom FE, Du Lac S, Ghosh A, Spitzer NC (2012) Fundamental neuroscience. Academic Press, Cambridge
2. Feher JJ (2017) Quantitative human physiology: an introduction. Academic Press, Cambridge
3. Rathelot J-A, Strick PL (2009) Subdivisions of primary motor cortex based on corticomotoneuronal cells. Proc Natl Acad Sci U S A 106(3):918–923
4. Purves D, Augustine GJ, Fitzpatrick D, Hall WC, LaMantia A-S, McNamara JO et al (2001) Neuroscience Glutamate. Sinauer Associates, Sunderland (MA)
5. Voogd J, Nieuwenhuys R, Huijzen CV (2008) The human central nervous system. Springer, New York
6. Yokoi A, Arbuckle SA, Diedrichsen J (2018) The role of human primary motor cortex in the production of skilled finger sequences. J Neurosci 38(6):1430–1442
7. Barthas F, Kwan AC (2017) Secondary motor cortex: where 'sensory' meets 'motor' in the rodent frontal cortex. Trends Neurosci 40(3):181–193
8. Morandell K, Huber D (2017) The role of forelimb motor cortex areas in goal directed action in mice. Sci Rep 7(1):15759
9. Nachev P, Kennard C, Husain M (2008) Functional role of the supplementary and pre-supplementary motor areas. Nat Rev Neurosci 9(11):856
10. Fontaine D, Capelle L, Duffau H (2002) Somatotopy of the supplementary motor area: evidence from correlation of the extent of surgical resection with the clinical patterns of deficit. Neurosurgery 50(2):297–305
11. Squire LR, Dronkers N, Baldo J (2009) Encyclopedia of neuroscience. Elsevier, Amsterdam
12. Akkal D, Dum RP, Strick PL (2007) Supplementary motor area and presupplementary motor area: targets of basal ganglia and cerebellar output. J Neurosci 27(40):10659–10673
13. Nachev P, Wydell H, O'neill K, Husain M, Kennard C (2007) The role of the pre-supplementary motor area in the control of action. NeuroImage 36:T155–TT63
14. Latash ML (2012) Fundamentals of motor control. Academic Press, Cambridge
15. Rizzolatti G, Fabbri-Destro M (2009) Premotor cortex in primates: dorsal and ventral. Encyclopedia Neurosci:935–945
16. Geyer S, Luppino G, Rozzi S (2012) Motor cortex. In The human nervous system (3rd edn). Elsevier. pp. 1003–1026. Number of pages 1428. https://doi.org/10.1016/C2009-0-02721-4
17. Rizzolatti G, Fogassi L, Gallese V (2002) Motor and cognitive functions of the ventral premotor cortex. Curr Opin Neurobiol 12(2):149–154
18. Pastor-Bernier A, Tremblay E, Cisek P (2012) Dorsal premotor cortex is involved in switching motor plans. Front Neuroeng 5:5
19. Haines DE, Mihailoff GA (2017) Fundamental neuroscience for basic and clinical applications E-book. Elsevier Health Sciences, Amsterdam

20. Solopchuk O, Alamia A, Zénon A (2016) The role of the dorsal premotor cortex in skilled action sequences. J Neurosci 36 (25):6599–6601

21. Siddiqui SV, Chatterjee U, Kumar D, Siddiqui A, Goyal N (2008) Neuropsychology of prefrontal cortex. Indian J Psychiatry 50 (3):202

22. Barbas H (2016) Executive functions: the prefrontal cortex: structure and anatomy. Reference Module in Biomedical Sciences, Elsevier, 2016, ISBN 9780128012383. https://doi.org/10.1016/B978-0-12-801238-3.04731-0

23. Diamond A (2001) Prefrontal cortex development and development of cognitive function. In: Smelser NJ, Baltes PB (eds) International encyclopedia of the Social & Behavioral Sciences. Pergamon, Oxford, pp 11976–11982

24. Banks SJ, Eddy KT, Angstadt M, Nathan PJ, Phan KL (2007) Amygdala-frontal connectivity during emotion regulation. Soc Cogn Affect Neurosci 2(4):303–312

25. Sturm VE, Haase CM, Levenson RW (2016) Chapter 22 - emotional dysfunction in psychopathology and neuropathology: neural and genetic pathways. In: Lehner T, Miller BL, State MW (eds) Genomics, circuits, and pathways in clinical neuropsychiatry. Academic Press, San Diego, pp 345–364

26. Rae CL, Hughes LE, Anderson MC, Rowe JB (2015) The prefrontal cortex achieves inhibitory control by facilitating subcortical motor pathway connectivity. J Neurosci 35 (2):786–794

27. Laubach M, Amarante LM, Swanson K, White SR (2018) What, if anything, is rodent prefrontal cortex? eNeuro 5(5):ENEURO.0315-18.2018

28. Nambu A (2009) Basal ganglia: physiological circuits. In: Squire LR (ed) Encyclopedia of neuroscience. Academic Press, Oxford, pp 111–117

29. Singer HS, Mink JW, Gilbert DL, Jankovic J (2016) Chapter 1—basal ganglia anatomy, biochemistry, and physiology. In: Singer HS, Mink JW, Gilbert DL, Jankovic J (eds) Movement Disorders in Childhood, 2nd edn. Academic Press, Boston, pp 3–12

30. Lanciego JL, Luquin N, Obeso JA (2012) Functional neuroanatomy of the basal ganglia. Cold Spring Harb Perspect Med 2(12): a009621-a

31. Watkins KE, Jenkinson N (2016) Chapter 8—The anatomy of the basal ganglia. In: Hickok G, Small SL (eds) Neurobiology of

language. Academic Press, San Diego, pp 85–94

32. Mink JW. Chapter 30—The basal ganglia. In: Squire LR, Berg D, Bloom FE, du Lac S, Ghosh A, Spitzer NC, editors. Fundamental neuroscience (4th edn). San Diego: Academic Press; 2013. p. 653-676

33. Nagano-Saito A, Martinu K, Monchi O (2014) Function of basal ganglia in bridging cognitive and motor modules to perform an action. Front Neurosci 8:187

34. Deppe M, Krämer J, Tenberge JG, Marinell J, Schwindt W, Deppe K et al (2016) Early silent microstructural degeneration and atrophy of the thalamocortical network in multiple sclerosis. Hum Brain Mapp 37(5):1866–1879

35. Bosch-Bouju C, Hyland BI, Parr-Brownlie LC (2013) Motor thalamus integration of cortical, cerebellar and basal ganglia information: implications for normal and parkinsonian conditions. Front Comput Neurosci 7:163

36. McFarland NR, Haber SN (2002) Thalamic relay nuclei of the basal ganglia form both reciprocal and nonreciprocal cortical connections, linking multiple frontal cortical areas. J Neurosci 22(18):8117–8132

37. Hooks BM, Mao T, Gutnisky DA, Yamawaki N, Svoboda K, Shepherd GMG (2013) Organization of cortical and thalamic input to pyramidal neurons in mouse motor cortex. J Neurosci 33(2):748–760

38. Gaidica M, Hurst A, Cyr C, Leventhal DK (2018) Distinct populations of motor thalamic neurons encode action initiation, action selection, and movement vigor. J Neurosci 38 (29):6563–6573

39. Miall RC (2013) Cerebellum: anatomy and function. In: Pfaff DW (ed) Neuroscience in the 21st century: from basic to clinical. Springer New York, New York, NY, pp 1149–1167

40. Bagnall M, du Lac S, Mauk M. Chapter 31—Cerebellum. In: Squire LR, Berg D, Bloom FE, du Lac S, Ghosh A, Spitzer NC, editors. Fundamental neuroscience (4th edn). San Diego: Academic Press; 2013. p. 677-696

41. Todorov DI, Capps RA, Barnett WH, Latash EM, Kim T, Hamade KC et al (2019) The interplay between cerebellum and basal ganglia in motor adaptation: a modeling study. PLoS One 14(4):e0214926

42. Sokolov AA, Miall RC, Ivry RB (2017) The cerebellum: adaptive prediction for movement and cognition. Trends Cogn Sci 21 (5):313–332

43. Popa LS, Ebner TJ (2019) Cerebellum, predictions and errors. Front Cell Neurosci 12:524

44. De Zeeuw CI, Ten Brinke MM (2015) Motor learning and the cerebellum. Cold Spring Harb Perspect Biol 7(9):a021683

45. Caligiore D, Pezzulo G, Baldassarre G, Bostan AC, Strick PL, Doya K et al (2017) Consensus paper: towards a systems-level view of cerebellar function: the interplay between cerebellum, basal ganglia, and cortex. Cerebellum 16 (1):203–229

46. Nógrádi A, Vrbová G (2006) Anatomy and physiology of the spinal cord. In: Transplantation of neural tissue into the spinal cord. Springer, New York, pp 1–23

47. Stifani N (2014) Motor neurons and the generation of spinal motor neurons diversity. Front Cell Neurosci 8:293

48. Hultborn H, Fedirchuk B (2009) Spinal motor neurons: properties. In: Squire LR (ed) Encyclopedia of neuroscience. Academic Press, Oxford, pp 309–319

49. Fuhr P, Borggrefe-Chappuis A, Schindler C, Kappos L (2001) Visual and motor evoked potentials in the course of multiple sclerosis. Brain 124(11):2162–2168

50. Van Schependom J (2017) Nagels G. targeting cognitive impairment in multiple sclerosis—the road toward an imaging-based biomarker. Front Neurosci 11:380

51. Glover GH (2011) Overview of functional magnetic resonance imaging. Neurosurg Clin N Am 22(2):133–139

52. da Silva FL (2013) EEG and MEG: relevance to neuroscience. Neuron 80(5):1112–1128

53. Burle B, Spieser L, Roger C, Casini L, Hasbroucq T, Vidal F (2015) Spatial and temporal resolutions of EEG: is it really black and white? A scalp current density view. Int J Psychophysiol 97(3):210–220

54. Chirapapaisan N, Laotaweerungsawat S, Chuenkongkaew W, Samsen P, Ruangvaravate N, Thuangtong A et al (2015) Diagnostic value of visual evoked potentials for clinical diagnosis of multiple sclerosis. Doc Ophthalmol 130(1):25–30

55. Kesar TM, Stinear JW, Wolf SL (2018) The use of transcranial magnetic stimulation to evaluate cortical excitability of lower limb musculature: challenges and opportunities. Restor Neurol Neurosci 36(3):333–348

56. Zhong J, Nantes JC, Holmes SA, Gallant S, Narayanan S, Koski L (2016) Abnormal functional connectivity and cortical integrity influence dominant hand motor disability in multiple sclerosis: a multimodal analysis. Hum Brain Mapp 37(12):4262–4275

57. Jaeger S, Paul F, Scheel M, Brandt A, Heine J, Pach D et al (2019) Multiple sclerosis–related fatigue: altered resting-state functional connectivity of the ventral striatum and dorsolateral prefrontal cortex. Mult Scler J 25(4):554–564

58. Leocani L, Colombo B, Magnani G, Martinelli-Boneschi F, Cursi M, Rossi P et al (2001) Fatigue in multiple sclerosis is associated with abnormal cortical activation to voluntary movement—EEG evidence. NeuroImage 13(6):1186–1192

59. Barratt EL, Tewarie PK, Clarke MA, Hall EL, Gowland PA, Morris PG et al (2017) Abnormal task driven neural oscillations in multiple sclerosis: a visuomotor MEG study. Hum Brain Mapp 38(5):2441–2453

60. Muthuraman M, Raethjen J, Koirala N, Anwar AR, Mideksa KG, Elble R et al (2018) Cerebello-cortical network fingerprints differ between essential, Parkinson's and mimicked tremors. Brain 141(6):1770–1781

61. Matthews W, Wattam-Bell J, Pountney E (1982) Evoked potentials in the diagnosis of multiple sclerosis: a follow up study. J Neurol Neurosurg Psychiatry 45(4):303–307

62. Giffroy X, Maes N, Albert A, Maquet P, Crielaard J-M, Dive D (2016) Multimodal evoked potentials for functional quantification and prognosis in multiple sclerosis. BMC Neurol 16(1):83

63. Kiiski HS, Riada SN, Lalor EC, Goncalves NR, Nolan H, Whelan R et al (2016) Delayed P100-like latencies in multiple sclerosis: a preliminary investigation using visual evoked spread spectrum analysis. PLoS One 11(1): e0146084

64. Fernández V, Valls-Sole J, Relova J, Raguer N, Miralles F, Dinca L et al (2013) Recommendations for the clinical use of motor evoked potentials in multiple sclerosis. Neurología 28 (7):408–416

65. Conte A, Li Voti P, Pontecorvo S, Quartuccio ME, Baione V, Rocchi L, Cortese A, Bologna M, Francia A, Berardelli A (2016) Attention-related changes in short-term cortical plasticity help to explain fatigue in multiple sclerosis. Mult Scler 22:1359–1366

66. Nantes JC, Zhong J, Holmes SA, Narayanan S, Lapierre Y, Koski L (2016) Cortical damage and disability in multiple sclerosis: relation to Intracortical inhibition and facilitation. Brain Stimul 9:566–573

67. Štětkářová I (2014) Evoked potentials in diagnosis and prognosis of multiple sclerosis. Clin Neurophysiol 125(5):e27

68. Treuting PM, Dintzis SM, Montine KS (2017) Comparative anatomy and histology: a mouse,

rat, and human atlas. Academic Press, Cambridge

69. Burrows DJ, McGown A, Jain SA, De Felice M, Ramesh TM, Sharrack B et al (2019) Animal models of multiple sclerosis: from rodents to zebrafish. Mult Scler J 25(3):306–324

70. Jonckers E, Delgado Y Palacios R, Shah D, Guglielmetti C, Verhoye M, Van der Linden A (2014) Different anesthesia regimes modulate the functional connectivity outcome in mice. Magn Reson Med 72(4):1103–1112

71. Schroeter A, Schlegel F, Seuwen A, Grandjean J, Rudin M (2014) Specificity of stimulus-evoked fMRI responses in the mouse: the influence of systemic physiological changes associated with innocuous stimulation under four different anesthetics. NeuroImage 94:372–384

72. Schlegel F, Schroeter A, Rudin M (2015) The hemodynamic response to somatosensory stimulation in mice depends on the anesthetic used: implications on analysis of mouse fMRI data. NeuroImage 116:40–49

73. Grandjean J, Schroeter A, Batata I, Rudin M (2014) Optimization of anesthesia protocol for resting-state fMRI in mice based on differential effects of anesthetics on functional connectivity patterns. NeuroImage 102:838–847

74. Bukhari Q, Schroeter A, Cole DM, Rudin M (2017) Resting State fMRI in mice reveals anesthesia specific signatures of brain functional networks and their interactions. Front Neural Circuits 11:5

75. Sirmpilatze N, Baudewig J, Boretius S (2019) Temporal stability of fMRI in medetomidine-anesthetized rats. Sci Rep 9:1): 1–1):13

76. Paasonen J, Stenroos P, Salo RA, Kiviniemi V, Gröhn O (2018) Functional connectivity under six anesthesia protocols and the awake condition in rat brain. NeuroImage 172:9–20

77. Juavinett AL, Bekheet G, Churchland AK (2019) Chronically implanted Neuropixels probes enable high-yield recordings in freely moving mice. Elife 8:e47188

78. Bermudez-Contreras E, Chekhov S, Sun J, Tarnowsky J, McNaughton BL, Mohajerani MH (2018) High-performance, inexpensive setup for simultaneous multisite recording of electrophysiological signals and mesoscale voltage imaging in the mouse cortex. Neurophotonics 5(2):025005

79. Wu B, Schonewille M (2018) Targeted electrophysiological recordings in vivo in the mouse cerebellum. In: Extracellular recording approaches. Springer, New York, pp 19–37

80. Polo-Castillo LE, Villavicencio M, Ramírez-Lugo L, Illescas-Huerta E, Moreno MG, Ruiz-Huerta L et al (2019) Reimplantable microdrive for long-term chronic extracellular recordings in freely moving rats. Front Neurosci 13:128

81. Hasegawa T, Fujimoto H, Tashiro K, Nonomura M, Tsuchiya A, Watanabe D (2015) A wireless neural recording system with a precision motorized microdrive for freely behaving animals. Sci Rep 5(1):7853

82. Jovalekic A, Cavé-Lopez S, Canopoli A, Ondracek JM, Nager A, Vyssotski AL et al (2017) A lightweight feedback-controlled microdrive for chronic neural recordings. J Neural Eng 14(2):026006

83. Inagaki S, Agetsuma M, Ohara S, Iijima T, Yokota H, Wazawa T et al (2019) Imaging local brain activity of multiple freely moving mice sharing the same environment. Sci Rep 9 (1):7460

84. Trevathan JK, Asp AJ, Nicolai EN, Trevathan JM, Kremer NA, Kozai TDY, et al. Calcium imaging in freely-moving mice during electrical stimulation of deep brain structures. bioRxiv 2020: 460220

85. Gulati S, Cao VY, Otte S (2017) Multi-layer cortical Ca^{2+} imaging in freely moving mice with prism probes and miniaturized fluorescence microscopy. J Vis Exp 124:e55579

86. Dombeck DA, Khabbaz AN, Collman F, Adelman TL, Tank DW (2007) Imaging large-scale neural activity with cellular resolution in awake, mobile mice. Neuron 56(1):43–57

87. Kerr JN, Nimmerjahn A (2012) Functional imaging in freely moving animals. Curr Opin Neurobiol 22(1):45–53

88. Ozbay BN, Futia GL, Ma M, Bright VM, Gopinath JT, Hughes EG et al (2018) Three dimensional two-photon brain imaging in freely moving mice using a miniature fiber coupled microscope with active axial-scanning. Sci Rep 8(1):1–14

89. Holtmaat A, Bonhoeffer T, Chow DK, Chuckowree J, De Paola V, Hofer SB et al (2009) Long-term, high-resolution imaging in the mouse neocortex through a chronic cranial window. Nat Protoc 4(8):1128

90. Krämer J, Brück W, Zipp F, Cerina M, Groppa S, Meuth SG (2019) Imaging in mice and men: pathophysiological insights into multiple sclerosis from conventional and advanced MRI techniques. Prog Neurobiol 182:101663

Chapter 11

Characterization of the Somatosensory System

Carsten H. Wolters, Marios Antonakakis, Asad Khan, Maria Carla Piastra, and Johannes Vorwerk

Abstract

For the characterization of the somatosensory system in the mammalian brain, somatosensory evoked potentials (SEP) and fields (SEF), SEP/SEF source analysis and targeted transcranial electric (TES) and magnetic (TMS) brain stimulation have become important techniques over the last decades. They were also shown to contribute to an improved diagnosis and therapy of neurological diseases such as multiple sclerosis (MS) and epilepsy. In this chapter, we will review the most recent advances of these modern methods and point out their translational potential for clinical use. We will start by describing the generation process of evoked responses with a focus on not only the high temporal resolution information contained in these data, but especially also the rich spatial information when using (combined) multichannel electroencephalography (EEG) and magnetoencephalography (MEG). This topographical information and the use of individualized realistic head models extracted from magnetic resonance imaging (MRI) data enables the reconstruction of the underlying sources in the brain using source analysis techniques. We will then discuss state-of-the-art individually optimized multichannel TES and TMS brain stimulation methods for a targeted manipulation of reconstructed sources in the brain. Finally, we will indicate how these modern noninvasive methods can be used to characterize the somatosensory system on an individual level and how they can be applied to gain a deeper understanding of the neural networks in MS and epilepsy.

Key words Somatosensory evoked potentials (SEP) and fields (SEF), Electroencephalography (EEG), Magnetoencephalography (MEG), Magnetic resonance imaging (MRI), Source analysis, Realistic head model, Optimized multi-sensor transcranial electric (TES) and magnetic stimulation (TMS), Multiple sclerosis (MS), Epilepsy

1 Introduction

Physiological recordings and brain stimulation in humans and animals play an essential role in developing an understanding of the mammalian brain. Signal recording technology includes noninvasive whole-brain monitoring through measurement of scalp potentials (EEG) and extracranial magnetic fields (MEG), but also invasive techniques such as for example stereo EEG (sEEG) or electrocorticography (ECoG) [1–4]. Somatosensory evoked potentials (SEP) and fields (SEF) are used not only to characterize

Sergiu Groppa and Sven G. Meuth (eds.), *Translational Methods for Multiple Sclerosis Research*, Neuromethods, vol. 166, https://doi.org/10.1007/978-1-0716-1213-2_11, © Springer Science+Business Media, LLC, part of Springer Nature 2021

the somatosensory system in healthy subjects [5–7] but also in clinical diagnostics, for example, in multiple sclerosis (MS) and epilepsy. Analysis of SEP/SEF data presents a host of challenges, from low level noise reduction and artifact rejection to sophisticated spatiotemporal source modeling and statistical inference. The multidisciplinary neuroscience research community has an ongoing need for state-of-the-art methods and tools to perform this analysis and to facilitate reproducible and large-scale research involving such electrophysiological data. Much progress especially with regard to open source tools has been made in the last decade [8–12].

An important method in the analysis of such electrophysiological data is source analysis, which has significantly contributed to both fundamental brain research and clinical applications [1, 3]. In source analysis, the primary current sources in the brain that are underlying the measured potentials or fields are reconstructed. Source analysis is used, for example, in presurgical epilepsy diagnosis, where its significant contribution has recently been shown in the largest epilepsy patient cohort study performed up to date [13].[1] While the latter study was based on single modality MEG, multimodal EEG (sEEG), MEG and magnetic resonance imaging (MRI) based source analysis methods have also been developed [5, 14–16].

Increasingly, quasi-noninvasive stimulation methods such as transcranial electric (TES) [17–19] or magnetic stimulation (TMS) [20, 21] are also being used for fundamental research and therapeutic intervention. First studies show that multichannel targeted, optimized, and individualized brain stimulation might explain variability in the outcome measures, as observed in many brain stimulation studies [22, 23].

In this chapter, we will describe the generation process of SEP/ SEF with a focus on the complementarity of both modalities (Subheading 2). We will discuss the most recent advances in individualized realistic head modeling (Subheading 3). This will enable combined EEG/MEG source analysis for an appropriate targeting followed by optimized multichannel TES and TMS brain stimulation (Subheading 4). Finally, we will describe the application of these modern methods for an individualized characterization of the somatosensory system (Subheading 5) and review translational possibilities for their use in the treatment of MS and epilepsy (Subheading 6).

---

[1] See also the scientific commentary to this work by Burgess https://doi.org/10.1093/brain/awz281

2   Evoked Responses and the Complementarity of Electroencephalography and Magnetoencephalography

In the last decades, advances in brain imaging technology have significantly deepened our knowledge about the complex neural systems in humans. While functional magnetic resonance imaging (fMRI) techniques provide information at high spatial resolution, EEG and MEG provide the important advantage of high temporal resolution, an aspect that is crucial not only for the characterization of the somatosensory system but also more generally for connectivity investigations in neuronal networks [1–3, 24]. In this section, we will study the generation process of SEP and SEF and work out the complementarity of EEG and MEG.

2.1   SEP and SEF Generation Process

This section is dedicated to the description of how evoked responses are generated using the example of SEP and SEF with a focus on not only the high temporal but also the rich spatial information contained in multichannel EEG and MEG. The latter gets most often much less attention in neurological textbooks about evoked responses [25].

Figures 1 and 2 visualize step-by-step the evoked responses generation process. In the experiment, the median nerve was stimulated at the right wrist (see the stimulation picture in Fig. 2a) with monophasic square-wave electrical pulses having a duration of 0.5 ms. The stimulus strength was increased until a clear movement of the thumb was visible. The data were acquired with a sampling rate of 1200 Hz and online filtered with a 300 Hz low pass filter. The overall duration of the experiment was 9 min, in which 1200 trials were measured. The stimulus onset asynchrony varied randomly (350–450 ms) to avoid habituation and to allow obtaining clear prestimulus intervals for signal-to-noise ratio (SNR) determination. The raw EEG/MEG recordings were filtered between 20 and 250 Hz [6] and a notch filter was used for the 50 Hz power line interferences. Subsequently, the preprocessed recordings were separated into equally large segments of 300 ms (100 ms pre- and 200 ms post-stimulus). Figure 1 shows the resulting EEG (left) and MEG (right) signals at sensors CP3 and MLP33, respectively, of all 1200 trials, where amplitudes were color-coded over time (top) and the resulting averaged SEP and SEF signals for the chosen two sensors (bottom) [26].

Figure 1 clearly shows that the potential and field deflections at 20 ms after the trigger, and also later components, do not only get visible in the evoked responses (bottom) but are clearly observable already in the single trials (top). However, as also indicated in Figure 1 (compare the amplitudes of noise in the pre-stimulus and signal in the post-stimulus interval in single trials and average), the averaging procedure can increase the SNR by $\sqrt{N}$ with N the

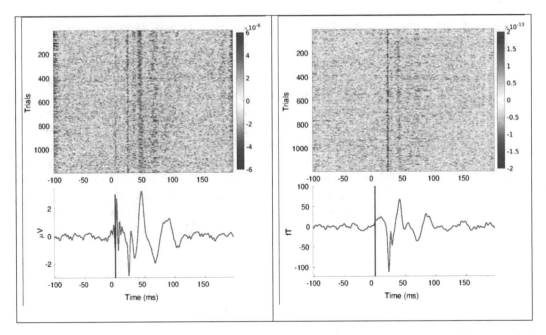

Fig. 1 EEG signal at sensor CP3 (left) and MEG signal at sensor MLP33 (right) of all trials with color-coded amplitudes over time (top) and the averaged signals at the chosen two sensors (bottom). The evoked responses start at 5 ms, indicated by the red triggers, whereas the first 5 ms only show the stimulation artifact (especially in the EEG) [26]

number of trials. In multi-channel EEG and MEG recordings, such signals over time are evoked at all sensors over the whole head surface, as shown in Figure 2b with the butterfly plots, in which all sensor signals were presented in single plots, SEF in green and SEP in blue (corrected by subtracting the 5 ms delay still contained in Figure 1). Fixing the time at 20 ms post-stimulus and using head surface splines [27] for amplitude interpolations, dipolar iso-potential (EEG) and -field (MEG) topographies of the so-called P20/N20 components in MEG (upper topography) and EEG (lower topography) can be visualized (Figure 2b). The orthogonality of the two topographies already points to their complementarity, discussed in the next section. The topographies contain the spatial information that is needed for a reconstruction of the underlying source activity (Subheading 4).

2.2 Complementarity of EEG and MEG

Due to their different sensitivity profiles, EEG and MEG contain complementary information. A first example for this complementarity has already been shown in Figure 2b, where the dipolar P20/N20 SEF and SEP topographies were orthogonal to each other. This aspect can later be exploited by a more stable reconstruction of the underlying mainly tangentially oriented P20/N20 source (*see* Subheading 5). In general, MEG can almost only measure quasi-tangentially oriented sources, while the EEG is sensitive

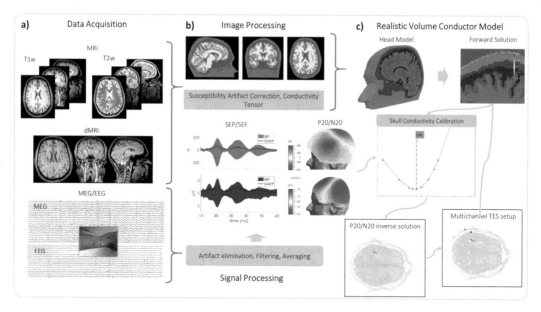

Fig. 2 Summary of an individualized SEP/SEF/MRI data based source analysis pipeline: (**a**) Data acquisition contains MRI datasets (T1w, T2w, and DTI) and combined SEP/SEF data elicited by somatosensory stimulation (e.g., electric-wrist, braille-tactile, or pneumato-tactile). (**b**) After preprocessing of the structural and functional data, (**c**) an individualized calibrated realistic FEM head model is generated and used for combined EEG/MEG source analysis of the somatosensory P20/N20 component as well as for targeted multi-channel TES optimization

to both quasi-radial and quasi-tangential neural generators and reveals a better depth resolution [1, 28]. However, MEG is much more sensitive to tangential source components and allows their reconstruction with much higher spatial resolution. Because of this complementarity, the combined analysis of EEG and MEG data is of increasing interest and it can outperform single modality EEG or MEG source analysis with regard to stability, spatial resolution and complete visualization of the time course of source activation [5, 15, 16, 28–32], as will also be studied for SEP/SEF in Subheading 5. Epileptic activity like for example interictal spike-wave complexes can often be seen in both MEG and EEG (helpful is then not only the orthogonality of their topographies). Even more importantly, trained epileptologists are often able to detect epileptic activity in only one of the two modalities, more in the MEG than in the EEG, which motivates the use of both modalities if available for the diagnosis [28, 33, 34]. However, combined EEG/MEG source analysis can only be successful if the different sensitivity profiles of EEG and MEG are taken into account with accurate forward solutions using realistic head models [5, 29, 30], as described in the next section.

3   Head Model Generation

Head models, needed for both source analysis and the simulation of brain stimulation, most often need to be built from MRI datasets. As shown in Figure 2a, b, generation pipelines for six-compartment (6C: skin, skull compacta, skull spongiosa, cerebrospinal fluid (CSF), brain grey and white matter) head models are based on T1-weighted- (T1w-) and T2w-MRI for tissue segmentation [5, 35–37] and on echo-planar imaging (EPI) diffusion-weighted- (dw-) MRI for brain conductivity anisotropy modeling [38–40]. More sophisticated approaches also model further compartments such as blood vessels [41] or dura [42]. Head model generation can then continue, for example, in the direction of surface-based tetrahedral [43] or (geometry-adapted) hexahedral [5, 44] volume conductor models, an example for the latter one being shown in Figure 2b, c. The importance of six compartment anisotropic (6CA) head modeling for reducing modeling errors caused by simplified head volume conductor modeling have been shown for source analysis [5], for connectivity investigations in source space [45] and for the forward problem in brain stimulation [46].

The second important aspect in a realistic head model are the tissue conductivity parameters. For the conductivity of the cerebrospinal fluid (CSF), a standard value of 1.79 S/m at body temperature ([47]: standard deviation of less than 2% for frequency range 10–10.000 Hz and age range 4.5 months to 70 years) can be assumed (*see* Figure 2b, c). The importance of CSF modeling has been shown in computer simulation studies for source analysis [48] and brain stimulation [46] and has been experimentally proven by [49].

The accurate modeling of other tissue conductivities is more demanding: Using sensitivity investigations for EEG/MEG [5, 50] and TES/TMS [51, 52], it has been shown that skull conductivity is the most influential tissue conductivity parameter for EEG and TES simulations, whereas it has hardly any influence on MEG and TMS. However, skull conductivity is known to vary strongly across individuals due to variations in age, disease state and environmental factors [5, 15, 29, 30, 53–57, 58]. Skull conductivity calibration has therefore been developed [5, 15, 55, 56, 58, 59], most of them based on simultaneously acquired SEP and SEF P20/N20 components to individually estimate conductivity parameters of the skull. An example for a calibration curve with skull conductivity on the *x*-axis and residual variance to the measured data on the *y*-axis and the indication of the individual calibration point is presented in Figure 2c. The additional SEP/SEF measurement only takes 7 min and can thus be easily acquired for each subject/patient in a first run of combined EEG/MEG data just before measuring the data of interest (e.g., epileptic activity).

White matter conductivity anisotropy ratios in the range of 1.96–3.25 together with significant inter-subject variabilities have recently been measured by Katoch et al., 2019 [60]. Another important step of the head model generation should thus be the measurement, registration, artifact correction and post-processing of dw-MRI for brain anisotropy modeling. dw-MRI can be measured using fast EPI sequences, which can be performed in less than 10 min of time. For this task, Ruthotto et al. 2012 [40] presented a new reversed gradient approach to correct for dw-MRI susceptibility artifacts and to accurately register dw-MRI data to the corresponding T1w- and T2w-MRI. In a second step, individual white matter conductivity anisotropy can then be derived using an effective medium approach [38, 39]. The importance of accurate white matter conductivity modeling has been shown for source analysis [5, 50, 61] and for brain stimulation [46, 52].

4 Forward and Inverse Problems in Bioelectromagnetism

This section deals with the inverse and forward problems that are involved in EEG and MEG source analysis, allowing for the identification of brain areas with appropriate spatial resolution and the investigation of their temporal dynamics.

4.1 Inverse Problem in EEG and MEG Source Analysis

Over the last decades, EEG and MEG source analysis have become a prominent technique for reconstructing neuronal networks with highest temporal and appropriate spatial resolution. The non-uniqueness of the inverse problem [62] implies that assumptions on the source model as well as anatomical and physiological a priori knowledge about the source region should be taken into account to obtain a unique solution. Therefore, different inverse approaches have been proposed. The classical dipole scan and fit approaches as well as multiple signal classification are often used [63–65], but they require as prior knowledge the number of active sources that are underlying the measured data. This prior knowledge is available for the reconstruction of the early transient SEP and SEF components (Subheading 4), but not in many other applications. It is thus an important advantage of beamforming [66–68] and (spatiotemporal) current density reconstruction (CDR) approaches [1–3, 69] that this parameter does not need to be predefined. In contrast to beamforming, CDR and hierarchical Bayesian modeling (HBM) [70, 71, 72] approaches can even reconstruct sources that are fully correlated in time. Beamformers and HBM can achieve higher spatial resolution with less localization bias than standard CDR approaches [1, 66, 71, 72].

4.2 Optimization Methods for Targeted Multichannel TES and TMS

Recently, multichannel brain stimulation hardware and optimization approaches for improved targeting have been proposed [73–76]. In these approaches (1) the target needs to be defined/reconstructed, for example using EEG/MEG source analysis or fMRI (*targeting*), (2) the injected current should be maximal at the target area (*intensity*), and (3) minimal at other areas (*focality*). Furthermore, the orientation of the injected current at the target location might play an important role, too (*directionality*) [77, 78]. The unification approach presented by Fernandez-Corazza et al. 2020 [79] allows for a good understanding of the nature of the TES optimization problem and a clear visualization and quantification of the intensity versus focality trade-off. In this context, the optimally focusing methods were proposed to be used for example in clinical situations where a series of stimulation sessions over a longer period of time is planned [75, 80], while the max intensity methods might be especially attractive in single stimulation experiments with short and few stimulation sessions, like used for the characterization of the somatosensory system [76] (*see* Fig. 2). Like EEG/MEG source analysis, the TES/TMS optimization methods rely on accurate solutions to the brain stimulation forward problem.

4.3 Bioelectromagnetic Forward Problems

The bioelectromagnetic forward problems are based on Maxwell's equations [1–3, 81, 82]. In EEG and MEG source analysis, the forward problem is defined as the simulation of the electric potentials and/or magnetic fields due to current sources in the brain. The TES/TMS forward problem consists of computing the induced current flow patterns in the brain for fixed sensor setups at the head surface. Both forward problems are related to each other by Helmholtz reciprocity [44, 83] and require a volume conductor model of the head. Realistic head models as described in Subheading 3 motivate the use of 3D approaches for the numerical solution such as the Finite Difference Method (FDM) [84, 85] or the Finite Element Method (FEM) [81, 86–91]. 3D approaches also enable a higher accuracy in solving intracranial inverse problems, for example, for sEEG data [14], and in performing deep brain stimulation (DBS) optimization [92–94]. Due to their flexibility in (1) the approximation of complex head geometries and tissue conductivity inhomogeneities and anisotropies [5, 46], as well as (2) source [43, 89, 95, 96] and (3) electrode modeling [97, 98] and (4) the approximation of the involved scalar and vector-valued functions [99–102], especially FEM has contributed to a more realistic and accurate solution of the bioelectromagnetic forward problems.

5   Application for the Characterization of the Somatosensory System

SEP/SEF and source analysis have been extensively used for the characterization of the somatosensory system and for localization of components in deep brain/thalamus, primary (SI) and secondary (SII) somatosensory cortices [5–7, 16, 72, 103–106]. In Antonakakis et al. 2019 [5], the effect of measurement modality (EEG, MEG, or combined EEG/MEG) and degree of realism in head modeling were shown to have a similar influence on P20/N20 source analysis as the stimulation types electric-wrist, braille-tactile, or pneumato-tactile index finger stimulation. The calibration results showed a considerable intersubject variability for skull conductivity, and it was furthermore demonstrated that combined EEG/MEG source reconstructions are less susceptible to forward modeling inaccuracies than single modality EEG or MEG with MEG being the superior modality for the localization and EEG for contributing the source orientation information.

Also brain stimulation methods could already contribute significantly to the characterization of the somatosensory system. TES [107, 108] and TMS [109–111] over the sensorimotor cortex were shown to induce long-lasting changes in SEP magnitude and in stimulus detection rates, and might thus be useful to investigate cortical sensory plasticity, for example, in MS. A recent study also showed strong interindividual differences in optimized multichannel stimulation of the somatosensory cortex [76] (see also Figure 2c). This study also took into account individual differences in localization and orientation of the target somatosensory sources [5], which are differentially stimulated by TES and TMS.

6   Translation to Clinical Applications in Multiple Sclerosis and Epilepsy

In the last section, we will focus on the translational aspects, that is, the contributions of the above described modern methodology to the diagnosis and therapy of neurological diseases such as epilepsy and multiple sclerosis (MS). This is not without mentioning that the described modern methodology is used as well in other areas of clinical research such as for example in psychiatry [112].

In presurgical epilepsy diagnosis, multimodal EEG (sEEG), MEG, and MRI source analysis methods and individualized calibrated head modeling have already been successfully applied and their superiority over standard approaches has been clearly demonstrated [14, 29, 30, 39, 58, 80]. In the context of the discussed trade-off between focality and intensity in TES optimization, Antonakakis et al. [80] proposed a targeting and optimally focusing multichannel TES approach to be used for patients who are unsuitable surgery candidates and for whom a series of stimulation sessions over a longer period of time are planned. In Holmes et al. 2019 [113], it was demonstrated that targeted spikes could be suppressed, using a safe protocol designed to induce long-term depression in epilepsy patients. It was, however, also discussed that spike suppression does not imply seizure suppression, which was left for future studies. Also in the review of Regner et al. 2018 [114], most studies reported positive and significant findings in translating tDCS to the treatment of epilepsy. On the other hand, it was also discussed that only a few preclinical findings could be translated into clinical research so far and that a successful preclinical study may not indicate success in a clinical study. Finally, in a recent large multicenter study in patients with refractory focal epilepsy, a significant decrease in seizure frequencies was shown in comparison to sham stimulation in a 14-day treatment using a standard bipolar TES montage with cathode placed over the epileptogenic focus [115]. In this study, a 2×20-min daily stimulation protocol was found to be superior to a 1×20-min one.

In MS, SEP/SEF and also visually evoked potentials (VEP) and fields (VEF) can be used to detect clinically silent foci and to objectify symptoms such as paresthesia or visual acuity disorders [25]. The methods furthermore help to locate sensitivity disorders in the peripheral or central nervous system or in spinal or supraspinal sections of the somatosensory nervous system and they can be used to proof the demyelinating character of the disease [25]. Multimodal evoked responses serve as a representative measure of the functional impairment, quantification, and prognosis in MS [116, 117]. Slight demyelinating processes can be identified by the analysis of 600 Hz high-frequency oscillations in MS patients in whom the P20/N20 SEP component remained unaffected [118]. Also the spatial information contained in multichannel recordings can be exploited as shown by Lascano et al. 2009 [119], where an objective topographic analysis increased the value of VEP (and to a lesser extent also SEP) as a surrogate marker for MS. In the MEG study of Dell'Acqua et al. 2010 [120], SEF source analysis was used to study the primary components of the sensorimotor network with focus on P20 and N30 in mildly disabled relapsing-remitting MS patients without sensory symptoms at the time of the investigation. Whereas a preserved S1 activation was demonstrated, the SEF morphology was found to be strikingly distorted, marking a disruption of primary somatosensory network patterning. For the treatment of tactile sensory deficit in MS it was shown that a 5-day application of standard bipolar TES is able to ameliorate tactile sensory loss with long-lasting beneficial effects [121]. Furthermore, the beneficial effect against MS fatigue of an individualized TES treatment targeting the bilateral primary somatosensory cortex was demonstrated by Cancelli et al. 2018 [122]. Finally, since it is not feasible for many individuals with MS to visit a clinic for treatment on a daily basis, Kasschau et al. 2016

[123] proposed remotely supervised self- or proxy-administration for home delivery of tDCS using specially designed equipment and a telemedicine platform.

7 Conclusion

In this book chapter, we reviewed the most recent advances in the methodology for source analysis based on measured electrophysiological signals such as EEG (sEEG, ECoG) and MEG as well as multichannel optimized and targeted electric (TES, DBS) or magnetic (TMS) brain stimulation methods. We, furthermore, showed how these innovative and mostly noninvasive new methodologies can be applied for the characterization of the human somatosensory system. In the last section, we presented the translational aspect for their use in neurological diseases such as epilepsy and MS. As a future outlook, the newest hardware developments such as OPM-MEGs [124] and multichannel TMS [125] will make the presented technologies even more accessible to a much larger neuroscientific and neurological clinical and research community.

Acknowledgments

This work was supported by the Deutsche Forschungsgemeinschaft, project WO1425/7-1, by the DFG priority program SPP1665, project WO1425/5-2 and by EU project ChildBrain (Marie Curie innovative training network, grant no. 641652). We thank Andreas Wollbrink, Karin Wilken, Hildegard Deitermann, Ute Trompeter, and Harald Kugel for their help with the acquisition of the EEG/MEG/MRI data.

References

1. Brette R, Destexhe A (2012) Handbook of neural activity measurement. Cambridge University Press, Cambridge. https://doi.org/10.1017/CBO9780511979958

2. Baillet S, Mosher JC, Leahy RM (2001) Electromagnetic brain mapping. IEEE Signal Process Mag 18(6):14–30. https://doi.org/10.1109/79.962275

3. Hämäläinen M, Hari R, Ilmoniemi R, Knuutila J, Lounasmaa O (1993) Magnetoencephalography—theory, instrumentation, and applications to noninvasive studies of the working human brain. Rev Mod Phys 65:413–497. https://doi.org/10.1103/RevModPhys.65.413

4. Pantev C, Hoke M, Lutkenhoner B, Lehnertz K (1989) Tonotopic organization of the auditory cortex: pitch versus frequency representation. Science 246(4929):486–488. https://doi.org/10.1126/science.2814476

5. Antonakakis M, Schrader S, Wollbrink A, Oostenveld R, Rampp S, Haueisen J, Wolters CH (2019) The effect of stimulation type, head modeling and combined EEG and MEG on the source reconstruction of the somatosensory P20/N20 component. Hum Brain Mapp 40(17):5011–5028. https://doi.org/10.1002/hbm.24754

6. Buchner H, Fuchs M, Wischmann HA et al (1994) Source analysis of median nerve and finger stimulated somatosensory evoked

potentials: multichannel simultaneous recording of electric and magnetic fields combined with 3D-MR tomography. Brain Topogr 6 (4):299–310

7. Hari R, Karhu J, Hämäläinen M, Knuutila J, Salonen O, Sams M, Vilkman V (1993) Functional organization of the human first and second somatosensory cortices: a neuromagnetic study. Eur J Neurosci 5(6):724–734

8. Nüßing A, Piastra MC, Schrader S, Miinalainen T, Brinck H, Wolters CH, Engwer C (2019), DUNEuro— A software toolbox for forward modeling in neuroscience. Preprint arXiv:1901.02874

9. Vorwerk J, Oostenveld R, Piastra MC, Magyari L, Wolters CH (2018) The FieldTrip-SimBio pipeline for EEG forward solutions. Biomed Eng Online 17:37. https://doi.org/10.1186/s12938-018-0463-y

10. Gramfort A, Luessi M, Larson E, Engemann DA, Strohmeier D, Brodbeck C, Parkkonen L, Hämäläinen MS (2014) MNE software for processing MEG and EEG data. NeuroImage 86:446–460. https://doi.org/10.1016/j.neuroimage.2013.10.027

11. Tadel F, Baillet S, Mosher JC, Pantazis D, Leahy RM (2011) Brainstorm: a user-friendly application for MEG/EEG analysis. Comput Intell Neurosci 8:879716. https://doi.org/10.1155/2011/879716

12. Oostenveld R, Fries P, Maris E, Schoffelen JM (2011) FieldTrip: open source software for advanced analysis of MEG, EEG, and invasive electrophysiological data. Comput Intell Neurosci 2011:1. https://doi.org/10.1155/2011/156869

13. Rampp S, Stefan H, Wu X, Kaltenhäuser M, Maess B, Schmitt F, Wolters CH, Hamer H, Kasper BS, Schwab S, Doerfler A, Blümcke I, Rössler K, Buchfelder M (2019) Magnetoencephalography for epileptic focus localization: a series of 1000 cases. Brain 142 (10):3059–3071. https://doi.org/10.1093/brain/awz231

14. Lanfer B, Röer C, Scherg M, Rampp S, Kellinghaus C, Wolters CH (2013) Influence of a silastic ECoG grid on EEG/ECoG based source analysis. Brain Topogr 26 (2):212–228. https://doi.org/10.1007/s10548-012-0251-0

15. Huang MX, Song T, Hagler DJ, Podgorny I, Jousmaki V et al (2007) A novel integrated MEG and EEG analysis method for dipolar sources. NeuroImage 37(3):731–748

16. Fuchs M, Wagner M, Wischmann HA, Köhler T, Theissen A, Drenckhahn R, Buchner H (1998) Improving source reconstructions by combining bioelectric and biomagnetic data. Electroencephalogr Clin Neurophysiol 107(2):93–111

17. Lefaucheur J-P, Antal A, Ayache SS et al (2017) Evidence-based guidelines on the therapeutic use of transcranial direct current stimulation (tDCS). Clin Neurophysiol 128 (1):56–92. https://doi.org/10.1016/j.clinph.2016.10.087

18. Antal A, Alekseichuk I, Bikson M et al (2017) Low intensity transcranial electric stimulation: safety, ethical, legal regulatory and application guidelines. Clin Neurophysiol 128 (9):1774–1809. https://doi.org/10.1016/j.clinph.2017.06.001

19. Bikson M, Grossman P, Thomas C et al (2016) Safety of transcranial direct current stimulation: evidence based update 2016. Brain Stimul 9(5):641–661. https://doi.org/10.1016/j.brs.2016.06.004

20. Lefaucheur JP, André-Obadia N, Antal A et al (2014) Evidence-based guidelines on the therapeutic use of repetitive transcranial magnetic stimulation (rTMS). Clin Neurophysiol 125(11):2150–2206. https://doi.org/10.1016/j.clinph.2014.05.021

21. Barker AT, Jalinous R, Freeston IL (1985) Non-invasive magnetic stimulation of the human motor cortex. Lancet 1:1106–1107

22. Kasten FH, Duecker K, Maack MC, Meiser A, Herrmann CS (2019) Integrating electric field modeling and neuroimaging to explain inter-individual variability of tACS effects. Nat Commun 10:5427. https://doi.org/10.1038/s41467-019-13417-6

23. Baltus A, Wagner S, Wolters CH, Herrmann CS (2018) Optimized auditory transcranial alternating current stimulation improves individual auditory temporal resolution. Brain Stimul 11(1):118–124

24. Bullmore E, Sporn O (2012) The economy of brain network organization. Nat Rev Neurosci 13:336–349. https://doi.org/10.1038/nrn3214

25. Stöhr M, Dichgans J, Büttner U, Hess CW (2005) Evozierte potenziale: SEP—VEP—AEP—EKP—MEP. Springer-Verlag, New York City, ISBN 978-3-540-26659-4

26. Piastra MC (2019), New finite element methods for solving the MEG and the combined MEG/EEG forward problem. Dissertation in mathematics, Fachbereich Mathematik und Informatik, Universität Münster, https://nbn-resolving.org/urn:nbn:de:hbz:6-53199662090(diss_piastra.pdf)

27. Perrin F, Pernier J, Bertrand O, Echallier JF (1989) Spherical splines for scalp potential and current density mapping. Electroencephalogr Clin Neurophysiol 72:184–187.

https://doi.org/10.1016/0013-4694(89) 90180-6

28. Piastra, M.C., Nüßing, A., Vorwerk, J., Clerc, M., Engwer, C., Wolters, C.H., A Comprehensive Study on EEG and MEG Sensitivity to Corticaland Sub-cortical Sources, Human Brain Mapping, pp.1-15 (2020). DOI:10.1002/hbm.25272

29. Aydin Ü, Rampp S, Wollbrink A, Kugel H, Cho J-H, Knösche TR, Grova C, Wellmer J, Wolters CH (2017) Zoomed MRI guided by combined EEG/MEG source analysis: a multimodal approach for optimizing presurgical epilepsy work-up and its application in a multi-focal epilepsy patient case study. Brain Topogr 30:417–433. https://doi.org/10. 1007/s10548-017-0568-9

30. Aydin Ü, Vorwerk J, Dümpelmann M, Küpper P, Kugel H, Heers M, Wellmer J, Kellinghaus C, Haueisen J, Rampp S, Stefan H, Wolters CH (2015) Combined EEG/MEG can outperform single modality EEG or MEG source reconstruction in presurgical epilepsy diagnosis. PLoS One 10: e0118753. https://doi.org/10.1371/jour nal.pone.0118753

31. Liu AK, Dale AM, Belliveau JW (2002) Monte Carlo simulation studies of EEG and MEG localization accuracy. Hum Brain Mapp 16:47–62

32. Baillet S, Garnero L, Marin G, Hugonin JP (1999) Combined MEG and EEG source imaging by minimization of mutual information. IEEE Trans Biomed Eng 46:522–534

33. Knake S, Halgren E, Shiraishi H, Hara K, Hamer HM, Grant PE et al (2006) The value of multichannel MEG and EEG in the presurgical evaluation of 70 epilepsy patients. Epilepsy Res 69(1):80–86

34. Iwasaki M, Pestana E, Burgess RC, Lüders HO, Shamoto H, Nakasato N (2005) Detection of epileptiform activity by human interpreters: blinded comparison between electroencephalography and magnetoencephalography. Epilepsia 46(1):59–68

35. Huang Y, Datta A, Bikson M, Parra LC (2019) Realistic volumetric-approach to simulate transcranial electric stimulation-ROAST-a fully automated open-source pipeline. J of Neur Eng 16(5):056006. https:// doi.org/10.1088/1741-2552/ab208d

36. Nielsen JD, Madsen KH, Puonti O, Siebner HR, Bauer C, Madsen CG, Saturnino GB et al (2018) Automatic skull segmentation from MR images for realistic volume conductor models of the head: assessment of the state-of-the-art. NeuroImage 174:587–598. https://doi.org/10.1016/j.neuroimage. 2018.03.001

37. Lanfer B (2014), Automatic generation of volume conductor models of the human head for EEG source analysis. Dissertation in Mathematics, Fachbereich Mathematik und Informatik, Universität Münster, http:// www.sci.utah.edu/~wolters/PaperWolters/ 2014/Lanfer_Dissertation_July2-2014.pdf

38. Tuch DS, Wedeen VJ, Dale AM, George JS, Belliveau JW (2001) Conductivity tensor mapping of the human brain using diffusion tensor MRI. Proc Natl Acad Sci 98:11697–11701. https://doi.org/10. 1073/pnas.171473898

39. Rullmann M, Anwander A, Dannhauer M, Warfield SK, Duffy FH, Wolters CH (2009) EEG source analysis of epileptiform activity using a 1 mm anisotropic hexahedra finite element head model. NeuroImage 44:399–410. https://doi.org/10.1016/j. neuroimage.2008.09.009

40. Ruthotto L, Kugel H, Olesch J, Fischer B, Modersitzki J, Burger M, Wolters CH (2012) Diffeomorphic susceptibility artifact correction of diffusion-weighted magnetic resonance images. Phys Med Biol 57:5715–5731. https://doi.org/10.1088/ 0031-9155/57/18/5715

41. Fiederer LDJ, Vorwerk J, Lucka F et al (2016) The role of blood vessels in high-resolution volume conductor head modeling of EEG. NeuroImage 128:193–120. https://doi. org/10.1016/j.neuroimage.2015.12.041

42. Ramon C, Garguilo P, Fridgeirsson EA, Haueisen J (2014) Changes in scalp potentials and spatial smoothing effects of inclusion of dura layer in human head models for EEG simulations. Front Neuroeng 7:32. https:// doi.org/10.3389/fneng.2014.00032

43. Miinalainen T, Rezaei A, Us D, Nüßing A, Engwer C, Wolters CH, Pursiainen S (2019) A realistic, accurate and fast source modeling approach for the EEG forward problem. NeuroImage 184:56–67

44. Wagner S, Lucka F, Vorwerk J, Herrmann CS, Nolte G, Burger M, Wolters CH (2016b) Using reciprocity for relating the simulation of transcranial current stimulation to the EEG forward problem. NeuroImage 140:163–173. https://doi.org/10.1016/j. neuroimage.2016.04.005

45. Cho J-H, Vorwerk J, Wolters CH, Knösche TR (2015) Influence of the head model on EEG and MEG source connectivity analysis. NeuroImage 110:60–77

46. Wagner S, Rampersad SM, Aydin Ü, Vorwerk J, Oostendorp TF, Neuling T, Herrmann CS, Stegeman DF, Wolters CH (2014) Investigation of tDCS volume conduction effects in a highly realistic head model. J

Neural Eng 11(1):016002. https://doi.org/10.1088/1741-2560/11/1/016002

47. Baumann SB, Wozny DR, Kelly SK, Meno FM (1997) The electrical conductivity of human cerebrospinal fluid at body temperature. IEEE Trans Biomed Eng 44:220–223. https://doi.org/10.1109/10.554770

48. Wendel K, Narra NG, Hannula M, Kauppinen P, Malmivuo J (2008) The influence of CSF on EEG sensitivity distributions of multilayered head models. IEEE Trans Biomed Eng 55:1454–1456. https://doi.org/10.1109/TBME.2007.912427

49. Rice JK, Rorden C, Little JS, Parra LC (2013) Subject position affects EEG magnitudes. NeuroImage 64:476–484. https://doi.org/10.1016/j.neuroimage.2012.09.041

50. Vorwerk J, Aydin Ü, Wolters CH, Butson CR (2019) Influence of head tissue conductivity uncertainties on EEG dipole reconstruction. Front Neurosci 13:531. https://doi.org/10.3389/fnins.2019.00531

51. Saturnino GB, Thielscher A, Madsen KH, Knösche TR, Weise K (2019) A principled approach to conductivity uncertainty analysis in electric field calculations. NeuroImage 188:821–834

52. Schmidt C, Wagner S, Burger M, van Rienen U, Wolters CH (2015) Impact of uncertain head tissue conductivity in the optimization of transcranial direct current stimulation for an auditory target. J Neural Eng 12:046028. https://doi.org/10.1088/1741-2560/12/4/046028

53. McCann H, Pisano G, Beltrachini L (2019) Variation in reported human head tissue electrical conductivity values. Brain Topogr 32 (5):825–858. https://doi.org/10.1007/s10548-019-00710-2

54. Azizollahi H, Darbas M, Diallo MM, El Badia A, Lohrengel S (2018) EEG in neonates: forward modeling and sensitivity analysis with respect to variations of the conductivity. Math Biosci Eng 15 (4):905–932. https://doi.org/10.3934/mbe.2018041

55. Baysal U, Haueisen J (2004) Use of a priori information in estimating tissue resistivities–application to human data in vivo. Physiol Meas 25:737–748

56. Goncalves S, de Munck JC, Verbunt JPA, Heethaar RM, da Silva FHL (2003) In vivo measurement of the brain and skull resistivities using an EIT-based method and the combined analysis of SEF/SEP data. IEEE Trans Biomed Eng 50:1124–1128. https://doi.org/10.1109/TBME.2003.816072.73

57. Akhtari M, Bryant HC, Mamelak AN, Flynn ER, Heller L et al (2002) Conductivities of three-layer live human skull. Brain Topogr 14:151–167. https://doi.org/10.1023/A:1014590923185

58. Antonakakis, M., Schrader, S., Aydin, Ü., Khan, A., Gross, J., Zervakis, M., Rampp, S., Wolters, C.H., Inter-Subject Variability of Skull Conductivity and Thickness in Calibrated Realistic Head Models, NeuroImage, 223:117353(2020). DOI: 10.1016/j.neuroimage.2020.117353

59. Akalin-Acar Z, Acar CE, Makeig S (2016) Simultaneous head tissue conductivity and EEG source location estimation. NeuroImage 124(Pt A):168–180. https://doi.org/10.1016/j.neuroimage.2015.08.032

60. Katoch N, Choi BK, Sajib SZK, Lee E, Kim HJ, Kwon OI, Woo EJ (2018) Conductivity tensor imaging of in vivo human brain and experimental validation using Giant vesicle suspension. IEEE Trans Med Imag 38(7). https://doi.org/10.1109/TMI.2018.2884440

61. Güllmar D, Haueisen J, Reichenbach JR (2010) Influence of anisotropic electrical conductivity in white matter tissue on the EEG-/MEG forward and inverse solution: a high-resolution whole head simulation study. NeuroImage 51(1):145–163. https://doi.org/10.1016/j.neuroimage.2010.02.014

62. Dassios G, Fokas A, Kariotou F (2005) On the non-uniqueness of the inverse MEG problem. Inverse Problems 21. https://doi.org/10.1088/0266-5611/21/2/L01

63. Scherg M, Von Cramon D (1986) Evoked dipole source potentials of the human auditory cortex. Electroencephalogr Clin Neurophysiol 65(5):344–360. https://doi.org/10.1016/0168-5597(86)90014-6

64. Mosher JC, Lewis PS, Leahy RM (1992) Multiple dipole modeling and localization from spatio-temporal MEG data. IEEE Trans Biomed Eng 39(6):541–557

65. Wolters CH, Beckmann RF, Rienäcker A, Buchner H (1999) Comparing regularized and non-regularized nonlinear dipole fit methods: a study in a simulated sulcus structure. Brain Topogr 12(1):3–18. https://doi.org/10.1023/A:1022281005608

66. Sekihara K, Sahani M, Nagarajan SS (2005) Localization bias and spatial resolution of adaptive and non-adaptive spatial filters for MEG source reconstruction. NeuroImage 25:1056–1067. https://doi.org/10.1016/j.neuroimage.2004.11.051

67. Gross J, Kujala J, Hämäläinen M, Timmermann L, Schnitzler A, Salmelin R (2001) Dynamic imaging of coherent sources: studying neural interactions in the human brain. Proc Natl Acad Sci U S A 98

(2):694–699. https://doi.org/10.1073/pnas.98.2.694

68. Neugebauer F, Möddel G, Rampp S, Burger M, Wolters CH (2017) The effect of head model simplification on beamformer source localization. Front Neurosci 11:625. https://doi.org/10.3389/fnins.2017.00625

69. Dannhauer M, Lämmel E, Wolters CH, Knösche TR (2013) Spatio-temporal regularization in linear distributed source reconstruction from EEG/MEG—a critical evaluation. Brain Topogr 26(2):229–246

70. Calvetti D, Hakula H, Pursiainen S, Somersalo E (2009) Conditionally Gaussian hypermodels for cerebral source localization. SIAM J Imaging Sci 2(3):879–909

71. Lucka F, Pursiainen S, Burger M, Wolters CH (2012) Hierarchical Bayesian inference for the EEG inverse problem using realistic FE head models: depth localization and source separation for focal primary currents. NeuroImage 61:1364–1382. https://doi.org/10.1016/j.neuroimage.2012.04.017

72. Rezaei, A., Antonakakis, M., Piastra, M.C., Wolters, C.H., Pursiainen, S., Parametrizing the Conditionally Gaussian Prior Model for Source Localization with Reference to the P20/N20 Component of Median Nerve SEP/SEF, Brain Sciences, 10(12), 934 (2020). DOI: 10.3390/brainsci10120934

73. Dmochowski JP, Datta A, Bikson M, Su Y, Parra LC (2011) Optimized multi-electrode stimulation increases focality and intensity at target. J Neural Eng 8:046011. https://doi.org/10.1088/1741-2560/8/4/046011

74. Ruffini G, Fox MD, Ripolles O, Miranda PC, Pascual-Leone A (2014) Optimization of multifocal transcranial current stimulation for weighted cortical pattern targeting from realistic modeling of electric fields. NeuroImage 89:216–225. https://doi.org/10.1016/j.neuroimage.2013.12.002

75. Wagner S, Burger M, Wolters CH (2016) An optimization approach for well-targeted transcranial direct current stimulation. SIAM J Appl Math 76(6):2154–2174. https://doi.org/10.1137/15M1026481

76. Khan A, Antonakakis M, Vogenauer N, Wollbrink A, Radecke JO, Schneider T, Nitsche M, Paulus W, Haueisen J, Wolters CH, (2019), Optimized multi-electrode tDCS targeting of human somatosensory network, 41st Int. Eng. in Medicine and Biology Conf. (EMBC), Berlin, July 23-27, KhanEtAl_EMBC_2019_1403.pdf

77. Creutzfeldt OD, Fromm GH, Kapp H (1962) Influence of transcortical d-c currents on cortical neuronal activity. Exp Neurol 5:436–452. https://doi.org/10.1016/0014-4886(62)90056-0

78. Krieg TD, Salinas FS, Narayana S, Fox PT, Mogul DJ (2013) PET-based confirmation of orientation sensitivity of TMS-induced cortical activation in humans. Brain Stimul 6:898–904. https://doi.org/10.1016/j.brs.2013.05.007

79. Fernández-Corazza M, Turovets S, Muravchik C (2020) Unification of optimal targeting methods in Transcranial Electrical Stimulation. NeuroImage 209:116403

80. Antonakakis M, Rampp S, Kellinghaus C, Wolters CH, Möddel G (2019), Individualized targeting and optimization of multichannel transcranial direct current stimulation in drug-resistant epilepsy, 19th annual IEEE Int. Conf. On bioinformatics and bioengineering (BIBE), Oct.28-30, Athens, Greece, http://www.sci.utah.edu/~wolters/PaperWolters/2019/AntonakakisEtAl_IEEEBIBE_2019.pdf

81. Miranda PC, Hallett M, Basser PJ (2003) The electric field induced in the brain by magnetic stimulation: a 3-D finite-element analysis of the effect of tissue heterogeneity and anisotropy. IEEE Trans Biomed Eng 50(9):1074–1085. https://doi.org/10.1109/TBME.2003.816079

82. Wolters CH, de Munck JC (2007), Volume conduction, encyclopedia of computational neuroscience, Scholarpedia, http://www.scholarpedia.org/article/Volume_conduction

83. Vallaghe S, Papadopoulo T, Clerc M (2009) The adjoint method for general EEG and MEG sensor-based lead field equations. Phys Med Biol 54:135–147. https://doi.org/10.1088/0031-9155/54/1/009

84. Cuartas Morales EC, Acosta-Medina CD, Castellanos-Dominguez G, Mantini D (2019) A finite-difference solution for the EEG forward problem in inhomogeneous anisotropic media. Brain Topogr 32(2):229–239. https://doi.org/10.1007/s10548-018-0683-2

85. Montes-Restrepo V, van Mierlo P, Strobbe G, Staelens S, Vandenberghe S, Hallez H (2014) Influence of skull modeling approaches on eeg source localization. Brain Topogr 27:95–11. https://doi.org/10.1007/s10548-013-0313-y

86. Weinstein D, Zhukov L, Johnson C (2000) Lead-field bases for electroencephalography source imaging. Ann Biomed Eng 28(9):1059–1065. https://doi.org/10.1114/1.1310220

87. Schimpf P, Ramon C, Haueisen J (2002) Dipole models for the EEG and MEG. IEEE Trans Biomed Eng 49(5):409–418

88. Gençer NG, Acar CE (2004) Sensitivity of EEG and MEG measurements to tissue conductivity. Phys Med Biol 49(5):701–717. https://doi.org/10.1088/0031-9155/49/5/004

89. Wolters CH, Köstler H, Möller C, Härdtlein J, Grasedyck L, Hackbusch W (2007a) Numerical mathematics of the subtraction approach for the modeling of a current dipole in EEG source reconstruction using finite element head models. SIAM J Sci Comput 30(1):24–45. https://doi.org/10.1137/060659053

90. Sadleir RJ, Vannorsdall TD, Schretlen DJ, Gordon B (2010) Transcranial direct current stimulation (tDCS) in a realistic head model. NeuroImage 51(4):1310–1318

91. Thielscher A, Antunes A, Saturnino GB (2015) Field modeling for transcranial magnetic stimulation: a useful tool to understand the physiological effects of TMS? Annu Int Conf IEEE Eng Med Biol Soc 2015:222–225. https://doi.org/10.1109/EMBC.2015.7318340

92. Jakobs M, Fomenko A, Lozano AM, Kiening KL (2019) Cellular, molecular, and clinical mechanisms of action of deep brain stimulation - a systematic review on established indications and outlook on future developments. EMBO Mol Med 11(4):e9575. https://doi.org/10.15252/emmm.201809575

93. Vorwerk J, Brock AA, Anderson DN, Rolston JD, Butson CR (2019c) A retrospective evaluation of automated optimization of deep brain stimulation parameters. J Neural Eng 16(6):064002

94. Deuschl G, Schade-Brittinger C, Krack P, Volkmann J, Schäfer H, Bötzel K, Daniels C, Deutschländer A, Dillmann U, Eisner W (2006) A randomized trial of deep-brain stimulation for Parkinson's disease. N Engl J Med 355(9):896–908. https://doi.org/10.1056/NEJMoa060281

95. Beltrachini L (2019) The analytical subtraction approach for solving the forward problem in EEG. J Neur Eng 16(5):056029. https://doi.org/10.1088/1741-2552/ab2694

96. Vorwerk J, Hanrath A, Wolters CH, Grasedyck L (2019a) The multipole approach for EEG forward modeling using the finite element method. NeuroImage 201:116039. https://doi.org/10.1016/j.neuroimage.2019.116039

97. Pursiainen S, Agsten B, Wagner S, Wolters CH (2018) Advanced boundary electrode modeling for tES and parallel tES/EEG. IEEE Trans Neural Syst Rehabil Eng 26:37–44. https://doi.org/10.1109/TNSRE.2017.2748930

98. von Ellenrieder N, Beltrachini L, Muravchik CH (2012) Electrode and brain modeling in stereo-EEG. Clin Neurophysiol 123(9):1745–1754. https://doi.org/10.1016/j.clinph.2012.01.019

99. Piastra MC, Nüßing A, Vorwerk J, Bornfleth H, Oostenveld R, Engwer C, Wolters CH (2018) The discontinuous galerkin finite element method for solving the MEG and the combined MEG/EEG forward problem. Front Neurosci 12:30. https://doi.org/10.3389/fnins.2018.00030

100. Vorwerk J, Engwer C, Pursiainen S, Wolters CH (2017) A mixed finite element method to solve the EEG forward problem. IEEE Trans Med Imaging 36(4):930–941. https://doi.org/10.1109/TMI.2016.2624634

101. Engwer C, Vorwerk J, Ludewig J, Wolters CH (2017) A discontinuous galerkin method to solve the EEG forward problem using the subtraction approach. SIAM J Sci Comput 39:B138–B164. https://doi.org/10.1137/15M1048392

102. Nüßing A, Wolters CH, Brinck H, Engwer C (2016) The unfitted discontinuous Galerkin method for solving the EEG forward problem. IEEE Trans Biomed Eng 63:2564–2575. https://doi.org/10.1109/TBME.2016.2590740

103. Haueisen J, Leistritz L, Süsse T, Curio G, Witte H (2007) Identifying mutual information transfer in the brain with differential-algebraic modeling: evidence for fast oscillatory coupling between cortical somatosensory areas 3b and 1. NeuroImage 37(1):130–136

104. Nakamura A, Yamada T, Goto A et al (1998) Somatosensory homunculus as drawn by MEG. NeuroImage 7(4 Pt 1):377–386

105. Curio G, Mackert BM, Burghoff M, Neumann J, Nolte G, Scherg M, Marx P (1997) Somatotopic source arrangement of 600 Hz oscillatory magnetic fields at the human primary somatosensory hand cortex. Neurosci Lett 234:131–134

106. Allison T, McCarthy G, Wood CC, Jones SJ (1991) Potentials evoked in human and monkey cerebral cortex by stimulation of the median nerve. Brain 114:2465–2503

107. Dieckhöfer A, Waberski T, Nitsche M, Paulus W, Buchner H, Gobbelé R (2006) Transcranial direct current stimulation applied over the somatosensory cortex – differential effect on low and high frequency SEPs. Clin Neurophysiol 117:2221–2227

108. Matsunaga K, Nitsche M, Tsuji S, Rothwell J (2004) Effect of transcranial DC sensorimotor cortex stimulation on somatosensory evoked potentials in humans. Clin Neurophysiol 115:456–460

109. Koch G, Franca M, Albrecht UV, Caltagiron C, Rothwell JC (2006) Effects of paired pulse TMS of primary somatosensory cortex on perception of a peripheral electrical stimulus. Exp Brain Res 172(3):416–424

110. McKay DR, Ridding MC, Miles TS (2003) Magnetic stimulation of motor and somatosensory cortices suppresses perception of ulnar nerve stimuli. Int J Psychophysiol 48 (1):25–33

111. Seyal M, Masuoka LK, Browne JK (1992) Suppression of cutaneous perception by magnetic pulse stimulation of the human brain. Electroencephalogr Clin Neurophysiol 85 (6):397–401

112. Gonzalez-Hernandez JA, Pita-Alcorta C, Wolters CH, Padron A, Finale A, Galan-Garcia L, Marot M, Lencer R (2015) Specificity and sensitivity of visual evoked potentials in the diagnosis of schizophrenia: rethinking VEPs. Schizophr Res 166:231–234. https://doi.org/10.1016/j.schres.2015.05.007

113. Holmes MD, Feng R, Wise MV, Ma C, Ramon C, Wu J, Luu P, Hou J, Pan L, Tucker DM (2019) Safety of slow-pulsed transcranial electrical stimulation in acute spike suppression. Ann Clin Transl Neurol 6 (12):2579–2585. https://doi.org/10.1002/acn3.50934

114. Regner GG, Pereira P, Leffa DT, de Oliveira C, Vercelino R, Fregni F, Torres ILS (2018) Preclinical to clinical translation of studies of transcranial direct-current stimulation in the treatment of epilepsy: a systematic review. Front Neurosci 12:189. https://doi.org/10.3389/fnins.2018.00189

115. Yang D et al (2020) Transcranial direct current stimulation reduces seizure frequency in patients with refractory focal epilepsy: a randomized, double-blind, sham-controlled, and three-arm parallel multicenter study. Brain Stimul 13(1):109–116. https://doi.org/10.1016/j.brs.2019.09.006

116. Hardmeier M, Leocani L, Fuhr P (2017) A new role for evoked potentials in MS? Repurposing evoked potentials as biomarkers for clinical trials in MS. Mult Scler 23 (10):1309–1319. https://doi.org/10.1177/1352458517707265

117. Giffroy X, Maes N, Albert A et al (2016) Multimodal evoked potentials for functional quantification and prognosis in multiple sclerosis. BMC Neurol 16:83. https://doi.org/10.1186/s12883-016-0608-1

118. Gobbelé R, Waberski TD, Dieckhöfer A, Kawohl W, Klostermann F, Curio G, Buchner H (2003) Patterns of disturbed impulse propagation in multiple SclerosisIdentified by low and high frequency somatosensory evoked potential components. J Clin Neurophysiol 20(4):283–290

119. Lascano AM, Brodbeck V, Lalive PH, Chofflon M, Seeck M, Michel CM (2009) Increasing the diagnostic value of evoked potentials in multiple sclerosis by quantitative topographic analysis of multichannel recordings. J Clin Neurophysiol 26(5):316–325. https://doi.org/10.1097/WNP.0b013e3181baac00

120. Dell'Acqua ML, Landi D, Zito G, Zappasodi F, Lupoi D, Rossini PM, Filippi MM, Tecchio F (2010) Thalamocortical sensorimotor circuit in multiple sclerosis: an integrated structural an electrophysiological assessment. Hum Brain Mapp 31:1588–1600. https://doi.org/10.1002/hbm.20961

121. Mori F, Nicoletti CG, Kusayanagi H, Foti C, Restivo DA, Marciani MG, Centonze D (2013) Transcranial direct current stimulation ameliorates tactile sensory deficit in multiple sclerosis. Brain Stimul 6:654e659

122. Cancelli A, Cottone C, Giordani A et al (2018) Personalized, bilateral whole-body somatosensory cortex stimulation to relieve fatigue in multiple sclerosis. Mult Scler J 24 (10):1366–1374. https://doi.org/10.1177/1352458517720528

123. Kasschau M, Reisner J, Sherman K, Bikson M, Datta A, Charvet LE (2016) Transcranial direct current stimulation is feasible for remotely supervised home delivery in multiple sclerosis. Neuromodulation 19:824–831. https://doi.org/10.1111/ner.12430

124. Labyt E, Corsi MC, Fourcault W, Palacios Laloy A, Bertrand F, Lenouvel F, Cauffet G, Le Prado M, Berger F, Morales S (2019) Magnetoencephalography with optically pumped 4He magnetometers at ambient temperature. IEEE Trans Med Imaging 38 (1):90–98. https://doi.org/10.1109/TMI.2018.2856367

125. Minusa S, Muramatsu S, Osanai H, Tateno T (2019) A multichannel magnetic stimulation system using submillimeter-sized coils: system development and experimental application to rodent brain in vivo. J Neural Eng 16 (6):066014. https://doi.org/10.1088/1741-2552/ab3187

Chapter 12

Translational Research in Neuroimmunology: Cognition

Maren Person and Miriam Becke

Abstract

Cognitive impairment affects up to 70% of patients with multiple sclerosis (MS). It may already be present in the early stages of disease and impedes with a patient's ability to maintain employment, take part in activities of daily life, and fully participate in society. Similar to the clinical heterogeneity of MS, there is great variability in cognitive symptoms among patients. Given the impact of such deficits on everyday functioning, an increasing interest in diagnosis and treatment of cognitive impairment has arisen.

   This chapter aims to present an overview of those cognitive domains most commonly affected in MS, common approaches to assess the respective cognitive functions as well as procedures which may facilitate cross-species study of cognition.

   Key words Cognition, Cognitive impairment, Multiple sclerosis

1   Introduction

Impairments of cognitive functions, including attention, memory, or executive functions, are now widely recognized as epiphenomena exerting far-reaching impacts on the quality of life of many. Whether the underlying cause is a neurological, psychiatric, or internist condition, cognitive deficits may impede a person's ability to partake in activities of daily living both inside and outside the workplace [1–5]. Indeed, cognitive impairment has been associated with loss of self-esteem, reduced involvement in social activities, and leaving the workforce [6–9]. Given the impact of such deficits on everyday functioning, the study of cognition has informed the search for treatments in various conditions and diseases including many mental health and neurological disorders. Should the development of a medication be deemed the goal of such a search for new interventions, the regulatory process mandates the study of nonhuman subjects before human volunteers are first exposed to the thus far unknown chemical entity. As these endeavors benefit from the assessment of cognitive abilities in both human and non-

Sergiu Groppa and Sven G. Meuth (eds.), *Translational Methods for Multiple Sclerosis Research*, Neuromethods, vol. 166, https://doi.org/10.1007/978-1-0716-1213-2_12, © Springer Science+Business Media, LLC, part of Springer Nature 2021

human subjects, the present chapter discusses procedures which may facilitate such cross-species study of cognition and its potential changes in neuroimmunological conditions.

2   Neuroimmunology and Cognition

Ever since Charcot held his *Lectures on the Diseases of the Nervous System* in 1877 and, in doing such, recognized cognitive impairment as part of neurological disorders, rising numbers of publications illustrate the research efforts aimed at elucidating changes in cognition among patients with these very conditions. As a result, impairments in domains including attention, memory, or executive functions are now widely acknowledged as accompanying symptoms in various neurological disorders of—among others—inflammatory, metabolic, or infective origin [10]. Owing to its common occurrence and early onset, the study of neuropsychological phenomena appears particularly pertinent to multiple sclerosis (MS). With an estimated 40–70% of patients with MS affected by changes in cognition [11], routine screening for cognitive impairment has been recommended and may be followed by more extensive neuropsychological testing. As such examination should be sensitive to the cognitive changes most commonly evident in MS, the following sections of this chapter will describe frequent findings of neuropsychological assessments undertaken in this patient sample and discuss tests available to examine the respective domains (*see* Table 1 for an overview).

Table 1
Overview of cognitive domains and respective cognitive measures commonly applied

Cognitive Domain	Common Measures	References
Attention Information processing speed	*SDMT* *TMT-A*	[16] for example [18]
Memory Verbal learning and memory Visual learning and memory	*RAVLT* *WMS-IV visual reproduction*	[18, 29, 33]
Executive functions Working memory Cognitive flexibility Verbal fluency	*PASAT* *TMT-B* *WLG*	[39] for example [18, 41]
Visuospatial functions Visuospatial learning and memory	*SPART*	[41]

3   Attention

Attention, first being defined by William James in 1890 [12] as "holding something before the mind to the exclusion of all else," is a broad term comprising several attentional mechanisms selectively guiding (*selective attention*), focusing (*sustained attention*), and dividing (*divided attention*) conscious awareness towards one or more sources of sensory input or information. A further aspect of attention is the rate at which the brain processes this incoming information (*processing speed*). Research on attentional processes in patients diagnosed with MS has focused mainly on the latter aspect of attention, as it is commonly found to be one of the first and most affected cognitive functions. Potagas and colleagues [13] report deficient information processing speed among 14% of examined patients, whereas Nilsson et al. [14] note such deficits for 22% of patients who had experienced a single episode of demyelinating and inflammatory disease of the central nervous system. Over the course of disease, a pronounced decline in processing speed has been observed, with the progressive subtypes of MS showing the most severe impairments [15]. Potagas and colleagues [13] observed prolonged processing speed among 28% of patients with remitting–relapsing MS (RRMS), 63% of patients with secondary progressive MS and 76% of patients with primary progressive MS.

3.1   Tests Measuring Attention

One of the most commonly tests applied to assess attentional capacity is the *Symbol Digit Modalities Test* (SDMT) [16], measuring processing speed of visual information. This task requires subjects to substitute symbols by the corresponding number as indicated by a reference key (*see* Fig. 1a). The number of symbols correctly substituted within 90 s forms the total score. Normative data subsequently allow for the translation of this total score into age- and education-corrected scores [17]. The SDMT has been reported to be the most sensitive task to detect cognitive dysfunction in MS that furthermore shows high correlations with MRI measures of atrophy, lesion burden as well as microstructural pathology [18]. Another frequently applied measure of processing speed, as well as visual-motor skills, is the *Trail Making Test Part A* (TMT-A) [19]. The examinee is presented a random allocation of encircled numbers, ranging from one to 25, and is instructed to connect all numbers in ascending order as quickly as possible (*see* Fig. 1b). Score is the time taken to correctly connect all numbers (for further information on administration and scoring, please refer to [17]).

A   Symbol Digit Modalities Test (SDMT)

/	O	%	-	\	+
1	2	3	4	5	6

O	-	\	%	/	+	%	O	+	/	%	-	\

B   Trail Making Test Part A (TMT-A)

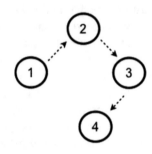

C   Trail Making Test Part B (TMT-B)

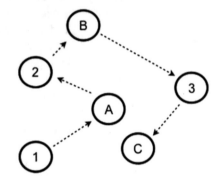

D   Paced Auditory Serial Addition Test (PASAT)

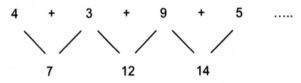

Note: These images do **not** reflect original test materials and serve for illustrative purposes only.

Fig. 1 Examples of common cognitive measures for processing speed (**a**, **b**), cognitive flexibility (**c**), and working memory (**d**)

4   Memory

Memory can be conceptualized as the process of encoding and storing information as well as experiences for a possible retrieval in the future. Impairments can occur at any of these stages of the memory process: encoding, consolidation or retrieval. Previous reports deemed memory impairments throughout the course of MS deficits in retrieval [20–22], whereas more recent studies suggest impairments in both encoding and retrieval [23–25]. Specifically, following DeLuca et al. [25], their sample of MS patients required more repetitions to attain a prespecified learning criterion, but performed as well as healthy controls on the subsequent tasks of recall and recognition. On the contrary, other authors observed an impaired recall performance even after successful learning [24]. Conflicting results are not limited to these different stages of the memory formation process, but have also been described for various types of material to be remembered. One may differentiate between memory for verbal information (e.g., words or digits) and visual information (e.g., geometric figures or spatial placement)—a distinction which is also evident in neuropsychological tests assessing memory function. Some studies suggest memory impairments in the early stages of MS to primarily affect the visual domain [13, 20, 26], whereas others propose memory deficits to be more prominent in the verbal domain [27]. As for the chronic subtypes of MS, memory deficits do not seem to be limited to one domain, with several studies suggesting most severe impairments among those suffering a progressive course of MS [13, 15].

4.1 Tests Assessing Memory

Measures of verbal learning and long-term memory typically require subjects to learn a list of words, as in the *Rey Auditory Verbal Learning Test* (RAVLT) [28, 29]. After an initial presentation of the target word list containing 15 words (List "A"), subjects are asked to immediately recall as many words as possible (*memory span*) and are then presented the same list for four more consecutive learning trials. After each trial, subjects are again asked to recall as many words as possible in any order. Interference effects are then determined by presenting a distractor list (List "B") in the same manner, followed by a free-recall of target List A. After a delay of 20–30 min, subjects are again asked to freely recall the previously learned target list. Finally, in a recognition trial, subjects are required to recognize the target words from a longer list of words, containing all target and distractor words as well as 15 semantically and phonetically similar words. Besides the original French and subsequent English versions of the RAVLT, the test has been translated into several different languages including, amongst others, German [30], Spanish [31] and Chinese [32] with normative data available.

Continuing with the assessment of visual learning and long-term memory, neuropsychological tests typically entail the presentation of a visual stimulus, such as five geometric figures during the *Visual Reproduction subscale* of the well-known *Wechsler Memory Scale—IV* (WMS-IV) [33]. A series of five designs is shown, one at a time, for 10 s each. After each presentation, the patient is requested to draw the respective design from memory. After a delay of about 20–30 min, the patient is again asked to draw all five designs from memory, followed by a recognition task wherein the original designs must be recognized among an array of six similar-looking designs.

5 Executive Functions

Executive functions are a set of cognitive processes necessary for adaptive, goal-directed behavior, and include, amongst others, working memory, cognitive flexibility and verbal fluency. Working memory has been described as a temporary storage of information, allowing for its active manipulation as necessary for complex cognitive tasks like learning, language comprehension or reasoning. Already during the early stages of MS, working memory has been found to be deficient in up to a fourth of MS patients [34]. Over the course of disease, the number of patients displaying impairments of working memory rises up to 47% among patients diagnosed with secondary progressive MS [13]. Generally, a better performance can be observed for the relapsing–remitting subtype of MS compared to the progressive subtypes [13, 15]. Cognitive flexibility, also referred to as set shifting, allows for the attentional shifting between multiple stimuli or sets of rules. Throughout the course of MS, cognitive flexibility seems to be impaired with MS patients consistently showing lower performance levels than healthy controls [34, 35]. Verbal fluency, requiring executive control over processes such as selective attention, selective inhibition, cognitive flexibility, and self-monitoring, refers to the conscious retrieval of relevant information from memory. Some authors report verbal fluency to be reduced during the early stages of MS [13, 36], whereas others find it to be intact [34]. However, substantial deficits in verbal fluency may become apparent especially during the later stages of MS, with most severe impairments among the progressive subtypes [13, 37, 38].

Generally, it has been argued that many tests assessing executive function encompass a substantial processing speed component, especially with respect to timed tasks. It is thus important to consider that formally deficient performance on a task intended to assess executive function might actually reflect prolonged processing speed [35].

5.1 Assessing Executive Functions

A frequently used paradigm assessing working memory is the *Paced Auditory Serial Addition Test* (PASAT) [39]. It is a sensitive measure of cognitive dysfunction in MS and was thus chosen as the cognitive component of the *Multiple Sclerosis Functional Composite* (MSFC) [40], a short outcome measure frequently used in clinical settings. Examinees are presented a series of 61 randomly selected digits and are required to add sequential pairs, so that each number is added to the one immediately preceding it, and verbally report the sum (*see* Fig. 1d). Digits are presented at a rate of 3 s per digit (PASAT3) in a first trial, and at a rate of 2 s per digit in a second trial (PASAT2). Scores are the total number of correct responses.

A common measure of cognitive flexibility is the *Trail Making Test Part B* (TMT-B) [19]. During this task, patients are presented encircled numbers as well as letters and are asked to connect all circles in ascending order, alternating between numbers and letters (i.e., 1-A-2-B-3-C etc.; *see* Fig. 1c). Performance is scored in terms of time taken to correctly connect all circles.

Verbal fluency is assessed by evaluating spontaneous production of words stemming from a given category, such as fruits and vegetables (*semantic verbal fluency*) or starting with a certain letter, such as letter 'S' (*lexical verbal fluency*) within a given time frame. The *Word List Generation Test* (WLG), a subtest of the *Brief Repeatable Battery of Neuropsychological Tests* (BRB-N) [41], is frequently used in English-speaking samples. However, various equivalents in other languages are available for the assessment of non-English speakers, including German [42] and Spanish [43].

6   Visuospatial Functions

Visuospatial functions refer to the ability to identify visual representations and their spatial relationships, such as distance, direction, and dimension. It can be measured in terms of a person's ability to perceive, reproduce or reconstruct, and manipulate visual representations. Visuospatial tasks are complex in nature, drawing on various spatial and executive processes. Little work has been done on visuospatial functions, despite an early study suggesting up to 26% of MS patients experience impairments of visual perceptual functions [44]. Most studies have instead focused on visuospatial memory and reports are in line with aforementioned findings on visual memory function. Especially among the progressive subtypes, visuospatial memory was found to be impaired, however, even during the early disease stages performance levels are below those of healthy controls [13, 14].

6.1   Tests of Visuospatial Functions

The *10/36 Spatial Recall Test* (SPART) [41] is a common measure of visuospatial learning and memory and part of the aforementioned *Brief Repeatable Battery of Neuropsychological Tests* (BRB-N). The examinee is presented a 6 × 6 checker board containing a random pattern of ten checkers for 10 s. After the initial presentation, the patient is required to immediately reproduce the pattern of checkers. This procedure is repeated for two more learning trials, followed by a delayed recall after 15 min.

7   Examining Cognition in Nonhuman Subjects

7.1   Barnes Maze

The Barnes maze, originally developed by Carol Banes in 1979 [45], is a dry-land based behavioral paradigm for assessing spatial navigational learning and memory in laboratory animals [46, 47]. It consists of a circular platform with several evenly spaced holes around the perimeter, of which one leads to an escape cage. Animals are placed in the center of the platform and aversive stimuli such as bright light are used to motivate animals to locate the escape hole. Latency to locate and enter the escape hole as well as errors in terms of incorrect holes checked before locating the escape hole are recorded. Furthermore, video recordings of their exploration allow for the determination of a search strategy (i.e., random, serial, or spatial). An improved performance over multiple trials is assumed to reflect learning and memory of the escape hole. In a reversal learning trial, the location of the escape hole is changed, requiring the animal to locate and remember the new location for subsequent trials, a task assumed to reflect cognitive flexibility [45]. The Barnes maze is most commonly used in rodent models [48]; however, it has been applied to investigate spatial learning and memory in other species as well, including corn snakes [49], cockroaches [50] and nonhuman primates [51]. In recent years, this paradigm proved to be a useful tool in the field of translational research, allowing for the assessment of disease-related deficits and drug effects on cognition in animal models. Hollinger and colleagues [52] investigated dose-dependent drug effects on cognitive performance in experimental autoimmune encephalomyelitis (EAE), a mouse model of MS that displays both physical and cognitive impairments. The drug investigated proved to be effective in alleviating cognitive impairment on the Barnes maze task.

8   CANTAB

The Cambridge Neuropsychological Test Automated Battery (CANTAB) [53] is a validated and widely used computerized measure of cognition and neurodevelopment in humans. It includes measures of attention and psychomotor speed, memory, executive functions as well as emotion and social cognition. The nonverbal nature of test administration and response recording via touch-sensitive computer monitors has facilitated their use to assess cognition in nonhuman primates. Due to structural and functional similarities between primates and humans, findings from experimental primate studies based on the same or similar cognitive tests as those used in humans translate more closely to human cognition than findings from studies including other laboratory animals. A qualitatively similar performance to that of humans on tasks of the

CANTAB has been observed for rhesus monkeys; moreover, evidence suggests performance of both humans and rhesus monkeys is affected by manipulations of the same parameters, such as a decreased level of memory performance with increased delays [54]. Hence, this approach offers new opportunities to the field of translational research in neuroimmunology by enhancing extrapolation of findings from primate studies to human clinical treatments and allowing for cross-species comparisons. Especially the characterization of changes in cognition following central nervous system manipulations, such as lesions of specific brain structures or networks, is of special interest.

9   Conclusion

Cognitive impairments are now widely recognized as epiphenomena occurring in up to 70% of MS patients. Besides the physical and emotional disease burden, these impairments further impact a patient's ability to partake in activities of daily life in and outside the workplace. Similar to the clinical heterogeneity of MS, there is great variability in cognitive dysfunction among patients with MS, however, a number of deficits have been repeatedly found. A core deficit seems to be prolonged information processing speed, underlying many other, especially higher order cognitive functions. Though, impairments of episodic memory, working memory, and verbal fluency have also been widely described. Frequent findings as well as a number of tests assessing the respective cognitive domains have been described in this chapter. Generally, there is accumulating evidence indicating progressive subtypes exhibit the greatest impairments of cognition, whereas the relapsing–remitting subtype present with less severe impairments. Nonetheless, the degree of dysfunction depends on many other factors, including the patient's previous level of cognitive performance, disease duration as well as the extent and location of neural damage. An approach to studying the effects of lesion location as well as treatment outcomes of newly developed medication includes cross-species studies of cognition. The Barnes maze is an early behavioural paradigm used to assess spatial learning and memory. It is frequently applied in experimental autoencephalomyelitis (EAE), a mouse model of MS that displays both physical and cognitive impairments. While using rodents in such preclinical studies is a time and cost-effective method, extrapolation of findings from those studies to human clinical treatments is limited. Findings from experimental studies investigating cognition in nonhuman primates, resembling humans functionally and structurally, translate more closely to human cognition. CANTAB, a computerized measure of cognition previously developed to assess cognition in humans has successfully been adapted for application in primate samples, allowing for the cross-species study of cognition.

References

1. Amato MP, Ponziani G, Pracucci G, Bracco L, Siracusa G, Amaducci L (1995) Cognitive impairment in early-onset multiple sclerosis: pattern, predictors, and impact on everyday life in a 4-year follow-up. Arch Neurol 52 (2):168–172

2. Benedict RHB, Wahlig E, Bakshi R, Fishman I, Munschauer F, Zivadinov R, Weinstock-Guttman B (2005) Predicting quality of life in multiple sclerosis: accounting for physical disability, fatigue, cognition, mood disorder, personality, and behavior change. J Neurol Sci 231 (1–2):29–34

3. Goverover Y, Chiaravalloti N, DeLuca J (2015) Brief international cognitive assessment for multiple sclerosis (BICAMS) and performance of everyday life tasks: actual reality. Mult Scler J 22(4):544–550

4. Morrow SA, Drake A, Zivadinov R, Munschauer F, Weinstock-Guttman B, Benedict RHB (2010) Predicting loss of employment over three years in multiple sclerosis: clinically meaningful cognitive decline. Clin Neuropsychol 24(7):1131–1145

5. Raggi A, Covelli V, Schiavolin S, Scaratti C, Leonardi M, Willems M (2016) Work-related problems in multiple sclerosis: a literature review on its associates and determinants. Disabil Rehabil 38(10):936–944

6. Hoogs M, Kaur S, Smerbeck A, Weinstock-Guttman B, Benedict RHB (2011) Cognition and physical disability in predicting health-related quality of life in multiple sclerosis. Int J MS Care 13(2):57–63

7. Kalmar JH, Gaudino EA, Moore NB, Halper J, DeLuca J (2008) The relationship between cognitive deficits and everyday functional activities in multiple sclerosis. Neuropsychology 22 (4):442–449

8. Lincoln NB, Radford KA (2008) Cognitive abilities as predictors of safety to drive in people with multiple sclerosis. Mult Scler J 14 (1):123–128

9. Rao SM, Leo GJ, Bernardin L, Unverzagt F (1991) Cognitive dysfunction in multiple sclerosis. I. Frequency, patterns, and prediction. Neurology 41(5):685–691

10. Larner AJ (2013) Neuropsychological neurology: the neurocognitive impairments of neurological disorders, 2nd edn. Cambridge University Press, Cambridge, U.K.

11. DeLuca GC, Yates RL, Beale H, Morrow SA (2015) Cognitive impairment in multiple sclerosis: clinical, radiologic and pathologic insights. Brain Pathol 25(1):79–98

12. James W (1890) Attention. In: The principles of psychology. Henry Holt and Company, New York, NY, USA, pp 402–458

13. Potagas C, Giogkaraki E, Koutsis G, Mandellos D, Tsirempolou E, Sfagos C, Vassilopoulos D (2008) Cognitive impairment in different MS subtypes and clinically isolated syndromes. J Neurol Sci 267(1–2):100–106

14. Nilsson PC, Rorsman I, Larsson EM, Norrving B, Sandberg-Wollheim M (2008) Cognitive dysfunction 24-31 years after isolated optic neuritis. Mult Scler 14 (7):913–918

15. Huijbregts SCJ, Kalkers NF, de Sonneville LMJ, de Groot V, Polman CH (2006) Cognitive impairment and decline in different MS subtypes. J Neurol Sci 245(1–2):187–194

16. Smith A (1973) Symbol digit modalities test. Western Psychological Services, Los Angeles, CA

17. Strauss E, Sherman EM, Spreen O (2006) A compendium of neuropsychological tests: administration, norms, and commentary. American Chemical Society, Washington, D.C.

18. Benedict RH, DeLuca J, Phillips G, LaRocca N, Hudson LD, Rudick R, Multiple Sclerosis Outcome Assessments Consortium (2017) Validity of the symbol digit modalities test as a cognition performance outcome measure for multiple sclerosis. Mult Scler 23 (5):721–733

19. Reitan RM, Wolfson D (1985) The Halstead Reitan neuropsychological test battery: theory and clinical interpretation. Neuropsychology Press, Tucson, AZ

20. Schulz D, Kopp B, Kunkel A, Faiss JH (2006) Cognition in the early stage of multiple sclerosis. J Neurol 253(8):1002–1010

21. Rao SM (1986) Neuropsychology of multiple sclerosis: a critical review. J Clin Exp Neuropsychol 8(5):503–542

22. Rao SM, Leo GJ, St. Aubin-Faubert P (1989) On the nature of memory disturbance in multiple sclerosis. J Clin Exp Neuropsychol 11 (5):699–712

23. Kocer B, Unal T, Nazliel B (2008) Evaluating sub-clinical cognitive dysfunction and event-related potentials (P300) in clinically isolated syndrome. Neurol Sci 29(6):435–444

24. Thornton AE, Raz N, Tucker KA (2002) Memory in multiple sclerosis: contextual encoding deficits. J Int Neuropsychol Soc 8 (3):395–409

25. DeLuca J, Gaudino EA, Diamond BJ, Christodoulou C, Engel RA (1998) Acquisition and storage deficits in multiple sclerosis. J Clin Exp Neuropsychol 20(3):376–390

26. Feinstein A, Kartsounis LD, Miller DH, Youl BD, Ron MA (1992) Clinically isolated lesions of the type seen in multiple sclerosis: a cognitive, psychiatric, and MRI follow up study. J Neurol Neurosurg Psychiatry 55(10):869–876

27. Glanz BI, Holland CM, Gauthier SA, Amunwa EL, Liptak Z, Houtchens MK et al (2007) Cognitive dysfunction in patients with clinically isolated syndromes or newly diagnosed multiple sclerosis. Mult Scler J 13 (8):1004–1010

28. Rey A (1964) L 'examen clinique en psychologie. Presses Universitaires de France, Paris

29. Schmidt M (1996) Rey auditory verbal learning test: a handbook. Western Psychological Services, Los Angeles, CA

30. Müller H, Hasse-Sander I, Horn R, Helmstädter C, Elger CE (1997) Rey auditory verbal learning test: structure of a modified German version. J Clin Psychol 53 (7):663–671

31. Miranda JP, Valencia RR (1997) English and Spanish versions of a memory test: word-length effects versus spoken duration effects. Hisp J Behav Sci 19(2):171–181

32. Lee TMC (2003) Normative data: neuropsychological measures for Hong Kong Chinese. Neuropsychology Laboratory, The University of Hong Kong, Hong Kong

33. Wechsler D (2009) Wechsler memory scale—fourth edition (WMS–IV) technical and interpretive manual. Pearson, San Antonio, TX

34. Feuillet L, Reuter F, Audoin B, Malikova I, Barrau K, Cherif AA, Pelletier J (2007) Early cognitive impairment in patients with clinically isolated syndrome suggestive of multiple sclerosis. Mult Scler J 13(1):124–127

35. Leavitt VM, Wylie G, Krch D, Chiaravalloti N, DeLuca J, Sumowski JF (2014) Does slowed processing speed account for executive deficits in multiple sclerosis? Evidence from neuropsychological performance and structural neuroimaging. Rehabil Psychol 59(4):422–428

36. Anhoque CF, Biccas Neto L, Domingues SCA, Teixeira AL, Domingues RB (2012) Cognitive impairment in patients with clinically isolated syndrome. Dement Neuropsychol 6 (4):266–269

37. Henry JD, Beatty WW (2006) Verbal fluency deficits in multiple sclerosis. Neuropsychologia 44(7):1166–1174

38. Huijbregts SCJ, Kalkers NF, de Sonneville LMJ, de Groot V, Reuling IEW, Polman CH (2004) Differences in cognitive impairment of relapsing remitting, secondary, and primary progressive MS. Neurology 63(2):335–339

39. Gronwall DM (1977) Paced auditory serial-addition task: a measure of recovery from concussion. Percept Mot Skills 44(2):367–373

40. Cutter GR, Baier ML, Rudick RA, Cookfair DL, Fischer JS, Petkau J, Ellison GW (1999) Development of a multiple sclerosis functional composite as a clinical trial outcome measure. Brain 122(5):871–882

41. Rao SM (1990) Neuropsychological screening battery for multiple sclerosis. National Multiple Sclerosis Society, New York, NY, USA

42. Aschenbrenner S, Tucha O, Lange KW (2001) Regensburger Wortflüssigkeits-Test. Hogrefe, Göttingen

43. Artiola L, Hermisollo D, Heaton R, Pardee RE (1999) Manual de normas y procedimientos para la batería neuropsicológica en español [manual of norms and procedures for the Spanish neuropsychological battery]. m Press, Tucson, AZ

44. Vleugels L, Lafosse C, van Nunen AN, Nachtergaele S, Ketelaer P, Charlier M, Vandenbussche E (2000) Visuoperceptual impairment in multiple sclerosis patients diagnosed with neuropsychological tasks. Mult Scler 6(4):241–254

45. Barnes CA (1979) Memory deficits associated with senescence: a neurophysiological and behavioral study in the rat. J Comp Physiol Psychol 93(1):74–104

46. Rosenfeld CS, Ferguson SA (2014) Barnes maze testing strategies with small and large rodent models. J Vis Exp 84:e51194

47. Harrison FE, Reiserer RS, Tomarken AJ, McDonald MP (2006) Spatial and nonspatial escape strategies in the Barnes maze. Learn Mem 13(6):809–819

48. Paul CM, Magda G, Abel S (2009) Spatial memory: theoretical basis and comparative review on experimental methods in rodents. Behav Brain Res 203(2):151–164

49. Holtzman DA, Harris TW, Aranguren G, Bostock E (1999) Spatial learning of an escape task by young corn snakes, Elaphe guttata guttata. Anim Behav 57(1):51–60

50. Brown S, Strausfeld N (2009) The effect of age on a visual learning task in the American cockroach. Learn Mem 16(3):210–223

51. Languille S, Aujard F, Pifferi F (2012) Effect of dietary fish oil supplementation on the exploratory activity, emotional status and spatial memory of the aged mouse lemur, a non-human primate. Behav Brain Res 235 (2):280–286

52. Hollinger KR, Alt J, Riehm AM, Slusher BS, Kaplin AI (2016) Dose-dependent inhibition of GCPII to prevent and treat cognitive impairment in the EAE model of multiple sclerosis. Brain Res 1635:105–112

53. CANTAB® [Cognitive assessment software]. Cambridge Cognition (2019). All rights reserved. www.cantab.com

54. Weed MR, Taffe MA, Polis I, Roberts AC, Robbins TW, Koob GF et al (1999) Performance norms for a rhesus monkey neuropsychological testing battery: acquisition and long-term performance. Cogn Brain Res 8 (3):185–201

Chapter 13

Models for Assessing Anxiety and Depression in Multiple Sclerosis: from Mouse to Man

Erik Ellwardt and Dirk Luchtman

Abstract

Multiple sclerosis (MS) patients often suffer from anxiety disorders and depression. An effective treatment would improve quality of life and would have a positive socioeconomic impact. However, research, diagnosis, and treatment of such mood disorders in MS have not been paid sufficient attention in the past. For that, validated and reliable models are necessary for a putative translation to the clinic. We here review animal models of anxiety and depression and summarize tools for anxiety/depression assessment in MS patients.

Key words Anxiety, Depression, Rodent, Multiple sclerosis, Mood disorder, Translation

1 Introduction

Mood disorders such as anxiety and depression are often observed in patients with multiple sclerosis (MS) [1]. The notion that mood disorders can occur even early on during the disease course and even precede motor or sensory symptoms has led to new interest in the field of neuropsychiatric disorders in MS [2]. Approximately one-third of MS patients are diagnosed during their disease course with an anxiety disorder [3, 4]. About the same percentage of patients suffers from depression [4, 5]. Neuropsychiatric symptoms and especially mood disorders in MS therefore contribute significantly to disease burden and display a relevant socioeconomic burden.

It is still under debate whether such mood disorders are just comorbidities of MS or part of the disease pathology and directly caused by demyelination or neurodegeneration. On the one hand, pain, chronic fatigue, alcohol abuse, and cognitive impairment, which can occur among MS patients, are contributing factors to the development of depression or anxiety [6]. On the other hand, pathophysiological changes that occur in MS, including lesion load and brain atrophy, can also contribute to the development of mood

Sergiu Groppa and Sven G. Meuth (eds.), *Translational Methods for Multiple Sclerosis Research*, Neuromethods, vol. 166, https://doi.org/10.1007/978-1-0716-1213-2_13, © Springer Science+Business Media, LLC, part of Springer Nature 2021

disorders [7–10]. Functional studies even suggest that connectivity between brain regions involved in emotional processing, such as prefrontal cortex and amygdala, is impaired in MS patients [11]. Further knowledge about the exact pathophysiology of anxiety and depression-related MS symptoms might shed light on overall disease pathology of MS and lead to new treatment approaches. For example, repetitive transcranial magnetic stimulation (rTMS) of impaired networks might help to restore network efficiency and improve clinical outcomes [12].

However, the correct diagnosis of such mood disorders is mandatory. This chapter will assess screening tools, especially questionnaires which are currently used for the diagnosis of anxiety or depression in MS. In addition, this chapter will outline experimental approaches in rodents to assess anxiety and depression like behavior which can be used to study new experimental therapeutic approaches.

2 Material and Methods for Measuring Anxiety and Depression in Rodents

The following tests are the most commonly used laboratory tests for assessing anxiety and depression in rodents. Though they can be applied to other animals, for simplicity and because mice are most commonly used in these tests, below we describe the setup used in mice. They can easily be combined with the animal model of MS, experimental autoimmune encephalomyelitis (EAE). However, because most of the tests require intact locomotion, they should be performed either in prodromal phases or in remission when no motor symptoms are present. Here, for example, the SJL mouse strain could be used due to its relapsing–remitting disease course following immunization [13]. As with most behavioral assessments, great care needs to be taken to a number of potentially confounding factors, including inherent mouse strain phenotypes (e.g., blindness, hyperactivity), time of day effects, day vs. night-phase testing, batch-related behavioral variance, and sample sizes, which are further discussed below. Failure to carefully evaluate these factors may result in nonvalid or nonreliable measurements.

2.1 Anxiety

Typical anxiety-related behaviors of rodents include, among others, pathological avoidance of exposed areas, increased grooming behavior, ultrasound vocalizations at frequencies of 20–30 kHz, limited interaction with new unfamiliar objects, sustained attention to stimuli, and an increased freezing response to threat or fearful situations [14]. In general, there are two types of anxiety. Normal, or "state," anxiety, is triggered through painful or nonpainful stimuli (e.g., highly illuminated areas), and is situation dependent but not generally present. Pathological, or "trait," anxiety is rather characterized by continuous anxiety-related behavior, even in the

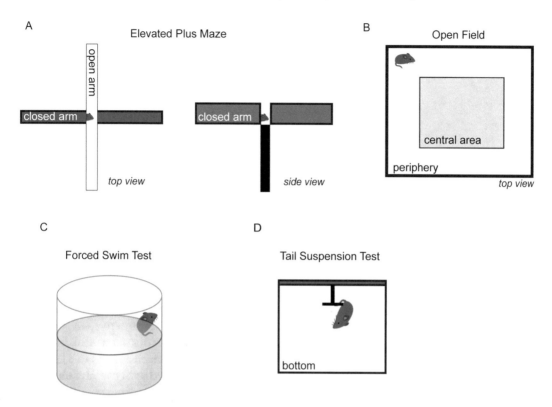

Fig. 1 Setup of frequently used animal tests for anxiety (EPM top view and side view, **a** and OF, **b**) and for depression like behavior (FST, **c** and TST, **d**)

absence of the aversive stimulus [15], and can be induced via genetic modification or disease models such as EAE [13]. There are several well-established tools to assess anxiety-related behavior (Fig. 1), the most common of which are the open field test, the elevated plus-maze, the elevated zero-maze, the staircase test, light/dark box exploration test or dark/light emergence test, and the mirrored chamber test. These exploration-based tests have the advantage that they do not require long training periods in advance. Other tests, for example punishment-associated tests (e.g., Geller–Seifter task), require longer training periods and often an elaborate experimental setup. All these behavioral tests have been used in preclinical testing, for example to assess the potency of anxiolytic drug candidates. We will here focus on exploratory tests due to their wide application.

2.1.1 Elevated Plus Maze (EPM)

The EPM is widely used in behavioral neuroscience experiments [16]. The rodent is placed on an elevated maze with four arms (e.g., 30 cm length for mice), two of them open (exposed) and two closed with surrounding walls (*see* Fig. 1a), configured in a plus shape which is elevated 40–50 cm above the floor (e.g., Maze Elevated Plus, Columbus Instruments, USA). Usually the animals prefer to stay in the closed arms but occasionally also explore the open arms. Standard dependent variables of the test include the number of entries into or time spent in the closed and open arms, but additional variables such as the numbers of rears and stretch-attend postures, grooming and freezing counts as well as metabolic parameters such as urination and defecation can all be evaluated to obtain an inclusive picture of the animals anxiety-related behavioral state. The animal is initially placed at the intersection of the closed and open arms, always facing the same direction. A video tracking system (e.g., Any-Maze, Stoelting, USA) is usually used to auto-matically detect the entries but additionally it is recommended that the experimenter also records entries and time spent in each area. An anxious mouse would rather stay in the closed arms and show fewer entries per minute, thus less exploration behavior. Usually 5 min recording time is sufficient.

2.1.2   Open Field (OF)

Besides the EPM the open field (OF) test is another very frequently used test to assess anxiety-related behavior quickly and without sophisticated equipment [17]. Either a square (*see* Fig. 1b, $420 \times 420 \times 420$ mm) or circular box (diameter 420 mm) can be used (e.g., open field from Maze Engineers, USA) [18]. Similar to the EPM, automated software combined with cameras should be used to objectively assess movement behavior of the animals [19]. Because of the larger movement surface of the OF, such programs may also be used to generate track patterns of the ani-mals' movement about the maze, which gives an immediate and easily presentable indication of their location tendencies. Apart from general assessment of locomotor activity (distance moved (cm) and mean as well as peak velocity (cm/s)) and the additional basic behavioral measures listed above for the EPM, a vital anxiety-related variable in the OF is the animal's avoidance of the central area. This can be measured by counting the number of crossings through this area and time duration spent in it. Normally animals tend to stick to the walls but frequently also explore the center areas. Anxious mice typically move less, spend more time in the periphery, and explore the central area less. Important to note here is that the anxiety provoking conditions of both the EPM and especially the OF depend on whether the tests are performed in the dark or in an illuminated environment. The center area of the OF is far more menacing to an anxious mouse if it is brightly illuminated than when it is situated in the dark. The dark condition could be used to test the animal's motivation to move, or lack thereof, rather than anxiety. While low motivation could be an indication of depression, there are more strict depression tests, as listed below.

2.1.3 Further Anxiety Tests

There are other less frequently used tests, which in some cases represent modified versions of the EPM and OF tests [14]. The *elevated zero maze* test, for example, is a circular elevated maze with open and closed areas. Similar to the EPM, time spent in the open and closed areas and number of entries can be assessed. The *light/dark box* is divided into two areas: a closed dark box (1/3 of total area) and an aversive open brightly illuminated area (2/3 of total area). Animals normally tend to stay within the dark box but naturally also explore the illuminated space occasionally [20]. The *mirrored chamber test* is a further but less frequently used test based on rodents' aversion to mirrors.

2.2 Depression

Typical symptoms of depression in humans include recurrent thoughts of death, feelings of worthlessness, and lack of ability to experience joy and pleasure (anhedonia). Since rodents cannot be interviewed for these symptoms, evaluation of depression-like symptoms has to be based instead on standardized objective measures, just as for anxiety. These include reduced locomotor activity, social isolation, reduced sexual motivation, impaired sleep architecture (requires rodent EEG recordings), reduced spatial memory, or abnormal weight loss. Typical depression models consist of stressful situations [21, 22], since stress is one of the major predisposing factors to depression. The forced swim test (FST) and the tail suspension test (TST) are two of the most widely used tests for evaluating therapeutic effects of antidepressant drug candidates. These tests are not time intensive and do not require elaborate or expensive apparatus and pretraining.

2.2.1 Forced Swim Test (FST)

For the FST, mice are placed into a cylinder filled with water without the ability to escape for 5 min (*see* Fig. 1c). Initially, mice will struggle to try to escape, but after a certain time they stay immobile and resign. The time spent immobile is measured and thought to be related to depressive behavior [23]. The test apparatus can be easily built by the experimenter or acquired commercially (e.g., Any-Maze, Stoelting, USA). However, it should be noted that there is some debate about whether the FST measures depression-like behavior or rather a reaction to an acute stressor [24].

2.2.2 Tail Suspension Test (TST)

The TST is a similar test to the FST, but in this case mice are hung upside-down by their tails so that they cannot move around [25]. Mice usually struggle to free themselves but after a certain time resign to their situation and stay immobile. The time spent immobile during the experiment, which normally lasts 6 min, is measured and is thought to be related to depression-like behavior just as in the FST. Both the FST and TST are tests that can be conducted quickly and without expensive equipment. Although there are some doubts about their interpretation concerning depression like behavior, they are frequently used for high through-put screening of antidepressant treatments or in genetic animal models.

2.2.3 Further Depression Tests

The *learned helplessness* model couples unpleasant stimuli (e.g., electric shocks) with the inability to escape [26]. At a later time point during the experiments, mice still receive the aversive stimuli, but are given the possibility to escape, which they usually fail to do. Antidepressant drugs normalize this behavior and reestablish the natural instinct to escape. The *chronic mild stress* (CMS) model consists of exposing mice to low-level stressors, such as wet bedding, food deprivation or constant lighting for several days [27], resulting in depression-like symptoms such as weight loss, reduced locomotor activity and less grooming behavior. The CMS is the most closely analogous to the human situation, where chronic mild stress is thought to be one of the major triggers of depression. However, the CMS usually lasts several days or weeks and therefore is more time-consuming and stressful to the animal than other measures such as the FST or TST.

2.3 Genetic Mouse Models of Anxiety and Depression

Recent advances in genetics and genetic engineering have led to a number of transgenic lines, mostly in mice, that mimic anxiety and/or depression. These mice can be used to investigate the direct causal relation of genes to symptoms and provide the opportunity to test promising drug candidates. For further reading, please *see* Table 1.

3   Notes for Animal Experiments

1. Statistical power analysis should be performed for animal number calculations; usually a lower effect size should be assumed due to high variability.

2. Larger samples require more time to complete the tests, as each animal often has to complete multiple trials and thus require careful planning.

3. Longer testing times warrant group and trial randomization to time of day, as different times of day may have a significant impact on the activity of the animals. When tests are repeated on different days, they should be performed during consistent day times.

4. It should be decided whether the tests should be performed in the day or night phase of the animals' sleep and wake cycle, as this may drastically impact the results.

Table 1
Genetic mouse models for anxiety and depression in rodents

Model	Target System (gene)	References
Depression	Norepinephrine system (NET, Adra2α, dopamine-b-hydroxylase)	[28, 29]
Depression	Serotonin system (5-HT$_{1A}$, 5-HT$_{1B}$, 5-HT$_{2C}$, 5–HT transporter)	[30, 31]
Depression	mGluR7 knockout	[32]
Depression	Substance P (Tachykinin NK1 receptor, Tac1 gene)	[33, 34]
Depression	Nicotinic β2	[35]
Depression	Phosphodiesterase 4D	[36]
Anxiety	Norepinephrine system (MAO, COMT, NET, VMAT2, Adra2α)	[37, 38]
Anxiety	Serotonin system (5-HT$_{1A}$, 5-HT$_{1B}$, 5–HT$_{2C}$, 5-HT$_{5A}$, 5–HT transporter)	[39]
Anxiety	GABA system (GAD65, β2 subunit, γ2-subunit)	[40, 41]
Anxiety	Corticotrophin releasing hormone (CRH) and hypothalamic–adrenal–pituitary (HPA) axis	[42]
Anxiety	Tac1	[34]

5. Animals should habituate to the testing environment for at least 7 days prior to any experiments and be regularly handled by the experimenter.

6. Different batches from the same or different vendors may display different behavior, despite the same testing conditions.

7. General testing conditions, including noise, movements and personnel should be kept consistent across trials.

8. Experimental apparatus should be carefully cleaned between each animal to remove odor cues.

9. If available: do not forget to turn on the video tracking software.

4 Material and Methods for Measuring Anxiety and Depression in Patients

4.1 Anxiety

Anxiety in humans is normally assessed using interviews and questionnaires, always in addition to a complete physical examination by a physician (overview *see* Table 2). We will focus here on the self-report tools, especially questionnaires, their sensitivity and specificity. The *hospital anxiety depression scale* (HADS) is a widely used screening tool, established in 1983, to assess both anxiety (HADS-A) and depression (HADS-D) in patients [43]. In contrast to what the name suggests, it is not only applicable in hospitals but also in outpatient clinics or at home. The questionnaire consists of 14 multiple-choice questions, 7 items relating to anxiety (HADS-A) and

Table 2
Frequently used anxiety and depression questionnaires for patients

Anxiety	Depression	References
HADS-A[a]	HADS-D[a]	[43]
Becks Anxiety inventory	Beck Depression inventory[a]	[45, 50]
State-trait anxiety[a]	Major Depression inventory	[47, 51]
HAM-A	HAM-D	[48, 52]
GAD-7	PHQ-9	[49, 53]
	Goldberg depression	[54]
	CES-D	[55]

[a]frequently used and best validated in MS patients

7 to depression (HADS-S). Each question has four possible answers that are scored from 0 to 3. A sum score of more than 10 points is pathological for each subscale (HADS-A and D). Although it rather screens general anxiety disorders and panic attacks and to a lesser extent phobic disorders, it is frequently used due to its good sensitivity and specificity compared to other self-reported questionnaires [44]. The *Becks anxiety inventory* [45], first published in 1988, consists of 21 items with a maximum score of 63 points (each item with 0 to 3 points) and measures panic disorders and general anxiety disorders. Scores from 16 points indicate moderate anxiety and above 25 points severe anxiety. The ability of this test to discriminate anxiety from depression is generally praised. The questionnaire takes no more than 10 min to complete, and it represents an alternative for anxiety screening in MS patients.

The *state-trait anxiety inventory* is a more complex self-report tool to detect anxiety [46, 47]. Trait anxiety represents a general feeling of anxiety which is constantly present, whereas state anxiety indicates an anxious response triggered by specific situations or events. This inventory consists of a total of 40 questions, with 20 items related to state and 20 to trait anxiety. Each answer can be scored on a 4-point scale (1 point relates to absence of anxiety and 4 points to pronounced symptoms) with a maximum of 80 points for both scales. Higher scoring translates to higher probability of anxiety.

Hamilton Anxiety Rating Scale (HAM-A, clinician rated), introduced already 1959 and *Generalized Anxiety Disorder 7* (GAD-7), is a seven-item screening tool for generalized anxiety disorders are further assessments which are often applied [48, 49].

5   Depression

As mentioned above, the *HADS-D* represents a good screening tool for subjects with depression [43]. The counterpart to the anxiety inventory developed by Beck is the *Beck Depression Inventory* which measures the severity of depression [50] using 21 questions (scoring from 0 to 3 points for each question just as for the anxiety test). It was developed in 1961 and since then there have been short forms and further developments of the original test. In contrast to other depression questionnaires it is not thought to be a screening tool for depression but rather assesses the current severity of a depressive disorder. In contrast, the *Patient Health Questionnaire* (PHQ-9) is a quick screening tool for depression consisting of nine questions which are directly related to the nine diagnostic criteria of major depression [53]. The test score directly represent the severity of the depression and therefore qualify for a standardized assessment of the disease course. Similarly, the *Major Depression Inventory*, developed by the World Health Organization, measures depression severity and serves as a screening tool [51]. Additional frequently used self-reported questionnaires are the *Goldberg Depression Questionnaire* [54] and the *Center for Epidemiological Studies Depression Scale* (CES-D) [55]. Similar to the HAM-A, the *Hamilton Depression Scale (HAM-D)* is used by the physician [52] and widely used in clinical practice.

6   Assessment in MS Patients

In principle, each of the abovementioned tools could be used to assess anxiety or depression in MS. The correct diagnosis of such mood disorders is vital, as they reduce quality of life and increase the risk of disease deterioration [56]. However, the HADS questionnaire, state-trait anxiety inventory, and the Becks depression inventory are most frequently used (*see* Table 2) for clinical and research purposes in MS patients [6, 10, 57, 58].

7   Experimental Approaches to Assess Anxiety and Depression in Humans

In addition to established screening tools like the aforementioned self-reported questionnaires and the physical examination, there are experimental imaging approaches to visualize alterations of networks involved in emotional processing [59, 60]. Passamonti et al. showed very elegantly that the connectivity between prefrontal cortex areas and amygdala, two key regions involved in mood disorders, is impaired during emotional processing in MS [11]. The knowledge of such "abnormal network communications" in MS patients could be used to apply personalized stimulations of impaired connectivity routes to improve patient symptoms [12]. Hence, both diagnosis and therapy of mood disorders such as depression and anxiety could profit from functional imaging studies in the future. However, validation in larger cohorts and randomized clinical trials are so far lacking and this research is a long way from being used in the clinic.

8   Conclusion

About one-third of MS patients suffer from anxiety and depression. Mood disorders can even precede motor or sensory symptoms [2] and represent a socioeconomic burden besides apparent disability (e.g., motoric deficits). To improve the situation for patients, we need reliable animal models to dissect the pathophysiology underlying such mood disorders and to test drug candidates. The murine system is commonly used due to the availability of standardized experimental tests, a multitude of genetic mouse lines and the general conditions applying to experimental mouse research, including easy housing and handling properties. For assessment of anxiety, exploration-based approaches such as the open field or the elevated plus maze are frequently used. To assess depression, the forced swim test or the tail suspension test can be applied, among others. All these tests have in common that they do not require sophisticated resources or long training periods and, importantly, can be easily applied in EAE, the murine model of MS, as long as the animal is free of motor impairments. One obvious drawback is, that it is not clear to what extent they measure what we call in patients "anxiety" or "depression" [24]. However, they all measure a certain level of stress resistance of the animals and are commonly used for preclinical testing of therapeutic interventions. In patients, several self-report questionnaires are available. For MS patients, the most widely used are the HADS, Becks depression inventory and the state-trait inventory to evaluate anxiety or depression [6, 57]. Thus, the validation of promising drug candidates or treatment interventions can be tested and translated from animal experiments to patients. Both the animal experiments and the questionnaires described here are easy applicable, time-efficient, and cost-effective methods. They are therefore good options for testing mood disorders in the context of MS.

Acknowledgments

We thank Rosalind Gilchrist for proofreading the manuscript. *Funding*: This work was supported by the German Research Council (DFG, CRC-TR-128).

References

1. Murphy R et al (2017) Neuropsychiatric syndromes of multiple sclerosis. J Neurol Neurosurg Psychiatry 88(8):697–708

2. Disanto G et al (2018) Prodromal symptoms of multiple sclerosis in primary care. Ann Neurol 83(6):1162–1173

3. Korostil M, Feinstein A (2007) Anxiety disorders and their clinical correlates in multiple sclerosis patients. Mult Scler 13(1):67–72

4. Kalron A, Aloni R, Allali G (2018) The relationship between depression, anxiety and cognition and its paradoxical impact on falls in multiple sclerosis patients. Mult Scler Relat Disord 25:167–172

5. Boeschoten RE et al (2017) Prevalence of depression and anxiety in multiple sclerosis: a systematic review and meta-analysis. J Neurol Sci 372:331–341

6. Feinstein A et al (2014) The link between multiple sclerosis and depression. Nat Rev Neurol 10(9):507–517

7. Bakshi R et al (2000) Brain MRI lesions and atrophy are related to depression in multiple sclerosis. Neuroreport 11(6):1153–1158

8. Zorzon M et al (2002) Depressive symptoms and MRI changes in multiple sclerosis. Eur J Neurol 9(5):491–496

9. Sanfilipo MP et al (2006) Gray and white matter brain atrophy and neuropsychological impairment in multiple sclerosis. Neurology 66(5):685–692

10. Rossi S et al (2017) Neuroinflammation drives anxiety and depression in relapsing-remitting multiple sclerosis. Neurology 89(13):1338–1347

11. Passamonti L et al (2009) Neurobiological mechanisms underlying emotional processing in relapsing-remitting multiple sclerosis. Brain 132(Pt 12):3380–3391

12. Hulst HE et al (2017) rTMS affects working memory performance, brain activation and functional connectivity in patients with multiple sclerosis. J Neurol Neurosurg Psychiatry 88(5):386–394

13. Ellwardt E et al (2018) Maladaptive cortical hyperactivity upon recovery from experimental autoimmune encephalomyelitis. Nat Neurosci 21(10):1392–1403

14. Cryan JF, Holmes A (2005) The ascent of mouse: advances in modelling human depression and anxiety. Nat Rev Drug Discov 4(9):775–790

15. Belzung C, Griebel G (2001) Measuring normal and pathological anxiety-like behaviour in mice: a review. Behav Brain Res 125(1-2):141–149

16. Walf AA, Frye CA (2007) The use of the elevated plus maze as an assay of anxiety-related behavior in rodents. Nat Protoc 2(2):322–328

17. Seibenhener ML, Wooten MC (2015) Use of the open field maze to measure locomotor and anxiety-like behavior in mice. J Vis Exp 96:e52434

18. Kraeuter AK, Guest PC, Sarnyai Z (2019) The open field test for measuring locomotor activity and anxiety-like behavior. Methods Mol Biol 1916:99–103

19. Samson AL et al (2015) MouseMove: an open source program for semi-automated analysis of movement and cognitive testing in rodents. Sci Rep 5:16171

20. Bourin M, Hascoet M (2003) The mouse light/dark box test. Eur J Pharmacol 463(1-3):55–65

21. Caspi A et al (2003) Influence of life stress on depression: moderation by a polymorphism in the 5-HTT gene. Science 301(5631):386–389

22. Menard C, Hodes GE, Russo SJ (2016) Pathogenesis of depression: insights from human and rodent studies. Neuroscience 321:138–162

23. Yankelevitch-Yahav R et al (2015) The forced swim test as a model of depressive-like behavior. J Vis Exp 97:52587

24. Molendijk ML, de Kloet ER (2015) Immobility in the forced swim test is adaptive and does not reflect depression. Psychoneuroendocrinology 62:389–391

25. Can A et al (2012) The tail suspension test. J Vis Exp 59:e3769

26. Vollmayr B, Gass P (2013) Learned helplessness: unique features and translational value of a cognitive depression model. Cell Tissue Res 354(1):171–178

27. Antoniuk S et al (2019) Chronic unpredictable mild stress for modeling depression in rodents: meta-analysis of model reliability. Neurosci Biobehav Rev 99:101–116

28. Haller J et al (2002) Behavioral responses to social stress in noradrenaline transporter knockout mice: effects on social behavior and depression. Brain Res Bull 58(3):279–284

29. Cryan JF et al (2004) Norepinephrine-deficient mice lack responses to antidepressant drugs, including selective serotonin reuptake inhibitors. Proc Natl Acad Sci U S A 101(21):8186–8191

30. Holmes A et al (2002) Evaluation of antidepressant-related behavioral responses in mice lacking the serotonin transporter. Neuropsychopharmacology 27(6):914–923

31. Heisler LK et al (1998) Elevated anxiety and antidepressant-like responses in serotonin 5-HT1A receptor mutant mice. Proc Natl Acad Sci U S A 95(25):15049–15054

32. Cryan JF et al (2003) Antidepressant and anxiolytic-like effects in mice lacking the group III metabotropic glutamate receptor mGluR7. Eur J Neurosci 17(11):2409–2417

33. Rupniak NM et al (2001) Comparison of the phenotype of NK1R−/− mice with pharmacological blockade of the substance P (NK1) receptor in assays for antidepressant and anxiolytic drugs. Behav Pharmacol 12(6-7):497–508

34. Bilkei-Gorzo A et al (2002) Diminished anxiety- and depression-related behaviors in mice with selective deletion of the Tac1 gene. J Neurosci 22(22):10046–10052

35. Caldarone BJ et al (2004) High-affinity nicotinic acetylcholine receptors are required for antidepressant effects of amitriptyline on behavior and hippocampal cell proliferation. Biol Psychiatry 56(9):657–664

36. Zhang HT et al (2002) Antidepressant-like profile and reduced sensitivity to rolipram in mice deficient in the PDE4D phosphodiesterase enzyme. Neuropsychopharmacology 27(4):587–595

37. Cases O et al (1995) Aggressive behavior and altered amounts of brain serotonin and norepinephrine in mice lacking MAOA. Science 268(5218):1763–1766

38. Xu F et al (2000) Mice lacking the norepinephrine transporter are supersensitive to psychostimulants. Nat Neurosci 3(5):465–471

39. Ramboz S et al (1998) Serotonin receptor 1A knockout: an animal model of anxiety-related disorder. Proc Natl Acad Sci U S A 95(24):14476–14481

40. Kash SF et al (1999) Increased anxiety and altered responses to anxiolytics in mice deficient in the 65-kDa isoform of glutamic acid decarboxylase. Proc Natl Acad Sci U S A 96(4):1698–1703

41. Sur C et al (2001) Loss of the major GABA (A) receptor subtype in the brain is not lethal in mice. J Neurosci 21(10):3409–3418

42. Heinrichs SC et al (1997) Anti-sexual and anxiogenic behavioral consequences of corticotropin-releasing factor overexpression are centrally mediated. Psychoneuroendocrinology 22(4):215–224

43. Zigmond AS, Snaith RP (1983) The hospital anxiety and depression scale. Acta Psychiatr Scand 67(6):361–370

44. Bjelland I et al (2002) The validity of the hospital anxiety and depression scale. An updated literature review. J Psychosom Res 52(2):69–77

45. Beck AT et al (1988) An inventory for measuring clinical anxiety: psychometric properties. J Consult Clin Psychol 56(6):893–897

46. Marteau TM, Bekker H (1992) The development of a six-item short-form of the state scale of the Spielberger state—trait anxiety inventory (STAI). Br J Clin Psychol 31(3):301–306

47. Spielberger CD, Gorsuch RL, Lushene RE (1968) State-trait anxiety inventory (STAI: Test Manual for Form X). Consulting Psychologists Press, Palo Alto, California

48. Hamilton M (1959) The assessment of anxiety states by rating. Br J Med Psychol 32(1):50–55

49. Spitzer RL et al (2006) A brief measure for assessing generalized anxiety disorder: the GAD-7. Arch Intern Med 166(10):1092–1097

50. Beck AT et al (1961) An inventory for measuring depression. Arch Gen Psychiatry 4:561–571

51. Bech P et al (2001) The sensitivity and specificity of the major depression inventory, using the present state examination as the index of diagnostic validity. J Affect Disord 66(2-3):159–164

52. Hamilton M (1960) A rating scale for depression. J Neurol Neurosurg Psychiatry 23:56–62

53. Spitzer RL, Kroenke K, Williams JB (1999) Validation and utility of a self-report version of PRIME-MD: the PHQ primary care study. Primary care evaluation of mental disorders. Patient health questionnaire. JAMA 282(18):1737–1744

54. Goldberg I (1993) Questions and answers about depression and its treatment: a consultation with a leading psychiatrist paperback. Charles Press Pubs, Philadelphia

55. Radloff LS (1977) The CES-D scale: a self-report depression scale for research in the general population. Appl Psychol Meas 1(3):385–401

56. Kowalec K et al (2017) Comorbidity increases the risk of relapse in multiple sclerosis: a prospective study. Neurology 89(24):2455–2461

57. Minden SL et al (2014) Evidence-based guideline: assessment and management of psychiatric disorders in individuals with MS: report of the guideline development Subcommittee of the

American Academy of neurology. Neurology 82(2):174–181

58. Honarmand K, Feinstein A (2009) Validation of the hospital anxiety and depression scale for use with multiple sclerosis patients. Mult Scler 15(12):1518–1524

59. Tahedl M et al (2018) Functional connectivity in multiple sclerosis: recent findings and future directions. Front Neurol 9:828

60. Louapre C et al (2014) Brain networks disconnection in early multiple sclerosis cognitive deficits: an anatomofunctional study. Hum Brain Mapp 35(9):4706–4717

Part IV

Exploring Translational Paths in MS with Brain Imaging: Small Animal to Human Imaging Tools

Chapter 14

Human Structural MRI

Menno M. Schoonheim

Abstract

Multiple sclerosis (MS) is a neuroinflammatory and neurodegenerative disease of the central nervous system featuring different forms of structural damage, such as lesions and atrophy. The extent and severity of this damage is highly heterogeneous between patients and requires different types of imaging modalities to visualize. In this chapter, different types of pathologies visible on conventional and advanced imaging methods are outlined. Recent concepts based on these findings are discussed, highlighting critical structures where damage is most predictive for disease progression, concluding with future perspectives.

Key words Multiple sclerosis, Pathology, MRI, Structural damage, Conventional, Advanced

1 Introduction

Multiple sclerosis (MS) [1] is a disease of the central nervous system featuring an autoimmune response against myelin, with extensive neuroinflammation and neurodegeneration. These changes have classically been described in the white matter only, which has led to the belief that MS is a only white matter disorder. This concept was driven by the lack of sufficient histopathological techniques allowing the visualization of grey matter damage in MS, but also by the lack of sufficiently advanced MRI sequences at the time. Interestingly, it is now known that MS can feature vastly expansive grey matter demyelination, with a very strong impact on patient symptoms [2]. These developments have now even led to the hypothesis that the primary cause of MS may not be driven by outside-in processes, that is, a primary cause within the immune system, but by inside-out processes, such as a primary neurodegenerative event [3]. As such, the past years have seen a push to further deepen our understanding of structural damage in MS, in order to grasp the disease fully.

This push has still mostly been constrained to research applications, however, as the radiological work-up of patients has remained similar for many years. In clinical practice, the

Sergiu Groppa and Sven G. Meuth (eds.), *Translational Methods for Multiple Sclerosis Research*, Neuromethods, vol. 166, https://doi.org/10.1007/978-1-0716-1213-2_14, © Springer Science+Business Media, LLC, part of Springer Nature 2021

visualization of grey matter pathology has long been excluded from diagnostic criteria and is still not part of routine clinical evaluations. Recently, however, this has changed, when grey matter lesions were included in the most recent diagnostic criteria for MS [4]. However, as visualizing grey matter lesions still requires a separate MR sequence, such as double inversion recovery (DIR) [5], it is unclear whether this change will lead to rapid changes to the clinical workup of patients.

In addition to grey matter lesions, additional advances in MS research have now highlighted the high potential for measures of atrophy [6]. Although still far from daily practice, it was shown that especially grey matter atrophy measures have some of the most strong correlations with clinical scores [7]. In addition, more advanced measures of white matter damage, such as diffusion tensor imaging (DTI) approaches and magnetization transfer imaging (MTI), have shown clear changes outside of lesional areas, that is, in the so-called normal appearing white matter (NAWM) [8]. The extent and severity of damage in these areas has also been shown to be highly clinically relevant, although still difficult or even impossible to measure in individual patients.

As such, the field has evolved drastically over the past few years. We can see more of the pathological changes in MS than ever before. Now, the next question is which patterns of damage we can identify that are especially relevant for patient care and/or future research. This chapter will outline some of the latest findings in MS on structural damage in the brain using MRI and their clinical relevance.

2   Conventional Structural MRI

In MS, the radiological protocol used for diagnosis typically includes 2D image sequences [4]. These include a T2-weighted sequence for the (hyperintense) visualization of white matter lesions, of which most will be visible at current field strengths. In addition, a T1-weighted sequence is used to identify active lesions, after gadolinium injection. This will allow the clinician to identify whiter matter lesions and whether they are old or new. These conventional imaging approaches are clinically very important, as they can identify the so-called dissemination in time and space, that is, multiple lesions in different locations at different time points. If both criteria are fulfilled, a diagnosis can be made [4]. Over the years these criteria have been developed further. Most recently this research line has led to the inclusion of grey matter lesions and CSF-based oligoclonal bands into the criteria [4].

In addition, T1-weighted sequences can identify hypointense lesions, that is, lesions that are more severely damaged. These so-called black holes [6] are associated with a more severe neurodegeneration (but not demyelination) within the lesion [9] but should ideally be confirmed longitudinally, to avoid misclassification due to edema. This longitudinal evolution has also been studied in the context of clinical trials, monitoring whether such lesions persist or not, which is strongly related to the severity of atrophy [10]. However, although information gained from T1-hypointense lesions is useful, the presence or absence of such lesions is currently not part of the diagnostic criteria.

It must be stressed, however, that while such lesional imaging is imperative for an early, sensitive and specific diagnosis, these measures do not commonly relate well with clinical functioning, the so-called clinic-radiological paradox [11]. In other words, these imaging modalities can diagnose patients well, but have limited value for prognosis [12]. Especially for more complex symptoms like cognitive dysfunction, a common and highly debilitating form of disability in MS, conventional lesion-based measures have been shown to correlate poorly [13].

As such, the MRI research field in MS has been pushed in recent years to find new ways to track pathological changes in vivo, explain more complex symptoms like cognition, and find better ways to predict progression in MS. This has led to a subsequent boom in the advanced MRI field [14].

3 Advanced Imaging of Lesions

The discovery of how to image grey matter lesions has solved some factors surrounding this paradox. Currently, the best way to visualize grey matter lesions at clinical field strengths is by suppressing the signal of CSF and white matter on T2-based sequences, resulting in a double-inversion recovery (DIR) sequence that can now be scored by clinicians with published consensus criteria [15]. Research using this technique has revealed that grey matter lesions are common in MS and seem to yield more power in explaining complex symptoms than white matter lesions do, and are now part of the diagnostic criteria [4]. Unfortunately, DIR-based imaging only identifies up to 20% of lesions, the so-called tip of the iceberg phenomenon, with no better technique currently available at 1.5 or 3T field strengths [16]. Especially missed are the smaller intracortical and ribbon-like subpial lesions that currently cannot be readily imaged at all, although these are also common, as seen in post-mortem pathology work [5]. Interestingly, some structures like the hippocampus are especially prone to grey matter lesions, which in some patients is entirely demyelinated [17]. Hippocampal lesions strongly relate to cognitive dysfunction [18, 19], and in some cases can even appear at disease onset [20].

Apart from grey matter lesions, the development of imaging techniques focused on white matter pathology has also continued. For instance, one of the main drawbacks of conventional MRI is that it is not quantitative. Conventional (clinical) scans can tell the clinician/researcher whether or not a lesion is present, and whether it is new or old. One of the latest technical achievements is the identification of the so-called central vein sign, which features the use of susceptibility-weighted approaches to identify a vein in the middle of an MS lesion, which seems to be quite specific for the disease and may help in diagnosis in the future [21]. These types of imaging approaches, however, can only identify whether or not a lesion may be present and specific for MS but does not provide information on how severe such a lesion would be. Using afore-mentioned conventional techniques, it is possible to distinguish black holes from other lesions, and detect how many lesions are especially destructive. As such, studies have used this information to look at the evolution of individual lesions.

For instance, several key papers have been published using magnetization transfer imaging [14]. This approach aims to quan-tify the myelin content within a certain area, such as a white matter lesion. Using this technique, it was identified that there is a large heterogeneity between lesions as well as between patients how these measures evolve over time. Some lesions will be highly destructive, indicating a continuous loss of myelin over time. Others, however, seem to regain signal, which may be indicative of remyelination [22]. This reparative potential of the brain may also underlie the seemingly altered appearance of some black hole lesions over time [23].

Additional techniques like diffusion tensor imaging (DTI) are able to measure the severity of damage as well, while MR spectros-copy (MRS) can quantify the content of specific metabolites in a given region, which can provide clues to what is going on in the tissue. Studies that have used such techniques have indicated that these may have the potential to detect the influx of immune cells as well as the initial structural damage during the formation of new lesions [24–26]. Most recently, the concept of slowly expanding "smouldering" lesions has gained a lot of attention [27]. These lesions are thought to represent chronic active lesions, that is, lesions that are no longer acute, but continuously feature destruc-tive processes, allowing them to slowly expand. In fact, such a measure of destructive lesional properties have now been shown to be predictive of subsequent clinical progression [28].

4   Normal-Appearing Tissue Damage

Outside of areas with focal lesional pathology, the integrity of the white matter has become increasingly important in recent years. On conventional MRI, there were already indications that there were additional areas of damage, as indicated by areas surrounding lesions, described as "diffusely abnormal" or "dirty" white matter, thought to reflect a dying back of axons starting at especially severe lesions [29]. Even further outside of lesions, in the normal-appearing white matter, additional changes can still be seen using quantitative MRI techniques [30]. This notion of diffuse and extensive patterns of damage that cannot be seen with conventional MRI has led to more advanced techniques, such as advanced processing techniques of DTI data. These techniques now enable the quantification of the extent and severity of damage, both within lesions as well as in the NAWM [31].

For DTI, tract-based spatial statistics (TBSS) is a commonly applied technique developed by FSL that circumvents a major problem in group comparisons using DTI data: registration errors and inter-individual variations that could result in spurious findings. This technique is therefore based on defining a so-called skeleton of major white matter bundles, determined by connecting areas with the highest fractional anisotropy (FA) in the images. These skeletons are then compared in a standard space environment, which allows for a reliable comparison of damage between groups, but only within the major white matter bundles. This approach has been extensively applied, and has for instance shown that the damage detectable with DTI far extends beyond lesional tissues into the normal-appearing white matter, with clear clinical correlations [32, 33].

5   Structural Disconnection

As it became clear that white matter damage extends far beyond focal lesions, the need to study individual tracts became clear. Tractography is a diffusion-based technique that aims to visualize specific white matter tracts of interest, identifying their anatomical location in the individual person, allowing the researcher to study tract integrity [34]. This technique has since been adopted into many software packages, each with a different approach. In MS, the technique has long been troubled by the poor potential for these techniques to pass through white matter lesions, overestimating structural disconnection. Nonetheless, tractography studies have shown clear patterns structural damage, for instance in the spinal cord, with strong correlations with disability [35]. As such, this information has led to a more holistic approach of the brain, looking at the brain as a network of interconnected systems [36]. This concept of network disconnection in MS [37] has coincided with a globally booming field of network neuroscience [38], aiming to unravel how connections in the brain lead to complex functions, like cognition.

In MS, the field of structural connectomics is still small, however, as most studies focus on functional connectivity. Early studies have related structural network changes to the presence of lesions, using cortical thickness-based structural networks, as tractography approaches at the time were not well-equipped to deal with lesions [39]. Such grey matter network studies also indicated a loss of network efficiency, related to cognitive dysfunction [40]. Newer processing pipelines have since shown that it is now possible to track across at least some of the lesions, again showing a less efficient network in MS. [41] This loss of efficiency might be driven by a more severe damage to long-distance connections [42], as well as network hubs [43]. A loss of such shortcuts in the brain would then result in a slower brain network, which would strongly affect brain functioning and hence, cognition. In addition, similar study designs have been used to show that the structural brain network is altered in its modularity, that is, the way subnetworks are interconnected, [44] which was related to the severity of atrophy [45]. Such changes in structural network topology are also related to a disconnect between brain structure and function [46], and may be predictive of conversion to MS [47]. And, most recently, it became very clear that analyzing network topology provides additional information, as a loss of structural network efficiency explained much more variability in disability compared to a TBSS-based summary of total damage [48]. Together, these findings indicate that white matter damage far exceeds lesional areas, but is non-random in MS, seemingly driven by network patterns, possibly leading up to the hypothesized "network collapse," after which clinical progression is unavoidable [49].

6   Atrophy

This network approach has also provided key insights into brain atrophy. Especially grey matter atrophy can be very severe in MS, both in the brain and spinal cord [50], featuring a rate of brain volume loss that can be up to four times higher than in healthy controls [51]. Atrophy has been shown to feature some of the strongest clinical correlations in MS, much more so than lesional pathology [7]. Histopathologically, cortical atrophy seems to be driven by a loss of axons and neurons, as well as a reduction in neuronal size, but not myelin content [52, 53]. The lack of a clear correlation between cortical demyelination and cortical atrophy has also been confirmed in-vivo, indicating that the two may be driven by different processes [54]. Further in-vivo evidence has since supported the hypothesis that structural disconnection seems to be the major driving force underlying atrophy in MS [55, 56].

Patterns if grey matter atrophy in MS have become much clearer in recent years [6], highlighting its non-random nature [57]. Atrophy seems to be worst in key network hubs like the

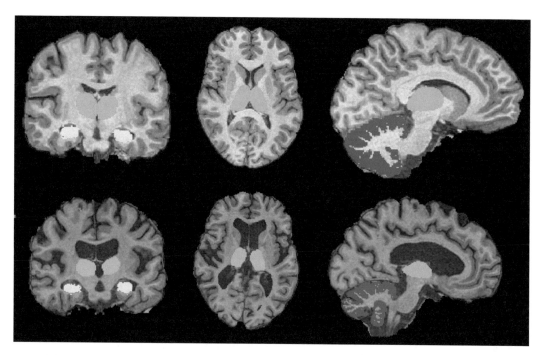

Fig. 1 Atrophy of cognitive network structures in multiple sclerosis. The thalamus (outlined in green), hippocampus (yellow), and cerebellum (blue) commonly feature pathological changes in MS, here outlined in a healthy female subject (top row) and cognitively impaired female MS patient (bottom row). Note that the cortex seems to be relatively spared in this patient, although cortical involvement is also strongly related to cognition

thalamus (*see* Fig. 1) [58, 59], which is affected very early in MS [60, 61] with strong relations to disability and cognitive impairment [13, 62]. Interestingly, thalamic lesions are not so commonly seen on MRI, which has sparked the hypothesis that thalamic atrophy might be the result of structural disconnections of thalamocortical tracts [63, 64], that is, acting like a "barometer" of all types of MS pathology in the brain [58]. Thalamic atrophy does not seem to accelerate over time [65]. This is different compared to the cortex, where atrophy seems to be centered around network hubs like the default-mode network [7, 57, 66], and does seems to accelerate in progressive MS [67]. Thalamic atrophy is predictive of disability progression and cognitive decline, especially in early stages of the disease, while cortical atrophy is especially predictive in later (progressive) stages [7, 68]. Spinal cord atrophy is also highly relevant for disability progression, and can now be readily measured in the upper cervical cord for research purposes [69].

The hippocampus has long been studied in MS and atrophy in this crucial structure is strongly linked to memory dysfunction [70] but also seems to be an important predictor of disability progression [7]. Similar to the thalamus, hippocampal atrophy has also been attributed to structural disconnection [71]. The thalamus is one of the few structures that has been studied extensively in MS, which has identified regional atrophy patterns [72], strongest in CA1 and subiculum [73–75], while CA2 and CA3 involvement has been specifically linked to depression [76]. The dentate has been reported to enlarge in MS, [74] while others did report CA1 atrophy [73]. Cerebellar atrophy is also important to note, as it is especially common in progressive phenotypes [77–79], with clear correlates with classical cerebellar symptoms such as tremor [80] and disability [81], but also more recent findings of relations with working memory and information processing speed [82, 83]. It should be noted that cerebellar pathology remains understudied in MS, however, as does the specific relation of its pathology with clinical progression over time [84].

7   Conclusion

Multiple sclerosis is a strongly heterogeneous disease, previously considered to mostly feature (random) white matter damage. However, as the potential of recent neuroimaging techniques has grown exponentially, specific and clinically relevant patterns of damage in MS have now clearly been identified. These patterns are mostly centered around the concept of structural network disconnection, most strongly affecting network hubs like the thalamus, with clear clinical consequences. It is now imperative that we continue this exploration beyond lesional pathology if we are to fully understand the complex dynamics of the different types of pathologies in MS. This journey will hopefully lead to an enrichment of neuro(radio)-logical practice in the near future, enabling clinicians to use additional metrics like thalamic volume to monitor patients and guide treatment decisions.

References

1. Thompson AJ, Baranzini SE, Geurts J, Hemmer B, Ciccarelli O (2018) Multiple sclerosis. Lancet 391(10130):1622–1636. https://doi.org/10.1016/S0140-6736(18)30481-1

2. Calabrese M, Magliozzi R, Ciccarelli O, Geurts JJ, Reynolds R, Martin R (2015) Exploring the origins of grey matter damage in multiple sclerosis. Nat Rev Neurosci 16(3):147–158. https://doi.org/10.1038/nrn3900

3. Stys PK, Zamponi GW, van Minnen J, Geurts JJ (2012) Will the real multiple sclerosis please stand up? Nat Rev Neurosci 13(7):507–514. https://doi.org/10.1038/nrn3275

4. Thompson AJ, Banwell BL, Barkhof F, Carroll WM, Coetzee T, Comi G, Correale J, Fazekas F, Filippi M, Freedman MS, Fujihara K, Galetta SL, Hartung HP, Kappos L, Lublin FD, Marrie RA, Miller AE, Miller DH, Montalban X, Mowry EM, Sorensen PS, Tintore M, Traboulsee AL, Trojano M, Uitdehaag BMJ, Vukusic S, Waubant E, Weinshenker BG, Reingold SC, Cohen JA (2018) Diagnosis of multiple sclerosis: 2017 revisions of the McDonald criteria. Lancet Neurol 17(2):162–173. https://doi.org/10.1016/S1474-4422(17)30470-2

5. Kilsdonk ID, Jonkman LE, Klaver R, van Veluw SJ, Zwanenburg JJ, Kuijer JP, Pouwels PJ, Twisk JW, Wattjes MP, Luijten PR, Barkhof F, Geurts JJ (2016) Increased cortical grey matter lesion detection in multiple

sclerosis with 7 T MRI: a post-mortem verification study. Brain 139(Pt 5):1472–1481. https://doi.org/10.1093/brain/aww037

6. Rocca MA, Comi G, Filippi M (2017) The role of T1-weighted derived measures of neurodegeneration for assessing disability progression in multiple sclerosis. Front Neurol 8:433. https://doi.org/10.3389/fneur.2017.00433

7. Eshaghi A, Prados F, Brownlee WJ, Altmann DR, Tur C, Cardoso MJ, De Angelis F, van de Pavert SH, Cawley N, De Stefano N, Stromillo ML, Battaglini M, Ruggieri S, Gasperini C, Filippi M, Rocca MA, Rovira A, Sastre-Garriga J, Vrenken H, Leurs CE, Killestein J, Pirpamer L, Enzinger C, Ourselin S, Wheeler-Kingshott C, Chard D, Thompson AJ, Alexander DC, Barkhof F, Ciccarelli O, group Ms (2018) Deep gray matter volume loss drives disability worsening in multiple sclerosis. Ann Neurol 83(2):210–222. https://doi.org/10.1002/ana.25145

8. Filippi M (2015) MRI measures of neurodegeneration in multiple sclerosis: implications for disability, disease monitoring, and treatment. J Neurol 262(1):1–6. https://doi.org/10.1007/s00415-014-7340-9

9. van Walderveen MA, Kamphorst W, Scheltens P, van Waesberghe JH, Ravid R, Valk J, Polman CH, Barkhof F (1998) Histopathologic correlate of hypointense lesions on T1-weighted spin-echo MRI in multiple sclerosis. Neurology 50(5):1282–1288

10. van den Elskamp IJ, Boden B, Dattola V, Knol DL, Filippi M, Kappos L, Fazekas F, Wagner K, Pohl C, Sandbrink R, Polman CH, Uitdehaag BM, Barkhof F (2010) Cerebral atrophy as outcome measure in short-term phase 2 clinical trials in multiple sclerosis. Neuroradiology 52 (10):875–881. https://doi.org/10.1007/s00234-009-0645-1

11. Barkhof F (2002) The clinico-radiological paradox in multiple sclerosis revisited. Curr Opin Neurol 15(3):239–245

12. Rocca MA, Battaglini M, Benedict RH, De Stefano N, Geurts JJ, Henry RG, Horsfield MA, Jenkinson M, Pagani E, Filippi M (2017) Brain MRI atrophy quantification in MS: from methods to clinical application. Neurology 88 (4):403–413. https://doi.org/10.1212/WNL.0000000000003542

13. Benedict RH, Zivadinov R (2011) Risk factors for and management of cognitive dysfunction in multiple sclerosis. Nat Rev Neurol 7 (6):332–342. https://doi.org/10.1038/nrneurol.2011.61

14. Enzinger C, Barkhof F, Ciccarelli O, Filippi M, Kappos L, Rocca MA, Ropele S, Rovira A, Schneider T, de Stefano N, Vrenken H,

Wheeler-Kingshott C, Wuerfel J, Fazekas F, Group Ms (2015) Nonconventional MRI and microstructural cerebral changes in multiple sclerosis. Nat Rev Neurol 11(12):676–686. https://doi.org/10.1038/nrneurol.2015.194

15. Geurts JJ, Roosendaal SD, Calabrese M, Ciccarelli O, Agosta F, Chard DT, Gass A, Huerga E, Moraal B, Pareto D, Rocca MA, Wattjes MP, Yousry TA, Uitdehaag BM, Barkhof F, Group MS (2011) Consensus recommendations for MS cortical lesion scoring using double inversion recovery MRI. Neurology 76(5):418–424. https://doi.org/10.1212/WNL.0b013e31820a0cc4

16. Seewann A, Vrenken H, Kooi EJ, van der Valk P, Knol DL, Polman CH, Pouwels PJ, Barkhof F, Geurts JJ (2011) Imaging the tip of the iceberg: visualization of cortical lesions in multiple sclerosis. Mult Scler 17 (10):1202–1210. https://doi.org/10.1177/1352458511406575

17. Geurts JJ, Bo L, Roosendaal SD, Hazes T, Daniels R, Barkhof F, Witter MP, Huitinga I, van der Valk P (2007) Extensive hippocampal demyelination in multiple sclerosis. J Neuropathol Exp Neurol 66(9):819–827. https://doi.org/10.1097/nen.0b013e3181461f54

18. Roosendaal SD, Moraal B, Vrenken H, Castelijns JA, Pouwels PJ, Barkhof F, Geurts JJ (2008) In vivo MR imaging of hippocampal lesions in multiple sclerosis. J Magn Reson Imaging 27(4):726–731. https://doi.org/10.1002/jmri.21294

19. Roosendaal SD, Moraal B, Pouwels PJ, Vrenken H, Castelijns JA, Barkhof F, Geurts JJ (2009) Accumulation of cortical lesions in MS: relation with cognitive impairment. Mult Scler 15(6):708–714. https://doi.org/10.1177/1352458509102907

20. Coebergh JA, Roosendaal SD, Polman CH, Geurts JJ, van Woerkom TC (2010) Acute severe memory impairment as a presenting symptom of multiple sclerosis: a clinical case study with 3D double inversion recovery MR imaging. Mult Scler 16(12):1521–1524. https://doi.org/10.1177/1352458510383302

21. Sinnecker T, Clarke MA, Meier D, Enzinger C, Calabrese M, De Stefano N, Pitiot A, Giorgio A, Schoonheim MM, Paul F, Pawlak MA, Schmidt R, Kappos L, Montalban X, Rovira A, Evangelou N, Wuerfel J, Group MS (2019) Evaluation of the central vein sign as a diagnostic imaging biomarker in multiple sclerosis. JAMA Neurol 76(12):1446–1456. https://doi.org/10.1001/jamaneurol.2019.2478

22. van Waesberghe JH, van Walderveen MA, Castelijns JA, Scheltens P, Lycklama a Nijeholt GJ, Polman CH, Barkhof F (1998) Patterns of lesion development in multiple sclerosis: longitudinal observations with T1-weighted spin-echo and magnetization transfer MR. AJNR Am J Neuroradiol 19(4):675–683

23. Barkhof F, Bruck W, De Groot CJ, Bergers E, Hulshof S, Geurts J, Polman CH, van der Valk P (2003) Remyelinated lesions in multiple sclerosis: magnetic resonance image appearance. Arch Neurol 60(8):1073–1081. https://doi.org/10.1001/archneur.60.8.1073

24. Klauser AM, Wiebenga OT, Eijlers AJ, Schoonheim MM, Uitdehaag BM, Barkhof F, Pouwels PJ, Geurts JJ (2018) Metabolites predict lesion formation and severity in relapsing-remitting multiple sclerosis. Mult Scler 24(4):491–500. https://doi.org/10.1177/1352458517702534

25. Ontaneda D, Sakaie K, Lin J, Wang X, Lowe MJ, Phillips MD, Fox RJ (2014) Identifying the start of multiple sclerosis injury: a serial DTI study. J Neuroimaging 24(6):569–576. https://doi.org/10.1111/jon.12082

26. Filippi M, Rocca MA, Martino G, Horsfield MA, Comi G (1998) Magnetization transfer changes in the normal appearing white matter precede the appearance of enhancing lesions in patients with multiple sclerosis. Ann Neurol 43 (6):809–814. https://doi.org/10.1002/ana.410430616

27. Elliott C, Wolinsky JS, Hauser SL, Kappos L, Barkhof F, Bernasconi C, Wei W, Belachew S, Arnold DL (2019) Slowly expanding/evolving lesions as a magnetic resonance imaging marker of chronic active multiple sclerosis lesions. Mult Scler 25(14):1915–1925. https://doi.org/10.1177/1352458518814117

28. Elliott C, Belachew S, Wolinsky JS, Hauser SL, Kappos L, Barkhof F, Bernasconi C, Fecker J, Model F, Wei W, Arnold DL (2019) Chronic white matter lesion activity predicts clinical progression in primary progressive multiple sclerosis. Brain 142(9):2787–2799. https://doi.org/10.1093/brain/awz212

29. Seewann A, Vrenken H, van der Valk P, Blezer EL, Knol DL, Castelijns JA, Polman CH, Pouwels PJ, Barkhof F, Geurts JJ (2009) Diffusely abnormal white matter in chronic multiple sclerosis: imaging and histopathologic analysis. Arch Neurol 66(5):601–609. https://doi.org/10.1001/archneurol.2009.57

30. Vrenken H, Geurts JJ, Knol DL, Polman CH, Castelijns JA, Pouwels PJ, Barkhof F (2006) Normal-appearing white matter changes vary with distance to lesions in multiple sclerosis. AJNR Am J Neuroradiol 27(9):2005–2011

31. Schoonheim MM, Vigeveno RM, Rueda Lopes FC, Pouwels PJ, Polman CH, Barkhof F, Geurts JJ (2014) Sex-specific extent and severity of white matter damage in multiple sclerosis: implications for cognitive decline. Hum Brain Mapp 35(5):2348–2358. https://doi.org/10.1002/hbm.22332

32. Hulst HE, Steenwijk MD, Versteeg A, Pouwels PJ, Vrenken H, Uitdehaag BM, Polman CH, Geurts JJ, Barkhof F (2013) Cognitive impairment in MS: impact of white matter integrity, gray matter volume, and lesions. Neurology 80(11):1025–1032. https://doi.org/10.1212/WNL.0b013e31828726cc

33. Welton T, Kent D, Constantinescu CS, Auer DP, Dineen RA (2015) Functionally relevant white matter degradation in multiple sclerosis: a tract-based spatial meta-analysis. Radiology 275(1):89–96. https://doi.org/10.1148/radiol.14140925

34. Ciccarelli O, Catani M, Johansen-Berg H, Clark C, Thompson A (2008) Diffusion-based tractography in neurological disorders: concepts, applications, and future developments. Lancet Neurol 7(8):715–727. https://doi.org/10.1016/S1474-4422(08)70163-7

35. Ciccarelli O, Wheeler-Kingshott CA, McLean MA, Cercignani M, Wimpey K, Miller DH, Thompson AJ (2007) Spinal cord spectroscopy and diffusion-based tractography to assess acute disability in multiple sclerosis. Brain 130 (Pt 8):2220–2231. https://doi.org/10.1093/brain/awm152

36. Gorgoraptis N, Wheeler-Kingshott CA, Jenkins TM, Altmann DR, Miller DH, Thompson AJ, Ciccarelli O (2010) Combining tractography and cortical measures to test system-specific hypotheses in multiple sclerosis. Mult Scler 16(5):555–565. https://doi.org/10.1177/1352458510362440

37. Fleischer V, Radetz A, Ciolac D, Muthuraman M, Gonzalez-Escamilla G, Zipp F, Groppa S (2019) Graph theoretical framework of brain networks in multiple sclerosis: a review of concepts. Neuroscience 403:35–53. https://doi.org/10.1016/j.neuroscience.2017.10.033

38. Bassett DS, Sporns O (2017) Network neuroscience. Nat Neurosci 20(3):353–364. https://doi.org/10.1038/nn.4502

39. He Y, Dagher A, Chen Z, Charil A, Zijdenbos A, Worsley K, Evans A (2009) Impaired small-world efficiency in structural cortical networks in multiple sclerosis associated with white matter lesion load. Brain 132(Pt 12):3366–3379. https://doi.org/10.1093/brain/awp089

40. Rimkus CM, Schoonheim MM, Steenwijk MD, Vrenken H, Eijlers AJ, Killestein J, Wattjes MP, Leite CC, Barkhof F, Tijms BM (2019) Gray matter networks and cognitive impairment in multiple sclerosis. Mult Scler 25(3):382–391. https://doi.org/10.1177/1352458517751650

41. Llufriu S, Rocca MA, Pagani E, Riccitelli GC, Solana E, Colombo B, Rodegher M, Falini A, Comi G, Filippi M (2019) Hippocampal-related memory network in multiple sclerosis: a structural connectivity analysis. Mult Scler 25(6):801–810. https://doi.org/10.1177/1352458518771838

42. Meijer KA, Steenwijk MD, Douw L, Schoonheim MM, Geurts JJG (2020) Long-range connections are more severely damaged and relevant for cognition in multiple sclerosis. Brain 143(1):150–160. https://doi.org/10.1093/brain/awz355

43. Solana E, Martinez-Heras E, Casas-Roma J, Calvet L, Lopez-Soley E, Sepulveda M, Sola-Valls N, Montejo C, Blanco Y, Pulido-Valdeolivas I, Andorra M, Saiz A, Prados F, Llufriu S (2019) Modified connectivity of vulnerable brain nodes in multiple sclerosis, their impact on cognition and their discriminative value. Sci Rep 9(1):20172. https://doi.org/10.1038/s41598-019-56806-z

44. Fleischer V, Groger A, Koirala N, Droby A, Muthuraman M, Kolber P, Reuter E, Meuth SG, Zipp F, Groppa S (2017) Increased structural white and grey matter network connectivity compensates for functional decline in early multiple sclerosis. Mult Scler 23(3):432–441. https://doi.org/10.1177/1352458516651503

45. Radetz A, Koirala N, Kramer J, Johnen A, Fleischer V, Gonzalez-Escamilla G, Cerina M, Muthuraman M, Meuth SG, Groppa S (2020) Gray matter integrity predicts white matter network reorganization in multiple sclerosis. Hum Brain Mapp 41(4):917–927. https://doi.org/10.1002/hbm.24849

46. Koubiyr I, Besson P, Deloire M, Charre-Morin J, Saubusse A, Tourdias T, Brochet B, Ruet A (2019) Dynamic modular-level alterations of structural-functional coupling in clinically isolated syndrome. Brain 142(11):3428–3439. https://doi.org/10.1093/brain/awz270

47. Tur C, Eshaghi A, Altmann DR, Jenkins TM, Prados F, Grussu F, Charalambous T, Schmidt A, Ourselin S, Clayden JD, Wheeler-Kingshott C, Thompson AJ, Ciccarelli O, Toosy AT (2018) Structural cortical network reorganization associated with early conversion to multiple sclerosis. Sci Rep 8(1):10715. https://doi.org/10.1038/s41598-018-29017-1

48. Pardini M, Yaldizli O, Sethi V, Muhlert N, Liu Z, Samson RS, Altmann DR, Ron MA, Wheeler-Kingshott CA, Miller DH, Chard DT (2015) Motor network efficiency and disability in multiple sclerosis. Neurology 85(13):1115–1122. https://doi.org/10.1212/WNL.0000000000001970

49. Schoonheim MM, Meijer KA, Geurts JJ (2015) Network collapse and cognitive impairment in multiple sclerosis. Front Neurol 6:82. https://doi.org/10.3389/fneur.2015.00082

50. Lansley J, Mataix-Cols D, Grau M, Radua J, Sastre-Garriga J (2013) Localized grey matter atrophy in multiple sclerosis: a meta-analysis of voxel-based morphometry studies and associations with functional disability. Neurosci Biobehav Rev 37(5):819–830. https://doi.org/10.1016/j.neubiorev.2013.03.006

51. De Stefano N, Stromillo ML, Giorgio A, Bartolozzi ML, Battaglini M, Baldini M, Portaccio E, Amato MP, Sormani MP (2016) Establishing pathological cut-offs of brain atrophy rates in multiple sclerosis. J Neurol Neurosurg Psychiatry 87(1):93–99. https://doi.org/10.1136/jnnp-2014-309903

52. Popescu V, Klaver R, Voorn P, Galis-de Graaf Y, Knol DL, Twisk JW, Versteeg A, Schenk GJ, Van der Valk P, Barkhof F, De Vries HE, Vrenken H, Geurts JJ (2015) What drives MRI-measured cortical atrophy in multiple sclerosis? Mult Scler 21(10):1280–1290. https://doi.org/10.1177/1352458514562440

53. Klaver R, Popescu V, Voorn P, Galis-de Graaf Y, van der Valk P, de Vries HE, Schenk GJ, Geurts JJ (2015) Neuronal and axonal loss in normal-appearing gray matter and subpial lesions in multiple sclerosis. J Neuropathol Exp Neurol 74(5):453–458. https://doi.org/10.1097/NEN.0000000000000189

54. van de Pavert SH, Muhlert N, Sethi V, Wheeler-Kingshott CA, Ridgway GR, Geurts JJ, Ron M, Yousry TA, Thompson AJ, Miller DH, Chard DT, Ciccarelli O (2016) DIR-visible grey matter lesions and atrophy in multiple sclerosis: partners in crime? J Neurol Neurosurg Psychiatry 87(5):461–467. https://doi.org/10.1136/jnnp-2014-310142

55. Steenwijk MD, Daams M, Pouwels PJ, Balk LJ, Tewarie PK, Killestein J, Uitdehaag BM, Geurts JJ, Barkhof F, Vrenken H (2014) What explains gray matter atrophy in long-standing multiple sclerosis? Radiology 272(3):832–842. https://doi.org/10.1148/radiol.14132708

56. Steenwijk MD, Daams M, Pouwels PJ, L JB, Tewarie PK, Geurts JJ, Barkhof F, Vrenken H (2015) Unraveling the relationship between regional gray matter atrophy and pathology in connected white matter tracts in long-standing multiple sclerosis. Hum Brain Mapp 36 (5):1796–1807. https://doi.org/10.1002/hbm.22738

57. Steenwijk MD, Geurts JJ, Daams M, Tijms BM, Wink AM, Balk LJ, Tewarie PK, Uitdehaag BM, Barkhof F, Vrenken H, Pouwels PJ (2016) Cortical atrophy patterns in multiple sclerosis are non-random and clinically relevant. Brain 139(Pt 1):115–126. https://doi.org/10.1093/brain/awv337

58. Kipp M, Wagenknecht N, Beyer C, Samer S, Wuerfel J, Nikoubashman O (2015) Thalamus pathology in multiple sclerosis: from biology to clinical application. Cell Mol Life Sci 72 (6):1127–1147. https://doi.org/10.1007/s00018-014-1787-9

59. Minagar A, Barnett MH, Benedict RH, Pelletier D, Pirko I, Sahraian MA, Frohman E, Zivadinov R (2013) The thalamus and multiple sclerosis: modern views on pathologic, imaging, and clinical aspects. Neurology 80(2):210–219. https://doi.org/10.1212/WNL.0b013e31827b910b

60. Schoonheim MM, Popescu V, Rueda Lopes FC, Wiebenga OT, Vrenken H, Douw L, Polman CH, Geurts JJ, Barkhof F (2012) Subcortical atrophy and cognition: sex effects in multiple sclerosis. Neurology 79 (17):1754–1761. https://doi.org/10.1212/WNL.0b013e3182703f46

61. Zivadinov R, Havrdova E, Bergsland N, Tyblova M, Hagemeier J, Seidl Z, Dwyer MG, Vaneckova M, Krasensky J, Carl E, Kalincik T, Horakova D (2013) Thalamic atrophy is associated with development of clinically definite multiple sclerosis. Radiology 268 (3):831–841. https://doi.org/10.1148/radiol.13122424

62. Bisecco A, Rocca MA, Pagani E, Mancini L, Enzinger C, Gallo A, Vrenken H, Stromillo ML, Copetti M, Thomas DL, Fazekas F, Tedeschi G, Barkhof F, Stefano ND, Filippi M, Network M (2015) Connectivity-based parcellation of the thalamus in multiple sclerosis and its implications for cognitive impairment: a multicenter study. Hum Brain Mapp 36(7):2809–2825. https://doi.org/10.1002/hbm.22809

63. Schoonheim MM, Geurts JJG (2019) What causes deep gray matter atrophy in multiple sclerosis? AJNR Am J Neuroradiol 40 (1):107–108. https://doi.org/10.3174/ajnr.A5942

64. van de Pavert SH, Muhlert N, Sethi V, Wheeler-Kingshott CA, Ridgway GR, Geurts JJ, Ron M, Yousry TA, Thompson AJ, Miller DH, Chard DT, Ciccarelli O (2015) DIR-visible grey matter lesions and atrophy in multiple sclerosis: partners in crime? J Neurol Neurosurg Psychiatry 87(5):461–467. https://doi.org/10.1136/jnnp-2014-310142

65. Azevedo CJ, Cen SY, Khadka S, Liu S, Kornak J, Shi Y, Zheng L, Hauser SL, Pelletier D (2018) Thalamic atrophy in multiple sclerosis: a magnetic resonance imaging marker of neurodegeneration throughout disease. Ann Neurol 83(2):223–234. https://doi.org/10.1002/ana.25150

66. Eshaghi A, Marinescu RV, Young AL, Firth NC, Prados F, Jorge Cardoso M, Tur C, De Angelis F, Cawley N, Brownlee WJ, De Stefano N, Laura Stromillo M, Battaglini M, Ruggieri S, Gasperini C, Filippi M, Rocca MA, Rovira A, Sastre-Garriga J, Geurts JJG, Vrenken H, Wottschel V, Leurs CE, Uitdehaag B, Pirpamer L, Enzinger C, Ourselin S, Gandini Wheeler-Kingshott CA, Chard D, Thompson AJ, Barkhof F, Alexander DC, Ciccarelli O (2018) Progression of regional grey matter atrophy in multiple sclerosis. Brain 141(6):1665–1677. https://doi.org/10.1093/brain/awy088

67. Eijlers AJC, Dekker I, Steenwijk MD, Meijer KA, Hulst HE, Pouwels PJW, Uitdehaag BMJ, Barkhof F, Vrenken H, Schoonheim MM, Geurts JJG (2019) Cortical atrophy accelerates as cognitive decline worsens in multiple sclerosis. Neurology 93(14):e1348–e1359. https://doi.org/10.1212/WNL.0000000000008198

68. Eijlers AJC, van Geest Q, Dekker I, Steenwijk MD, Meijer KA, Hulst HE, Barkhof F, Uitdehaag BMJ, Schoonheim MM, Geurts JJG (2018) Predicting cognitive decline in multiple sclerosis: a 5-year follow-up study. Brain 141 (9):2605–2618. https://doi.org/10.1093/brain/awy202

69. Moccia M, Prados F, Filippi M, Rocca MA, Valsasina P, Brownlee WJ, Zecca C, Gallo A, Rovira A, Gass A, Palace J, Lukas C, Vrenken H, Ourselin S, CAM GW-K, Ciccarelli O, Barkhof F, Group MS (2019) Longitudinal spinal cord atrophy in multiple sclerosis using the generalized boundary shift integral. Ann Neurol 86(5):704–713. https://doi.org/10.1002/ana.25571

70. Rocca MA, Barkhof F, De Luca J, Frisen J, Geurts JJG, Hulst HE, Sastre-Garriga J, Filippi M, Group MS (2018) The hippocampus in multiple sclerosis. Lancet Neurol 17 (10):918–926. https://doi.org/10.1016/S1474-4422(18)30309-0

71. Koenig KA, Sakaie KE, Lowe MJ, Lin J, Stone L, Bermel RA, Beall EB, Rao SM, Trapp BD, Phillips MD (2014) Hippocampal volume is related to cognitive decline and fornicial diffusion measures in multiple sclerosis. Magn Reson Imaging 32(4):354–358. https://doi.org/10.1016/j.mri.2013.12.012

72. Sicotte NL, Kern KC, Giesser BS, Arshanapalli A, Schultz A, Montag M, Wang H, Bookheimer SY (2008) Regional hippocampal atrophy in multiple sclerosis. Brain 131(Pt 4):1134–1141. https://doi.org/10.1093/brain/awn030

73. Planche V, Koubiyr I, Romero JE, Manjon JV, Coupe P, Deloire M, Dousset V, Brochet B, Ruet A, Tourdias T (2018) Regional hippocampal vulnerability in early multiple sclerosis: dynamic pathological spreading from dentate gyrus to CA1. Hum Brain Mapp 39(4):1814–1824. https://doi.org/10.1002/hbm.23970

74. Rocca MA, Longoni G, Pagani E, Boffa G, Colombo B, Rodegher M, Martino G, Falini A, Comi G, Filippi M (2015) In vivo evidence of hippocampal dentate gyrus expansion in multiple sclerosis. Hum Brain Mapp 36(11):4702–4713. https://doi.org/10.1002/hbm.22946

75. Longoni G, Rocca MA, Pagani E, Riccitelli GC, Colombo B, Rodegher M, Falini A, Comi G, Filippi M (2015) Deficits in memory and visuospatial learning correlate with regional hippocampal atrophy in MS. Brain Struct Funct 220(1):435–444. https://doi.org/10.1007/s00429-013-0665-9

76. Gold SM, Kern KC, O'Connor MF, Montag MJ, Kim A, Yoo YS, Giesser BS, Sicotte NL (2010) Smaller cornu ammonis 2-3/dentate gyrus volumes and elevated cortisol in multiple sclerosis patients with depressive symptoms. Biol Psychiatry 68(6):553–559. https://doi.org/10.1016/j.biopsych.2010.04.025

77. Mesaros S, Rovaris M, Pagani E, Pulizzi A, Caputo D, Ghezzi A, Bertolotto A, Capra R, Falautano M, Martinelli V, Comi G, Filippi M (2008) A magnetic resonance imaging voxel-based morphometry study of regional gray matter atrophy in patients with benign multiple sclerosis. Arch Neurol 65(9):1223–1230. https://doi.org/10.1001/archneur.65.9.1223

78. Anderson VM, Fox NC, Miller DH (2006) Magnetic resonance imaging measures of brain atrophy in multiple sclerosis. J Magn Reson Imaging 23(5):605–618. https://doi.org/10.1002/jmri.20550

79. Anderson VM, Fisniku LK, Altmann DR, Thompson AJ, Miller DH (2009) MRI measures show significant cerebellar gray matter volume loss in multiple sclerosis and are associated with cerebellar dysfunction. Mult Scler 15(7):811–817. https://doi.org/10.1177/1352458508101934

80. Boonstra F, Noffs G, Perera T, Jokubaitis V, Vogel A, Evans A, Moffat B, Butzkueven H, Kolbe S, van der Walt A (2018) Functional neuroplasticity in response to cerebellothalamic injury underpins the clinical presentation of tremor in multiple sclerosis. Mult Scler J 24:419–419

81. Cordani C, Meani A, Esposito F, Valsasina P, Colombo B, Pagani E, Preziosa P, Comi G, Filippi M, Rocca MA (2019) Imaging correlates of hand motor performance in multiple sclerosis: a multiparametric structural and functional MRI study. Mult Scler 26(2):233–244. https://doi.org/10.1177/1352458518822145

82. Moroso A, Ruet A, Lamargue-Hamel D, Munsch F, Deloire M, Coupe P, Ouallet JC, Planche V, Moscufo N, Meier DS, Tourdias T, Guttmann CR, Dousset V, Brochet B (2016) Posterior lobules of the cerebellum and information processing speed at various stages of multiple sclerosis. J Neurol Neurosurg Psychiatry 88(2):146–151. https://doi.org/10.1136/jnnp-2016-313867

83. D'Ambrosio A, Pagani E, Riccitelli GC, Colombo B, Rodegher M, Falini A, Comi G, Filippi M, Rocca MA (2017) Cerebellar contribution to motor and cognitive performance in multiple sclerosis: an MRI sub-regional volumetric analysis. Mult Scler 23(9):1194–1203. https://doi.org/10.1177/1352458516674567

84. Parmar K, Stadelmann C, Rocca MA, Langdon D, D'Angelo E, D'Souza M, Burggraaff J, Wegner C, Sastre-Garriga J, Barrantes-Freer A, Dorn J, BMJ U, Montalban X, Wuerfel J, Enzinger C, Rovira A, Tintore M, Filippi M, Kappos L, Sprenger T, Group Ms (2018) The role of the cerebellum in multiple sclerosis-150 years after Charcot. Neurosci Biobehav Rev 89:85–98. https://doi.org/10.1016/j.neubiorev.2018.02.012

Chapter 15

Human Functional MRI

Paolo Preziosa, Paola Valsasina, Massimo Filippi, and Maria A. Rocca

Abstract

This chapter reviews the basic principles, main acquisition, and postprocessing techniques of functional magnetic resonance imaging (fMRI) applied to study multiple sclerosis (MS).

First, we describe the blood-oxygenation level dependent (BOLD) effect and the principal analysis techniques used to process fMRI data, including those acquired during the performance of active tasks and those acquired at resting state.

Subsequently, we summarize how the different fMRI techniques have contributed to investigate MS pathophysiology, by demonstrating that functional reorganization occurs as a consequence of structural damage accumulation in MS patients and can contribute, at least in the earliest phases of the disease, to limit the clinical consequences of widespread structural abnormalities. We discuss also how the failure or exhaustion of central nervous system adaptive properties might be among the factors responsible for the accumulation of irreversible clinical disability and cognitive impairment.

The identification of MS-related adaptive and maladaptive reorganization is an attractive goal to understand the mechanisms of action of pharmacologic and rehabilitative treatments and to develop novel therapeutic strategies able to promote individual adaptive capacity.

Key words Multiple sclerosis; Functional MRI, Active task, Resting state, Brain networks

1 Introduction

Over the past decades, magnetic resonance imaging (MRI) has been extensively applied to improve the understanding of multiple sclerosis (MS) pathophysiology, monitor disease evolution, and evaluate the effects of pharmacologic and rehabilitative treatments [1–4]. Structural MRI techniques are providing an accurate mapping of central nervous system (CNS) macroscopic and microscopic brain tissue abnormalities occurring in MS patients, thus advancing our knowledge of the mechanisms responsible for the accumulation of irreversible disability and cognitive impairment [5–7]. Despite this, the magnitude of the correlations between structural MRI measures and clinical findings remains suboptimal [8, 9]. This might be explained, at least partially, by the presence of

Sergiu Groppa and Sven G. Meuth (eds.), *Translational Methods for Multiple Sclerosis Research*, Neuromethods, vol. 166, https://doi.org/10.1007/978-1-0716-1213-2_15, © Springer Science+Business Media, LLC, part of Springer Nature 2021

functional brain plasticity and a possible variable effectiveness of reparative and recovery mechanisms following the progressive accumulation of structural tissue damage.

Brain plasticity is a well-known feature of the human brain, which is likely to have different substrates (including increased axonal expression of sodium channels, synaptic changes, increased recruitment of parallel existing pathways or "latent" connections, and reorganization of distant sites) [10].

Cortical reorganization has been suggested as a potential contributor to the recovery or the maintenance of function in the presence of irreversible white matter (WM) and gray matter (GM) damage [11, 12]. Functional MRI (fMRI) techniques based on measurement of changes in the blood-oxygenation level dependent (BOLD) signal [13, 14] are able to provide indirect measures of neuronal activity, thus representing a powerful tool to measure brain plasticity.

In this chapter, we review the basic principles of fMRI techniques and their application to MS. First, we describe the BOLD effect and present the main analysis techniques used to process fMRI data, both those acquired during the performance of an active task and those acquired at resting state (RS). Then, we summarize the main results of observational and treatment studies applying fMRI techniques to investigate brain plasticity and functional reorganization occurring in patients with MS.

2   Basic Principles of Functional MRI Techniques

FMRI is a noninvasive technique that allows to study CNS function and to define altered patterns of activation and/or functional connectivity (FC) caused by injury or disease. This is possible because the activity of neuronal cells requires energy, which is produced through chemical reactions involving glucose and oxygen consumption. Local increases in neuronal activity result in a rise of cerebral blood flow (CBF) and cerebral blood volume (CBV). However, the increase of blood flow is greater than oxygen consumption, thus determining an increased ratio between oxygenated and deoxygenated hemoglobin, which enhances the MRI signal [14]. This phenomenon is known as BOLD effect and is one of the basic principles underlying findings from fMRI experiments.

Acquisition sequences weighted for the $T2^*$ decay are particularly sensitive to the BOLD effect [13]. Clearly, the acquisition of an fMRI experiment has to be fast, to allow for the temporal correlation of the physiological changes with the stimulus. Therefore, $T2^*$-weighted echo-planar imaging (EPI) sequences, which allow for a complete brain coverage within a few seconds thanks to their fast readout of the entire k-space, are the most used approach to acquire fMRI data. The recent introduction of simultaneous multi-slice imaging techniques with multi-band excitation [15] has allowed for a further boosting of data acquisition.

As the BOLD effect is reflected in an increased signal on MRI scans at the site of neuronal activation, post-processing of fMRI data should rely on the comparison of signal intensities between two sets of images, one acquired at rest and one acquired during a given task. However, such a signal intensity change is relatively small (about 2–3% of the total signal) [16]. As a consequence, an fMRI experiment requires an adequate setup, with a proper alternation of stimuli and resting conditions and a proper number of repetitions, to obtain a sufficient statistical power.

3   Methods to Investigate Task-Related Functional Activity in MS

3.1   Analysis of Active Task fMRI

3.1.1   The General Linear Model

Several techniques can be applied to analyze fMRI datasets acquired during the performance of an active task. Among them, the most popular and robust approach is the general linear model (GLM) [17] (Fig. 1a). This is based on a model-driven analysis depending on a priori hypotheses about the temporal intensity change expected to occur in the data. GLM is a univariate approach, that is, it characterizes the activity in each voxel, separately, by performing voxel-wise statistical analyses in parallel. To measure the magnitude of the BOLD signal, fMRI data at each voxel are modelled as a linear combination of explanatory variables and a residual error term. Explanatory variables can be related to the signal evoked by a specific task or can be due to noise, such as that originated from physiological effects or from scanner-related factors. Then, a statistical parametric map (SPM) [18] is created for each explanatory variable, reflecting significance of that variable in the data. SPM maps are interpreted as spatially extended statistical processes that behave according to the theory of Gaussian fields [19]. This enables the statistical characterization of regionally specific responses using appropriate statistical tests (e.g., F tests or T tests). Some methods of correction for multiple comparisons (e.g., family-wise error [20] or false discovery rate [21]) are usually applied to SPM maps to minimize the risk of obtaining false-positive fMRI activations.

3.1.2   Functional and Effective Connectivity

While GLM is a post-processing technique aimed at assessing functional segregation (i.e., which specialized brain areas respond to a task), fMRI experiments can also be used to investigate functional integration, defined as the interaction among specialized brain areas activated by a given task. Functional integration is investigated by functional and effective connectivity analysis. Functional connectivity (FC) is defined as the temporal correlation between spatially remote neurophysiological events [22]. As such, FC is a way to assess how strong are the associations between observed time series

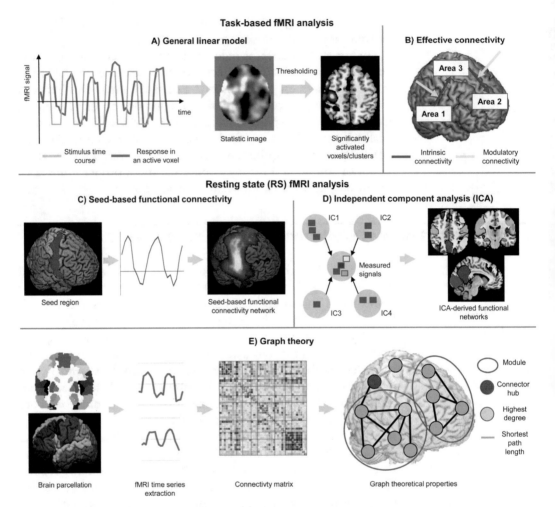

Fig. 1 Schematic representation of the main post-processing techniques used for the analysis of task-based and resting state functional MRI (fMRI) data. Top row: in the general linear model (panel **a**), active fMRI time series at each voxel are modelled as a linear combination of explanatory variables and a residual error term. Statistical parametric maps are created, reflecting significance of each explanatory variable in the data. Effective connectivity (panel **b**) investigates causal relationships between the activity detected in one brain region and the activity detected in all remaining brain regions included in a model. Changes of intrinsic connectivity by external factors (i.e., modulatory connectivity) can also be investigated. Middle row: the most popular approaches to analyze resting state functional connectivity are seed-based correlation analysis (panel **c**) and independent component analysis (panel **d**), which are able to detect the main large-scale functional networks of the brain. Graph theoretical analysis (panel **e**) describes the brain as a collection of nodes (brain regions) and edges (functional connections). Graph analysis quantifies local and global connectivity properties, such as degree, path length, and modularity. Hubs (i.e., highly connected regions ensuring efficient communication between distinct functional systems) can be also identified

but does not investigate the causal relationships between them. This is the topic of effective connectivity (Fig. 1b), which is defined as the influence that one neural system exerts over another [23, 24]. Several techniques have been proposed to perform effective connectivity analyses. Psychophysiological interactions (PPI) explore how connectivity between one region and the rest of brain changes according to the experimental or psychological context [25]. Other methods, such as dynamic causal modelling (DCM) [26], structural equation modelling (SEM) [27], or Bayes network analysis [28], investigate causal relationships between the activity detected in one brain region and the activity detected in all remaining brain regions included in a model, and analyze how intrinsic connectivity changes by external manipulations.

4 Resting State fMRI

4.1 Basic Principles

In the last two decades, the presence of signal fluctuations due to the BOLD effect has been ascertained not only in active fMRI scans, when subjects are performing a specific task, but also in fMRI data acquired at RS, that is, while subjects are lying still in the scanner. One of the features that makes RS fMRI particularly attractive is that it is a task-free approach, thus providing the unique opportunity to perform fMRI studies in patients who may have difficulties with task instructions and execution.

At rest, these signal fluctuations seem to occur at a low frequency (<0.1 Hz) [29] and they have been demonstrated to be synchronous in spatially distinct but functionally related brain regions [30]. Several studies found a high temporal correlation at rest in multiple brain areas resembling specific neuroanatomical systems, such as the motor, visual, auditory, language circuits and the dorsal and ventral attention systems [30–32]. It has been suggested that these RS networks (RSN) may reflect an intrinsic property of brain organization that serves to stabilize brain function. Particularly relevant is the intrinsic FC of the default mode network (DMN) [33], a medial cortical network involving several regions of the parietal and frontal lobes (medial prefrontal cortex, rostral anterior cingulate cortex, posterior cingulate cortex, precuneus and lateral parietal cortex). The DMN has been found to be active at rest, and to be deactivated when subjects perform attention-demanding, goal-oriented tasks [33, 34]. FC of the DMN has been shown to be vulnerable to aging and to a variety of neurological and psychiatric conditions [35, 36].

4.2 Methods to Investigate RS FC in MS

There are multiple ways to analyze RS fMRI data, and each approach has different implications in terms of type of information that can be extracted from the data [37]. As it is the case for active fMRI, RS fMRI analysis approaches can be broadly grouped into two categories: functional segregation techniques, which rely on the analysis of RS fMRI focusing on local function of specific brain regions, and functional integration techniques, which rely on FC analysis looking at the brain as an integrated network [38]. It is important to note that BOLD-related RS fMRI fluctuations occur in a frequency range (<0.1 Hz) contaminated by several sources of noise, including micro-head movements, cardiac and respiratory cycles, and scanner-related drifts [39]. Therefore, before proceeding with any analysis, it is important to minimize any source of spurious signal, by performing regression of motion parameters [40] and/or volume scrubbing [41], regression of signals of WM and cerebrospinal fluid, from cardiac/respiratory traces [42] or independent component analysis (ICA)-based denoising [43, 44], as well as detrending and band-pass filtering.

4.2.1 Local Methods

Methods commonly used for functional segregation analysis include amplitude of low-frequency fluctuations (ALFF) [45], fractional-ALFF, and regional homogeneity (ReHo) [46]. These methods reflect different aspects of regional neural activity. ALFF is measured by extracting the RS fMRI time series from each voxel and transforming it to the frequency domain with a fast Fourier transform. The average square root of the spectrum in the range where RS fluctuations are known to occur is taken as a measure of amplitude of the RS regional activity [45]. ReHo assumes that the time course of a given voxel is temporally similar to that of its neighbors [46]. The Kendall's coefficient of concordance (KCC) is used to measure the temporal similarity among a time series and those of its nearest neighbors in a voxel-wise way, and the spatial map of KCC is taken as a measure of regional RS activity.

4.2.2 Functional Integration Methods

Because the brain is more appropriately studied as an integrated network rather than isolated clusters, the excitement for stand-alone functional segregation methods has gradually receded in favor of functional integration methods, which measure the degree of synchrony of BOLD RS fMRI time series between spatially distinct brain regions. To assess functional integration features, commonly used computational methods include FC density analysis, seed-based FC analysis, ICA, and graph theoretical analysis.

FC density analysis attempts to identify highly connected functional hubs but it does not indicate which regions are connected [47]. Seed-based FC, also called region of interest (ROI)-based FC [48] (Fig. 1c), extracts the average RS fMRI time series from a region of interest and correlates it with the signal time courses of any other voxel in the brain. This allows to obtain a cross-correlation map, showing how strong is the connection between the seed region and the rest of the brain. Results depend from the a priori selection of seeds, which is often based on a hypothesis or prior results.

Another popular approach is ICA (Fig. 1d), where the observed fMRI time series are considered a linear mixture of unknown, spatially independent sources of signal. Each source is contributing to the data with an unknown time profile [49]. ICA employs mathematical algorithms to decompose RS fMRI data from the whole brain into spatially and temporally independent components [50, 51]. By investigating multiple simultaneous voxel-to-voxel interactions, it represents a powerful technique to perform group-level analyses of large-scale RSN and their changes in different conditions.

In graph theoretical analysis (Fig. 1e), the brain is described as a graph, which consists of a collection of nodes (i.e., brain regions) and edges (i.e., structural and functional connections) [52, 53]. After calculating bivariate RS FC between each pair of brain nodes, several metrics can be used to quantify local and global connectivity properties, such as degree, clustering, efficiency, and modularity [52]. Graph analysis has revealed important features of brain organization, such as an efficient "small-world" architecture (which combines a high level of segregation with a high level of global efficiency) and distributed hubs, that is, highly connected network regions ensuring efficient communication between distinct functional systems [52, 53].

4.2.3 Dynamic RS FC

One of the main assumptions of classical methods of RS FC assessment, described in the previous paragraphs, is that connectivity is static across the entire examination, that is, it can be assessed by calculating the mean correlation between whole-length RS fMRI time series [30]. However, as widely evident by neurophysiological techniques, brain FC is highly variable at a very fast time-scale. The functioning human brain during any state of wakefulness repeatedly changes between different combinations of cognitive, sensory, motor, attentional, and emotional tasks [54]. The study of temporal fluctuations of FC occurring within RS fMRI sessions has been defined as dynamic FC (dFC) [55, 56]. Several analysis strategies have been applied to quantify time-varying RS FC. Some methods aim at capturing variations in inter-regional associations between pairs of brain areas [57, 58], while others try to detect changing patterns of temporal synchrony at a multivariate level [59]. One of the most popular methods for dFC analysis, which is based on the use of the so-called sliding windows [57, 58], belongs to the first category, since it relies on the calculation of a series of pairwise correlation coefficients over small shifting segments of fMRI time series. Despite the great variability of available pipelines, dFC analysis usually requires the performance of the following steps: (1) selection of a set of ROIs in the brain; (2) assessment of time-varying correlations among the selected ROIs; and (3) extraction of features summarizing to what extent connectivity is dynamic, for example, the standard deviation (or variance) of sliding window correlation time series [58], or time of permanence/number of transitions in recurring FC states [57].

5   Functional MRI to Understand MS Pathophysiology

5.1   Active Task fMRI

Functional brain abnormalities have been consistently demonstrated in MS patients using different active paradigms. fMRI changes have been detected following the formation of acute symptomatic lesions (e.g., along the motor pathways), but also independently from an acute relapse and in clinically stable patients. Such fMRI changes occur early in the disease course, already in patients with a clinically isolated syndrome (CIS) suggestive of MS and those with pediatric MS, and tend to vary over the course of the disease.

By evaluating functional brain reorganization following an acute relapse due to the development of a symptomatic lesion, several studies demonstrated dynamic modifications of motor system recruitment, according to the time of the investigation and the degree of clinical recovery [60–62]. For instance, by evaluating MS patients with an acute motor relapse due to the formation of a pseudotumoral lesion along the motor pathways, a longitudinal study showed that, at the onset of the relapse, increased recruitment of motor areas located in the contralesional (healthy) hemisphere was present upon movement of the impaired upper limb [61]. After 1 year, patients with good clinical recovery had relateralization of motor network recruitment to the previously affected hemisphere, whereas those patients with poor clinical recovery continued to show recruitment of motor areas in the contralesional hemisphere. The recruitment of contralesional parallel existing pathways may be a protective phenomenon contributing to replace functional activity of the affected hemisphere early after the acute insult, whereas the restoration of function in motor areas of the affected hemisphere over time may be classified as an adaptive mechanism fostering clinical recovery. Conversely, the persistence of contralesional activation over time is likely to represent a maladaptive mechanism, contributing to poor clinical recovery.

Concerning abnormalities of fMRI activity in MS patients with the main disease clinical phenotypes, studies evaluating active motor tasks support the notion of a "natural history" of brain plasticity in this condition, with functional reorganization being variable along disease course [63, 64] (Fig. 2a). In CIS [65–67] and in patients with a mild clinical disability [63], including pediatric [68] or benign MS [69], the preservation of clinical functions is typically associated with an increased recruitment of regions "normally" devoted to the performance of a specific task and with the preservation of a focused and strictly lateralized movement-associated pattern of cortical activations. The accumulation of initial levels of disability is subsequently associated with a bilateral activation of these regions.

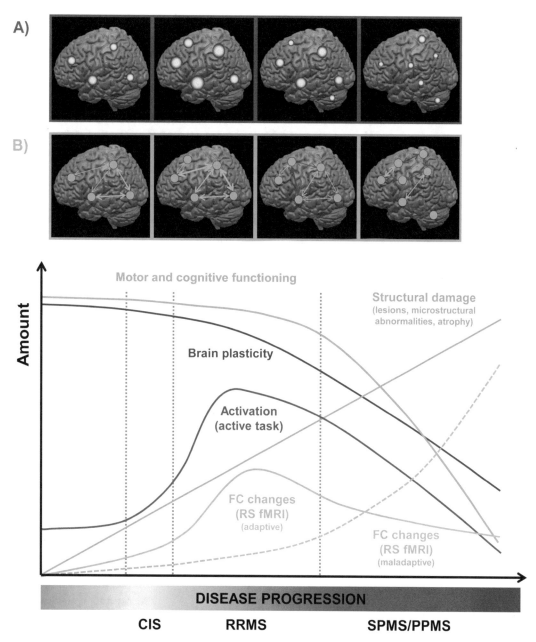

Fig. 2 Schematic representation of the main task-based and resting state (RS) functional MRI (fMRI) findings in multiple sclerosis (MS). At the beginning of the disease, in patients with a clinically isolated syndrome (CIS) and in early relapsing-remitting MS, the preservation of motor and cognitive functions is associated with (**a**) an increased recruitment of regions "normally" involved in a specific task and (**b**) increased functional connectivity (FC). In late RRMS and in progressive MS forms (i.e., primary progressive MS [PPMS] and secondary progressive MS [SPMS]), which are characterized by the accumulation of a severe structural damage, there is a progressive exhaustion or inefficiency of the adaptive brain properties, with (**a**) reduced activations of regions typically involved in a specific function, combined with a more widespread recruitment of additional areas usually recruited in normal people to perform novel/complex tasks and increased heterogeneous and aberrant activations. Additionally, (**b**) abnormal FC, associated with clinical dysfunction (and thus maladaptive), can be observed

In patients with progressive MS, several findings support a maladaptive role of functional abnormalities [70–72]. A progressive exhaustion or inefficiency of adaptive brain properties determine a collapse of functional network competencies that can be among the factors contributing to clinical worsening [63, 64]. In line with this, in patients with secondary progressive (SP) MS, reduced activation of "classical" regions of the sensorimotor network is combined with a widespread recruitment of additional areas usually recruited in normal people to perform novel/complex tasks as well as aberrant activations of many areas, located within the frontal, parietal, and temporal lobes [63, 64].

The correlations found between measures of abnormal activations and quantitative MR metrics of disease burden suggest that functional reorganization might play an adaptive role to limit the clinical consequences of disease-related structural damage [71, 73]; however, after reaching a given threshold, functional modifications are not able to further compensate structural damage accumulation, supporting a plateauing effect.

Brain fMRI modifications have also been assessed to explain the presence and severity of cognitive impairment [27, 74–89]. Several cognitive domains have been investigated in MS patients using active fMRI tasks, including working memory, attention, episodic memory, planning, and emotional processing [6].

In CIS [76, 77, 90] [74], RRMS [81, 84, 91], and benign MS [88] patients, cognitive-related brain activations are typically characterized by an increased recruitment of critical brain areas and activation of several regions located bilaterally in the frontotemporoparietal lobes and cerebellum during the performance of different cognitive tasks, especially with increased complexities.

Two studies addressed the relevance for cognitive performance of longitudinal modifications of cognitive network recruitment. In the first study, a higher activation in the prefrontal cortices over 1 year contributed to stability of Paced Auditory Serial Addition Test (PASAT) performance in early RRMS [92], whereas in the other, worsening of cognitive performance in RRMS was associated with modification of recruitment of parietal regions during a go/no-go paradigm [93]. These findings, combined with the associations reported between abnormal fMRI activation and measures of brain structural damage, suggest that such functional mechanisms might contribute to attenuate the negative effect of MS-related tissue damage on cognition [6].

However, increased cortical recruitment cannot continue indefinitely. In patients with progressive forms of the disease and cognitive impairment, reduced activations of several crucial brain regions such as frontoparietal and cingulate cortices have been demonstrated during simple cognitive tasks. Such abnormalities of recruitment tend to increase with increasing task complexity, suggesting the presence of a reduced functional reserve to preserve cognitive functions and the loss or exhaustion of adaptive mechanisms [6, 80, 86, 91, 94].

5.2 Resting State fMRI

RS FC abnormalities have been consistently demonstrated in MS patients with the main clinical phenotypes, already from disease onset.

5.2.1 Resting State FC

CIS [95], RRMS [96, 97] and pediatric MS patients [98] are characterized by a higher RS FC in the sensorimotor network compared to healthy controls (HC) (Fig. 2b). Conversely, in patients with progressive MS and more severe disability, a reduced sensorimotor network RS FC has been shown, suggesting that when functional reorganization reaches a plateau a further accumulation of structural damage causes an exhaustion of functional plasticity that contributes to disability progression [96, 99–102].

The analysis of deep GM RS FC has shown that MS patients had typically an increased deep GM RS FC [103–107]. A recent study showed a temporal evolution of such RS FC abnormalities [104]: patients with late RRMS and SPMS showed increased connectivity of the deep GM with other deep GM structures and the cortex, whereas within-cortex connectivity was decreased in SPMS. Deep GM and cortical RS FC abnormalities were associated with clinical disability and cognitive dysfunction, suggesting that an increased deep GM RS FC and cortical network collapse may explain the severe and progressive clinical deterioration characteristic of progressive MS forms [104].

Notably, a decreased RS FC between the thalami and several brain regions has also been described [107, 108], possibly due to the different pattern of structural connectivity of the main thalamic subregions, which may result in opposite RS FC abnormalities [109].

The analysis of cerebellar network has found that a higher cerebellar RS FC was associated with a shorter disease duration, a lower T2 lesion load and better motor performances, suggesting an adaptive mechanism contributing to preserved clinical function, whereas decreased cerebellar RS FC was correlated with longer disease duration, higher T2 lesion load and cognitive impairment, suggesting a maladaptive mechanism [110, 111].

Modifications of RS FC of networks involved in specific cognitive processes have been also demonstrated, especially in the DMN. An increased RS FC in the DMN has been found in CIS [95] and RRMS patients [112]. Conversely, a reduced DMN RS FC was found to be more severe in SPMS [113], to contribute to clinical disability and cognitive impairment, both in RRMS [100, 114] and progressive MS patients [115], and to correlate with the severity of structural damage [100, 115]. While adult patients with cognitive impairment had reduced RS FC in the anterior node of the DMN (in frontal lobe regions), pediatric MS patients with cognitive impairment showed decreased RS FC in the posterior regions of the DMN [98], suggesting that MS onset during childhood might impair network maturational trajectories.

Since brain functioning requires high level of integration between functionally relevant networks, functional interactions among the main RS networks have been also explored in MS patients. RS FC abnormalities within and between large-scale neuronal networks have been demonstrated both in adult and pediatric MS patients. Such RS abnormalities have been related to the extent of WM lesions and the severity of disability and cognitive impairment [98, 100, 116]. Interestingly, compared to adult-onset patients, pediatric-onset MS patients had reduced long-range RS FC between DMN and secondary visual network and more severe structural damage, suggesting a detrimental effect of MS at earlier onset in promoting a higher vulnerability to unfavorable clinical outcome in the long term [117].

5.2.2 Graph Theory Analysis

Graph theory analysis is a promising approach to detect functional changes, characterized by both global and local network measure alterations.

Only a few studies have applied graph theory analysis to investigate RS network functional alterations in MS patients [118–122]. Studies evaluating the earliest stages of MS have so far provided conflicting results [118, 121, 122]. CIS patients had a decreased whole-brain network efficiency, which was less pronounced than that observed in RRMS patients [121, 122]. Conversely, at regional analysis, alterations in nodal efficiency and RS FC detected in CIS patients were similar to those found in MS patients, supporting the hypothesis that regional network degeneration is already present in CIS [121, 122].

By combining structural and functional network analysis in CIS and MS patients, another study showed that CIS patients had only structural network abnormalities, while in MS patients both the structural and functional networks were altered, suggesting that structural network damage may precede functional network modifications [118].

A study including a large cohort of relapse-onset MS patients and HC [120] found that global network properties were abnormal in MS patients compared to HC, and helped to distinguish cognitively impaired MS patients from HC, while they did not allow the main MS clinical phenotypes to be differentiated. Regional analyses of brain hubs (i.e., highly connected regions with a central position in network organization) showed that, compared to HC, MS patients had a loss of hubs in the superior frontal gyrus, precuneus and anterior cingulum in the left hemisphere; a different lateralization of basal ganglia hubs (mostly located in the left hemisphere in HC, and in the right hemisphere in MS patients); and the formation of hubs, not seen in HC, in the left temporal pole and cerebellar lobules IV and V. This modification of regional network properties contributed to explain cognitive impairment and MS phenotypic variability. The role of network functional abnormalities for cognitive deficits in MS patients has been confirmed by other studies [119, 123, 124]. In early MS patients, diminished functional integration between separate functional modules contributed to explain worse performance in a dual task [123]. Another study demonstrated a decreased network efficiency correlated with reduced visuospatial memory in male MS patients only, suggesting gender differences in the pathological substrates of cognitive impairment [119]. A recent study showed an increased centrality of the DMN in MS patients with cognitive impairment, hypothesizing that this shifted balance toward the DMN could represent an overall DMN dysfunction underlying cognitive impairment [124].

5.2.3 Dynamic FC

Although heterogeneous dFC methodologies have been applied and regional patterns of dFC findings are quite variable across studies, dFC abnormalities have been consistently found in MS patients compared to HC [125–130].

CIS patients showed a reduced dFC in the functional networks involved by clinical onset, suggesting a correspondence between dFC dysfunction and clinical symptoms at disease onset [130]. During the first 2 years from disease onset, a progressive increase of dFC occurred in CIS patients and it was associated with a lower modification of WM lesion volume [130].

In RRMS, complex patterns of increased and decreased dFC have been shown [126, 128, 129]. A stronger dFC was found in temporoparietal regions [126, 128], subcortical and visual/cognitive networks [129], and between the salience network and the ascending nociceptive pathway [126]. A reduction of dFC, both widespread [125] or involving specific regions, such as subcortical, sensorimotor networks and amygdala [128, 129], was also detected. Another study showed a connectivity disorganization that was increased for regions involved in motor, executive control, spatial coordinating, and memory systems compared to HC [127].

In MS patients, a higher disability was associated with more disorganized connectivity in the bilateral supplementary motor area and right precentral operculum [127], as well as with a more rigid (less fluid) global connectivity [125]. Decreased dFC and reduced global dynamism were also associated with more severe tissue damage, including WM lesion load [131], microstructural tissue damage [127], and brain atrophy [129].

Although studies aimed to investigate dFC in progressive MS phenotypes are still missing, all these results suggest that, in the earliest phases of the disease, the occurrence of a progressive increase of dFC oscillations could represent a compensatory mechanism for disease-related damage. Later on, a globally more "static"

FC configuration is observed, associated with a loss of coordination and flexibility among brain regions [125, 127–129], which may be compensated by local increased fluctuations between specific areas [125, 132, 133].

When applied to investigate cognition, worse performances were associated to reduced dFC between subcortical and DMN areas, as well as to reduced global dynamism [129]. By evaluating specific cognitive-relevant systems (i.e., the hippocampal network, the DMN, and the attention network), dFC changes contributed to explain the performance of MS patients at visuospatial memory [125, 132, 133], information processing speed [125, 132, 133] and attention tasks [131]. In particular, better cognitive scores were correlated with a higher network dynamism [125, 132], whereas a lower hippocampal dFC contributed to explain maintained memory performances [133].

6   Functional MRI to Monitor Pharmacologic and Rehabilitative Treatments

6.1 Pharmacologic Treatments

At present, the application of fMRI to investigate the influence of different pharmacologic treatments has been limited in MS and mainly focused on symptomatic therapies rather than on disease-modifying drugs.

By investigating the execution of a right-hand motor task, one study showed that abnormal activation of frontothalamic areas, which are part of the sensorimotor network, could contribute to explain fatigue induced by interferon beta 1a administration in RRMS patients [134]. A more recent study showed that the positive effects promoted by interferon beta 1a could be associated with normalization of brain activity during a visuomotor task possibly through a reduction of MS-related inflammatory activity and creation of a more favorable environment [135].

Although cannabis may promote positive effects on spasticity, it also exerts detrimental effects on cognitive functions. According to recent studies, cannabis use is not associated with RS FC or structural MRI modifications, but it determines a more diffuse activation at fMRI during the N-Back task and increases the amount of errors [136–138].

Cholinesterase inhibitors have been proposed to limit cognitive dysfunction in MS patients. Using cognitive active tasks (i.e., Stroop, N-back, or a modified version of the paced visual serial addition test), several studies showed that rivastigmine can modulate functional recruitment of frontoparietal regions involved in cognitive functions [79, 139, 140], although no concomitant improvement of cognitive performance was detected.

By applying high-frequency (≥5 Hz) stimulation on the right dorsolateral prefrontal cortex (DLPFC), another study showed that repetitive transcranial magnetic stimulation (rTMS) could promote an improvement of working memory performances associated with a normalization of frontal activations and an increased FC between the right DLPFC and the right caudate nucleus and bilateral (para)-cingulate gyrus [141].

Potassium-channel blockers, such as 3,4-diaminopyridine and 4-aminopyridine have been suggested to improve fatigue and motor function in MS patients. Based on this, an fMRI study showed that 3,4-diaminopyridine modulated the recruitment of the sensorimotor network by increasing the activation of the ipsilateral sensorimotor cortex and supplementary motor area during a simple motor task in MS patients [142].

A recent study evaluating fMRI activation at the time of strong urgency, showed that OnabotulinumtoxinA, a drug used to improve neurogenic overactive bladder, may promote beneficial effects not only through a direct action on detrusor muscle, but also through the modulation of the activity of brain regions involved in the sensation and process of urinary urgency [143].

6.2 Rehabilitative Treatments

Motor rehabilitation has been consistently demonstrated to promote beneficial effects on MS patients' well-being, possibly through neuroplasticity mechanisms able to restore functions and competence of dysfunctional brain networks [10, 144].

Several studies suggested that, despite a severe structural damage, functional brain plasticity can still be present in MS patients and could be enhanced by task-dependent and target-selected motor rehabilitation procedures.

By applying fMRI during active tasks or at rest, modifications of recruitment, FC or dFC have been consistently found following motor rehabilitation [4, 144, 145]. These functional changes mainly characterize brain regions and networks involved in sensorimotor functions (i.e., the primary motor and somatosensory cortices, supplementary motor area, premotor cortex, thalamus, and cerebellum) [4, 144].

By applying action observation therapy (AOT) in MS patients with upper limb motor impairment, recent investigations showed that, compared to conventional rehabilitative approach, AOT can promote several functional changes which are correlated with clinical improvements at functional motor scales [145, 146]. AOT induced a greater activation of bilateral inferior frontal gyrus and left insula during a manipulation task with the right hand, it promoted a higher RS FC in left cerebellum and right inferior frontal gyrus of the motor network, a higher RS FC in right cerebellum and calcarine sulcus of the mirror neuron system [146] and also a larger dFC increase in sensorimotor and cognitive networks [145].

Interestingly, a recent study showed that treadmill walking exercise promoted an increase in thalamocortical RS FC that was correlated with an improvement of information processing speed, thus confirming the hypothesis that motor training could improve not only motor disability, but also cognitive dysfunction in MS patients [147].

Both task-related and RS fMRI studies have consistently demonstrated that cognitive rehabilitation results in an improvement of the rehabilitated function (i.e., attention, memory, and executive functions) and that these improvements are somehow mediated by brain plasticity, with modification of recruitment and/or FC occurring in functionally related brain regions and networks such as the anterior and posterior cingulate cortex, precuneus, DLPFC, thalamus, and cerebellum.

Two studies, investigating rehabilitation of attentive functions, showed an increased cerebellar and superior parietal lobule activation during PASAT fMRI task [148, 149].

Memory rehabilitation in MS patients has been shown to promote activation changes mainly in frontal and temporal regions during memory and learning tasks [150], but also to induce an increased RS FC in the hippocampal memory network and DMN [151]. Similarly, working memory training promoted increased performance in working memory and alertness in pediatric MS patients, which were associated with an increased working memory network activation and enhancement of inter-network connectivity [152].

Using both task-related and RS fMRI to study functional brain changes after three-months of computer-assisted cognitive rehabilitation of attention and executive functions, a pivotal study showed a significant improvement of these functions which was associated with an enhanced recruitment of brain networks/regions implicated in cognitive functions (i.e., the DLPFC bilaterally, and the posterior cingulate cortex and/or precuneus), an increase or at least stable RS FC of the DMN, salience processing network and executive function network as well as an increased FC of the anterior cingulate cortex with frontoparietal regions [153, 154]. Interestingly, the positive effects of cognitive rehabilitation on cognitive tests were still present 6 months after the termination of cognitive rehabilitation [155] and RS fMRI measures evaluated during the rehabilitation phase of the study were the only predictors of these positive effects at 6 months, suggesting that cognitive rehabilitation may act by optimizing cognitive network recruitment, resulting in a generalized functional improvement [155].

A recent study showed that a video game-based cognitive rehabilitation program may promote changes in thalamic RS FC with cognitively relevant brain regions, possibly representing a functional substrate for cognitive improvement [156].

7   Conclusions

The application of different fMRI techniques to study MS patients has consistently demonstrated how fMRI can provide important insights into the pathophysiological mechanisms of cortical reorganization in MS patients.

As a consequence, fMRI holds promise to improve our understanding of the processes occurring after acute relapses and lesion formations, and of the substrates associated with the accumulation of irreversible disability and cognitive impairment. fMRI studies suggest that, together with the accumulation of structural damage, neuroplasticity may occur in MS. Available data support the concept that brain adaptive responses may have an important role in compensating for irreversible tissue damage, such as demyelination and neuro-axonal loss. However, after a certain level of structural damage accumulation, a faster rate of disability progression might be a function of the progressive failure of the adaptive capacity and the occurrence of maladaptive functional changes.

Finally, fMRI might provide a relevant contribution to investigate the mechanisms underlying the beneficial effects of both pharmacologic and rehabilitative treatments.

Conflict of Interest Statement

Paolo Preziosa received speakers' honoraria from Biogen Idec, Novartis, and ExceMED.

Paola Valsasina received speakers' honoraria from Biogen Idec, Novartis, and ExceMED.

Massimo Filippi is Editor-in-Chief of the Journal of Neurology and Associate Editor of Human Brain Mapping; received compensation for consulting services and/or speaking activities from Almiral, Alexion, Bayer, Biogen, Celgene, Eli Lilly, Genzyme, Merck-Serono, Novartis, Roche, Sanofi, Takeda, and Teva Pharmaceutical Industries; and receives research support from Biogen Idec, Merck-Serono, Novartis, Roche, Teva Pharmaceutical Industries, Italian Ministry of Health, Fondazione Italiana Sclerosi Multipla, and ARiSLA (Fondazione Italiana di Ricerca per la SLA).

Maria A. Rocca received speaker honoraria from Bayer, Biogen, Bristol Myers Squibb, Celgene, Genzyme, Merck Serono, Novartis, Roche, and Teva, and receives research support from the MS Society of Canada and Fondazione Italiana Sclerosi Multipla.

References

1. Filippi M, Bruck W, Chard D et al (2019) Association between pathological and MRI findings in multiple sclerosis. Lancet Neurol 18:198–210

2. Filippi M, Preziosa P, Rocca MA (2014) Magnetic resonance outcome measures in multiple sclerosis trials: time to rethink? Curr Opin Neurol 27:290–299

3. Filippi M, Preziosa P, Rocca MA (2017) Microstructural MR imaging techniques in multiple sclerosis. Neuroimaging Clin N Am 27:313–333

4. Rocca MA, Preziosa P, Filippi M (2019) Application of advanced MRI techniques to monitor pharmacologic and rehabilitative treatment in multiple sclerosis: current status and future perspectives. Expert Rev Neurother 19:835–866

5. Enzinger C, Barkhof F, Ciccarelli O et al (2015) Nonconventional MRI and microstructural cerebral changes in multiple sclerosis. Nat Rev Neurol 11:676–686

6. Rocca MA, Amato MP, De Stefano N et al (2015) Clinical and imaging assessment of cognitive dysfunction in multiple sclerosis. Lancet Neurol 14:302–317

7. Rocca MA, Battaglini M, Benedict RH et al (2017) Brain MRI atrophy quantification in MS: from methods to clinical application. Neurology 88:403–413

8. Filippi M, Rocca MA (2004) Magnetization transfer magnetic resonance imaging in the assessment of neurological diseases. J Neuroimaging 14:303–313

9. Filippi M, Rocca MA, Comi G (2003) The use of quantitative magnetic-resonance-based techniques to monitor the evolution of multiple sclerosis. Lancet Neurol 2:337–346

10. Zatorre RJ, Fields RD, Johansen-Berg H (2012) Plasticity in gray and white: neuroimaging changes in brain structure during learning. Nat Neurosci 15:528–536

11. Rocca MA, Filippi M (2006) Functional MRI to study brain plasticity in clinical neurology. Neurol Sci 27(Suppl 1):S24–S26

12. Rocca MA, Filippi M (2007) Functional MRI in multiple sclerosis. J Neuroimaging 17 (Suppl 1):36S–41S

13. Ogawa S, Lee TM, Kay AR et al (1990) Brain magnetic resonance imaging with contrast dependent on blood oxygenation. Proc Natl Acad Sci U S A 87:9868–9872

14. Ogawa S, Menon RS, Tank DW et al (1993) Functional brain mapping by blood oxygenation level-dependent contrast magnetic resonance imaging. A comparison of signal characteristics with a biophysical model. Biophys J 64:803–812

15. Feinberg DA, Moeller S, Smith SM et al (2010) Multiplexed echo planar imaging for sub-second whole brain FMRI and fast diffusion imaging. PLoS One 5:e15710

16. Buxton RB, Wong EC, Frank LR (1998) Dynamics of blood flow and oxygenation changes during brain activation: the balloon model. Magn Reson Med 39:855–864

17. Friston KJ, Holmes AP, Poline JB et al (1995) Analysis of fMRI time-series revisited. NeuroImage 2:45–53

18. Flandin G, Friston KJ (2008) Statistical parametric mapping. Scholarpedia 3(4):6232

19. Adler RJ (1981) The geometry of random fields. Wiley, New York

20. Worsley KJ, Marrett S, Neelin P et al (1996) A unified statistical approach for determining significant signals in images of cerebral activation. Hum Brain Mapp 4:58–73

21. Chumbley J, Worsley K, Flandin G et al (2010) Topological FDR for neuroimaging. NeuroImage 49:3057–3064

22. Friston KJ, Frith CD, Liddle PF et al (1993) Functional connectivity: the principal-component analysis of large (PET) data sets. J Cereb Blood Flow Metab 13:5–14

23. Friston KJ (2011) Functional and effective connectivity: a review. Brain Connect 1:13–36

24. Stephan KE, Friston KJ (2010) Analyzing effective connectivity with functional magnetic resonance imaging. Wiley Interdiscip Rev Cogn Sci 1:446–459

25. Friston KJ, Buechel C, Fink GR et al (1997) Psychophysiological and modulatory interactions in neuroimaging. NeuroImage 6:218–229

26. Friston KJ, Harrison L, Penny W (2003) Dynamic causal modelling. NeuroImage 19:1273–1302

27. Au Duong MV, Boulanouar K, Audoin B et al (2005) Modulation of effective connectivity inside the working memory network in patients at the earliest stage of multiple sclerosis. NeuroImage 24:533–538

28. Ramsey JD, Hanson SJ, Glymour C (2011) Multi-subject search correctly identifies causal connections and most causal directions in the DCM models of the Smith et al. simulation study. NeuroImage 58:838–848

29. Cordes D, Haughton VM, Arfanakis K et al (2001) Frequencies contributing to

functional connectivity in the cerebral cortex in "resting-state" data. AJNR Am J Neuroradiol 22:1326–1333

30. Biswal BB, Mennes M, Zuo XN et al (2010) Toward discovery science of human brain function. Proc Natl Acad Sci U S A 107:4734–4739

31. Damoiseaux JS, Rombouts SA, Barkhof F et al (2006) Consistent resting-state networks across healthy subjects. Proc Natl Acad Sci U S A 103:13848–13853

32. Smith SM, Fox PT, Miller KL et al (2009) Correspondence of the brain's functional architecture during activation and rest. Proc Natl Acad Sci U S A 106:13040–13045

33. Raichle ME, MacLeod AM, Snyder AZ et al (2001) A default mode of brain function. Proc Natl Acad Sci U S A 98:676–682

34. Fox MD, Raichle ME (2007) Spontaneous fluctuations in brain activity observed with functional magnetic resonance imaging. Nat Rev Neurosci 8:700–711

35. Greicius MD, Srivastava G, Reiss AL et al (2004) Default-mode network activity distinguishes Alzheimer's disease from healthy aging: evidence from functional MRI. Proc Natl Acad Sci U S A 101:4637–4642

36. Fox MD, Greicius M (2010) Clinical applications of resting state functional connectivity. Front Syst Neurosci 4:19

37. Lv H, Wang Z, Tong E et al (2018) Resting-state functional MRI: everything that nonexperts have always wanted to know. AJNR Am J Neuroradiol 39:1390–1399

38. Tononi G, Sporns O, Edelman GM (1994) A measure for brain complexity: relating functional segregation and integration in the nervous system. Proc Natl Acad Sci U S A 91:5033–5037

39. Lund TE, Madsen KH, Sidaros K et al (2006) Non-white noise in fMRI: does modelling have an impact? NeuroImage 29:54–66

40. Van Dijk KR, Sabuncu MR, Buckner RL (2012) The influence of head motion on intrinsic functional connectivity MRI. NeuroImage 59:431–438

41. Whitfield-Gabrieli S, Thermenos HW, Milanovic S et al (2009) Hyperactivity and hyperconnectivity of the default network in schizophrenia and in first-degree relatives of persons with schizophrenia. Proc Natl Acad Sci U S A 106:1279–1284

42. Glover GH, Li TQ, Ress D (2000) Image-based method for retrospective correction of physiological motion effects in fMRI: RETROICOR. Magn Reson Med 44:162–167

43. Griffanti L, Salimi-Khorshidi G, Beckmann CF et al (2014) ICA-based artefact removal and accelerated fMRI acquisition for improved resting state network imaging. NeuroImage 95:232–247

44. Behzadi Y, Restom K, Liau J et al (2007) A component based noise correction method (CompCor) for BOLD and perfusion based fMRI. NeuroImage 37:90–101

45. Yang H, Long XY, Yang Y et al (2007) Amplitude of low frequency fluctuation within visual areas revealed by resting-state functional MRI. NeuroImage 36:144–152

46. Zang Y, Jiang T, Lu Y et al (2004) Regional homogeneity approach to fMRI data analysis. NeuroImage 22:394–400

47. Tomasi D, Volkow ND (2010) Functional connectivity density mapping. Proc Natl Acad Sci U S A 107:9885–9890

48. Biswal B, Yetkin FZ, Haughton VM et al (1995) Functional connectivity in the motor cortex of resting human brain using echo-planar MRI. Magn Reson Med 34:537–541

49. McKeown MJ, Sejnowski TJ (1998) Independent component analysis of fMRI data: examining the assumptions. Hum Brain Mapp 6:368–372

50. Calhoun VD, Adali T, Pearlson GD et al (2001) A method for making group inferences from functional MRI data using independent component analysis. Hum Brain Mapp 14:140–151

51. Beckmann CF, DeLuca M, Devlin JT et al (2005) Investigations into resting-state connectivity using independent component analysis. Philos Trans R Soc Lond Ser B Biol Sci 360:1001–1013

52. Rubinov M, Sporns O (2010) Complex network measures of brain connectivity: uses and interpretations. NeuroImage 52:1059–1069

53. Bullmore E, Sporns O (2009) Complex brain networks: graph theoretical analysis of structural and functional systems. Nat Rev Neurosci 10:186–198

54. Tagliazucchi E, van Someren EJW (2017) The large-scale functional connectivity correlates of consciousness and arousal during the healthy and pathological human sleep cycle. NeuroImage 160:55–72

55. Hutchison RM, Womelsdorf T, Allen EA et al (2013) Dynamic functional connectivity: promise, issues, and interpretations. NeuroImage 80:360–378

56. Calhoun VD, Miller R, Pearlson G et al (2014) The chronnectome: time-varying connectivity networks as the next frontier in fMRI data discovery. Neuron 84:262–274

57. Allen EA, Damaraju E, Plis SM et al (2014) Tracking whole-brain connectivity dynamics

in the resting state. Cereb Cortex 24:663–676

58. Sakoglu U, Pearlson GD, Kiehl KA et al (2010) A method for evaluating dynamic functional network connectivity and task-modulation: application to schizophrenia. MAGMA 23:351–366

59. Liu X, Duyn JH (2013) Time-varying functional network information extracted from brief instances of spontaneous brain activity. Proc Natl Acad Sci U S A 110:4392–4397

60. Reddy H, Narayanan S, Matthews PM et al (2000) Relating axonal injury to functional recovery in MS. Neurology 54:236–239

61. Mezzapesa DM, Rocca MA, Rodegher M et al (2008) Functional cortical changes of the sensorimotor network are associated with clinical recovery in multiple sclerosis. Hum Brain Mapp 29:562–573

62. Zaaraoui W, Rico A, Audoin B et al (2010) Unfolding the long-term pathophysiological processes following an acute inflammatory demyelinating lesion of multiple sclerosis. Magn Reson Imaging 28:477–486

63. Rocca MA, Colombo B, Falini A et al (2005) Cortical adaptation in patients with MS: a cross-sectional functional MRI study of disease phenotypes. Lancet Neurol 4:618–626

64. Enzinger C, Pinter D, Rocca MA et al (2016) Longitudinal fMRI studies: exploring brain plasticity and repair in MS. Mult Scler 22:269–278

65. Filippi M, Rocca MA, Mezzapesa DM et al (2004) Simple and complex movement-associated functional MRI changes in patients at presentation with clinically isolated syndromes suggestive of multiple sclerosis. Hum Brain Mapp 21:108–117

66. Rocca MA, Mezzapesa DM, Falini A et al (2003) Evidence for axonal pathology and adaptive cortical reorganization in patients at presentation with clinically isolated syndromes suggestive of multiple sclerosis. NeuroImage 18:847–855

67. Pantano P, Mainero C, Lenzi D et al (2005) A longitudinal fMRI study on motor activity in patients with multiple sclerosis. Brain 128:2146–2153

68. Rocca MA, Absinta M, Ghezzi A et al (2009) Is a preserved functional reserve a mechanism limiting clinical impairment in pediatric MS patients? Hum Brain Mapp 30:2844–2851

69. Rocca MA, Ceccarelli A, Rodegher M et al (2010) Preserved brain adaptive properties in patients with benign multiple sclerosis. Neurology 74:142–149

70. Rocca MA, Gavazzi C, Mezzapesa DM et al (2003) A functional magnetic resonance imaging study of patients with secondary progressive multiple sclerosis. NeuroImage 19:1770–1777

71. Filippi M, Rocca MA, Falini A et al (2002) Correlations between structural CNS damage and functional MRI changes in primary progressive MS. NeuroImage 15:537–546

72. Rocca MA, Matthews PM, Caputo D et al (2002) Evidence for widespread movement-associated functional MRI changes in patients with PPMS. Neurology 58:866–872

73. Rocca MA, Falini A, Colombo B et al (2002) Adaptive functional changes in the cerebral cortex of patients with nondisabling multiple sclerosis correlate with the extent of brain structural damage. Ann Neurol 51:330–339

74. Staffen W, Mair A, Zauner H et al (2002) Cognitive function and fMRI in patients with multiple sclerosis: evidence for compensatory cortical activation during an attention task. Brain 125:1275–1282

75. Au Duong MV, Audoin B, Boulanouar K et al (2005) Altered functional connectivity related to white matter changes inside the working memory network at the very early stage of MS. J Cereb Blood Flow Metab 25:1245–1253

76. Audoin B, Ibarrola D, Ranjeva JP et al (2003) Compensatory cortical activation observed by fMRI during a cognitive task at the earliest stage of MS. Hum Brain Mapp 20:51–58

77. Audoin B, Au Duong MV, Ranjeva JP et al (2005) Magnetic resonance study of the influence of tissue damage and cortical reorganization on PASAT performance at the earliest stage of multiple sclerosis. Hum Brain Mapp 24:216–228

78. Hillary FG, Chiaravalloti ND, Ricker JH et al (2003) An investigation of working memory rehearsal in multiple sclerosis using fMRI. J Clin Exp Neuropsychol 25:965–978

79. Parry AM, Scott RB, Palace J et al (2003) Potentially adaptive functional changes in cognitive processing for patients with multiple sclerosis and their acute modulation by rivastigmine. Brain 126:2750–2760

80. Penner IK, Rausch M, Kappos L et al (2003) Analysis of impairment related functional architecture in MS patients during performance of different attention tasks. J Neurol 250:461–472

81. Mainero C, Caramia F, Pozzilli C et al (2004) fMRI evidence of brain reorganization during attention and memory tasks in multiple sclerosis. NeuroImage 21:858–867

82. Sweet LH, Rao SM, Primeau M et al (2004) Functional magnetic resonance imaging of

working memory among multiple sclerosis patients. J Neuroimaging 14:150–157

83. Sweet LH, Rao SM, Primeau M et al (2006) Functional magnetic resonance imaging response to increased verbal working memory demands among patients with multiple sclerosis. Hum Brain Mapp 27:28–36

84. Wishart HA, Saykin AJ, McDonald BC et al (2004) Brain activation patterns associated with working memory in relapsing-remitting MS. Neurology 62:234–238

85. Chiaravalloti N, Hillary F, Ricker J et al (2005) Cerebral activation patterns during working memory performance in multiple sclerosis using FMRI. J Clin Exp Neuropsychol 27:33–54

86. Cader S, Cifelli A, Abu-Omar Y et al (2006) Reduced brain functional reserve and altered functional connectivity in patients with multiple sclerosis. Brain 129:527–537

87. Li Y, Chiaravalloti ND, Hillary FG et al (2004) Differential cerebellar activation on functional magnetic resonance imaging during working memory performance in persons with multiple sclerosis. Arch Phys Med Rehabil 85:635–639

88. Rocca MA, Valsasina P, Ceccarelli A et al (2009) Structural and functional MRI correlates of Stroop control in benign MS. Hum Brain Mapp 30:276–290

89. Lazeron RHC, Rombouts SARB, Scheltens P et al (2004) An fMRI study of planning-related brain activity in patients with moderately advanced multiple sclerosis. Mult Scler 10:549–555

90. Forn C, Rocca MA, Valsasina P et al (2012) Functional magnetic resonance imaging correlates of cognitive performance in patients with a clinically isolated syndrome suggestive of multiple sclerosis at presentation: an activation and connectivity study. Mult Scler 18:153–163

91. Rocca MA, Valsasina P, Hulst HE et al (2014) Functional correlates of cognitive dysfunction in multiple sclerosis: a multicenter fMRI study. Hum Brain Mapp 35:5799–5814

92. Audoin B, Reuter F, Duong MV et al (2008) Efficiency of cognitive control recruitment in the very early stage of multiple sclerosis: a one-year fMRI follow-up study. Mult Scler 14:786–792

93. Loitfelder M, Fazekas F, Koschutnig K et al (2014) Brain activity changes in cognitive networks in relapsing-remitting multiple sclerosis—insights from a longitudinal FMRI study. PLoS One 9:e93715

94. Rocca MA, Riccitelli G, Rodegher M et al (2010) Functional MR imaging correlates of

neuropsychological impairment in primary-progressive multiple sclerosis. AJNR Am J Neuroradiol 31:1240–1246

95. Roosendaal SD, Schoonheim MM, Hulst HE et al (2010) Resting state networks change in clinically isolated syndrome. Brain 133:1612–1621

96. Faivre A, Rico A, Zaaraoui W et al (2012) Assessing brain connectivity at rest is clinically relevant in early multiple sclerosis. Mult Scler 18:1251–1258

97. Dogonowski AM, Siebner HR, Soelberg Sorensen P et al (2013) Resting-state connectivity of pre-motor cortex reflects disability in multiple sclerosis. Acta Neurol Scand 128:328–335

98. Rocca MA, Absinta M, Amato MP et al (2014) Posterior brain damage and cognitive impairment in pediatric multiple sclerosis. Neurology 82:1314–1321

99. Rocca MA, Valsasina P, Leavitt VM et al (2018) Functional network connectivity abnormalities in multiple sclerosis: correlations with disability and cognitive impairment. Mult Scler 24:459–471

100. Rocca MA, Valsasina P, Martinelli V et al (2012) Large-scale neuronal network dysfunction in relapsing-remitting multiple sclerosis. Neurology 79:1449–1457

101. Basile B, Castelli M, Monteleone F et al (2014) Functional connectivity changes within specific networks parallel the clinical evolution of multiple sclerosis. Mult Scler 20:1050–1057

102. Janssen AL, Boster A, Patterson BA et al (2013) Resting-state functional connectivity in multiple sclerosis: an examination of group differences and individual differences. Neuropsychologia 51:2918–2929

103. Dogonowski AM, Siebner HR, Sorensen PS et al (2013) Expanded functional coupling of subcortical nuclei with the motor resting-state network in multiple sclerosis. Mult Scler 19:559–566

104. Meijer KA, Eijlers AJC, Geurts JJG et al (2018) Staging of cortical and deep grey matter functional connectivity changes in multiple sclerosis. J Neurol Neurosurg Psychiatry 89:205–210

105. Cui F, Zhou L, Wang Z et al (2017) Altered functional connectivity of striatal subregions in patients with multiple sclerosis. Front Neurol 8:129

106. Prosperini L, Fanelli F, Petsas N et al (2014) Multiple sclerosis: changes in microarchitecture of white matter tracts after training with a video game balance board. Radiology 273:529–538

107. De Giglio L, Tona F, De Luca F et al (2016) Multiple sclerosis: changes in thalamic resting-state functional connectivity induced by a home-based cognitive rehabilitation program. Radiology 280:202–211

108. Liu Y, Duan Y, Huang J et al (2015) Multimodal quantitative MR imaging of the thalamus in multiple sclerosis and neuromyelitis optica. Radiology 277:784–792

109. d'Ambrosio A, Hidalgo de la Cruz M, Valsasina P et al (2017) Structural connectivity-defined thalamic subregions have different functional connectivity abnormalities in multiple sclerosis patients: implications for clinical correlations. Hum Brain Mapp 38:6005–6018

110. Cirillo S, Rocca MA, Ghezzi A et al (2016) Abnormal cerebellar functional MRI connectivity in patients with paediatric multiple sclerosis. Mult Scler 22:292–301

111. Sbardella E, Upadhyay N, Tona F et al (2017) Dentate nucleus connectivity in adult patients with multiple sclerosis: functional changes at rest and correlation with clinical features. Mult Scler 23:546–555

112. Zhou F, Zhuang Y, Gong H et al (2014) Altered inter-subregion connectivity of the default mode network in relapsing remitting multiple sclerosis: a functional and structural connectivity study. PLoS One 9:e101198

113. Basile B, Castelli M, Monteleone F et al (2013) Functional connectivity changes within specific networks parallel the clinical evolution of multiple sclerosis. Mult Scler J 20:1050–1057

114. Bonavita S, Gallo A, Sacco R et al (2011) Distributed changes in default-mode resting-state connectivity in multiple sclerosis. Mult Scler 17:411–422

115. Rocca MA, Valsasina P, Absinta M et al (2010) Default-mode network dysfunction and cognitive impairment in progressive MS. Neurology 74:1252–1259

116. Huang MH, Zhou FQ, Wu L et al (2018) Synchronization within, and interactions between, the default mode and dorsal attention networks in relapsing-remitting multiple sclerosis. Neuropsychiatr Dis Treat 14:1241–1252

117. Giorgio A, Zhang J, Stromillo ML et al (2017) Pronounced structural and functional damage in early adult pediatric-onset multiple sclerosis with no or minimal clinical disability. Front Neurol 8:608

118. Shu N, Duan Y, Xia M et al (2016) Disrupted topological organization of structural and functional brain connectomes in clinically isolated syndrome and multiple sclerosis. Sci Rep 6:29383

119. Schoonheim MM, Hulst HE, Landi D et al (2012) Gender-related differences in functional connectivity in multiple sclerosis. Mult Scler 18:164–173

120. Rocca MA, Valsasina P, Meani A et al (2016) Impaired functional integration in multiple sclerosis: a graph theory study. Brain Struct Funct 221:115–131

121. Liu Y, Wang H, Duan Y et al (2017) Functional brain network alterations in clinically isolated syndrome and multiple sclerosis: a graph-based connectome study. Radiology 282:534–541

122. Abidin AZ, Chockanathan U, AM DS et al (2017) Using large-scale granger causality to study changes in brain network properties in the clinically isolated syndrome (CIS) stage of multiple sclerosis. Proc SPIE Int Soc Opt Eng 10137:101371B

123. Gamboa OL, Tagliazucchi E, von Wegner F et al (2014) Working memory performance of early MS patients correlates inversely with modularity increases in resting state functional connectivity networks. NeuroImage 94:385–395

124. Eijlers AJ, Meijer KA, Wassenaar TM et al (2017) Increased default-mode network centrality in cognitively impaired multiple sclerosis patients. Neurology 88:952–960

125. Lin SJ, Vavasour I, Kosaka B et al (2018) Education, and the balance between dynamic and stationary functional connectivity jointly support executive functions in relapsing-remitting multiple sclerosis. Hum Brain Mapp 39(12):5039–5049

126. Bosma RL, Kim JA, Cheng JC et al (2018) Dynamic pain connectome functional connectivity and oscillations reflect multiple sclerosis pain. Pain 159(11):2267–2276

127. Zhou F, Zhuang Y, Gong H et al (2016) Resting state brain entropy alterations in relapsing remitting multiple sclerosis. PLoS One 11:e0146080

128. Leonardi N, Richiardi J, Gschwind M et al (2013) Principal components of functional connectivity: a new approach to study dynamic brain connectivity during rest. NeuroImage 83:937–950

129. d'Ambrosio A, Valsasina P, Gallo A et al (2020) Reduced dynamics of functional connectivity and cognitive impairment in multiple sclerosis. Mult Scler 26(4):476–488

130. Rocca MA, Hidalgo de La Cruz M, Valsasina P et al (2020) Two-year dynamic functional

network connectivity in clinically isolated syndrome. Mult Scler 26(6):645–658

131. Huang M, Zhou F, Wu L et al (2019) White matter lesion loads associated with dynamic functional connectivity within attention network in patients with relapsing-remitting multiple sclerosis. J Clin Neurosci 65:59–65

132. van Geest Q, Douw L, van 't Klooster S et al (2018) Information processing speed in multiple sclerosis: relevance of default mode network dynamics. Neuroimage Clin 19:507–515

133. van Geest Q, Hulst HE, Meijer KA et al (2018) The importance of hippocampal dynamic connectivity in explaining memory function in multiple sclerosis. Brain Behav 8: e00954

134. Rocca MA, Agosta F, Colombo B et al (2007) fMRI changes in relapsing-remitting multiple sclerosis patients complaining of fatigue after IFNbeta-1a injection. Hum Brain Mapp 28:373–382

135. Tomassini V, d'Ambrosio A, Petsas N et al (2016) The effect of inflammation and its reduction on brain plasticity in multiple sclerosis: MRI evidence. Hum Brain Mapp 37:2431–2445

136. Romero K, Pavisian B, Staines WR et al (2015) Multiple sclerosis, cannabis, and cognition: a structural MRI study. Neuroimage Clin 8:140–147

137. Pavisian B, MacIntosh BJ, Szilagyi G et al (2014) Effects of cannabis on cognition in patients with MS: a psychometric and MRI study. Neurology 82:1879–1887

138. Pavisian B, Staines WR, Feinstein A (2015) Cannabis-induced alterations in brain activation during a test of information processing speed in patients with MS. Mult Scler J Exp Transl Clin 1:2055217315588223

139. Cader S, Palace J, Matthews PM (2009) Cholinergic agonism alters cognitive processing and enhances brain functional connectivity in patients with multiple sclerosis. J Psychopharmacol 23:686–696

140. Huolman S, Hamalainen P, Vorobyev V et al (2011) The effects of rivastigmine on processing speed and brain activation in patients with multiple sclerosis and subjective cognitive fatigue. Mult Scler 17:1351–1361

141. Hulst HE, Goldschmidt T, Nitsche MA et al (2017) rTMS affects working memory performance, brain activation and functional connectivity in patients with multiple sclerosis. J Neurol Neurosurg Psychiatry 88:386–394

142. Mainero C, Inghilleri M, Pantano P et al (2004) Enhanced brain motor activity in patients with MS after a single dose of 3,4-diaminopyridine. Neurology 62:2044–2050

143. Khavari R, Elias SN, Pande R et al (2019) Higher neural correlates in patients with multiple sclerosis and neurogenic overactive bladder following treatment with Intradetrusor injection of OnabotulinumtoxinA. J Urol 201:135–140

144. Prosperini L, Di Filippo M (2019) Beyond clinical changes: rehabilitation-induced neuroplasticity in MS. Mult Scler 25:1348–1362

145. Cordani C, Valsasina P, Preziosa P et al (2019) Action observation training promotes motor improvement and modulates functional network dynamic connectivity in multiple sclerosis. Mult Scler:1352458519887332

146. Rocca MA, Meani A, Fumagalli S et al (2019) Functional and structural plasticity following action observation training in multiple sclerosis. Mult Scler 25:1472–1487

147. Sandroff BM, Wylie GR, Sutton BP et al (2018) Treadmill walking exercise training and brain function in multiple sclerosis: preliminary evidence setting the stage for a network-based approach to rehabilitation. Mult Scler J Exp Transl Clin 4:2055217318760641

148. Cerasa A, Gioia MC, Valentino P et al (2013) Computer-assisted cognitive rehabilitation of attention deficits for multiple sclerosis: a randomized trial with fMRI correlates. Neurorehabil Neural Repair 27:284–295

149. Sastre-Garriga J, Alonso J, Renom M et al (2011) A functional magnetic resonance proof of concept pilot trial of cognitive rehabilitation in multiple sclerosis. Mult Scler 17:457–467

150. Chiaravalloti ND, Wylie G, Leavitt V et al (2012) Increased cerebral activation after behavioral treatment for memory deficits in MS. J Neurol 259:1337–1346

151. Leavitt VM, Wylie GR, Girgis PA et al (2014) Increased functional connectivity within memory networks following memory rehabilitation in multiple sclerosis. Brain Imaging Behav 8:394–402

152. Hubacher M, DeLuca J, Weber P et al (2015) Cognitive rehabilitation of working memory in juvenile multiple sclerosis-effects on cognitive functioning, functional MRI and network related connectivity. Restor Neurol Neurosci 33:713–725

153. Filippi M, Riccitelli G, Mattioli F et al (2012) Multiple sclerosis: effects of cognitive rehabilitation on structural and functional MR

imaging measures--an explorative study. Radiology 262:932–940

154. Parisi L, Rocca MA, Valsasina P et al (2014) Cognitive rehabilitation correlates with the functional connectivity of the anterior cingulate cortex in patients with multiple sclerosis. Brain Imaging Behav 8:387–393

155. Parisi L, Rocca MA, Mattioli F et al (2014) Changes of brain resting state functional connectivity predict the persistence of cognitive rehabilitation effects in patients with multiple sclerosis. Mult Scler 20:686–694

156. De Giglio L, Upadhyay N, De Luca F et al (2016) Corpus callosum microstructural changes associated with Kawashima Nintendo brain training in patients with multiple sclerosis. J Neurol Sci 370:211–213

Chapter 16

Functional Studies in Rodents

Lydia Wachsmuth and Cornelius Faber

Abstract

Anatomical MRI is well established in the diagnosis of MS in humans and in rodent models. In contrast, functional MRI is not routinely established in rodent models of MS. In this chapter, we introduce the major MRI methods and analysis approaches used to study brain function in healthy animals and in neuropathological disease models. We highlight differences in relation to human MRI, for example the hardware requirements for rodent MRI, and the specific issues associated with the need for anesthesia, regarding animal physiology and monitoring.

Key words Rodents, fMRI, BOLD, Stimulus-evoked fMRI, Resting state fMRI, Anesthesia

1 Introduction

Diagnosis and monitoring of disease progression of multiple sclerosis (MS) in humans and in animal models to date relies mainly on anatomical and structural magnetic resonance imaging (MRI) (Fig. 1 left). For clinical diagnosis in patients [1] contrast-enhanced T1 weighted (T1w) MRI and fluid attenuated inversion recovery (FLAIR) MRI are often complemented by double inversion recovery, T2 weighted (T2w), diffusion weighted (DW), susceptibility weighted, or other advanced MR sequences. In rodent models [2], commonly used MRI protocols for identification of lesions usually consist of T2w MRI for detection of edema and contrast-enhanced T1w MRI for detection of blood–brain barrier leakage. Contrast based on magnetization transfer and diffusion tensor imaging (DTI) is employed to identify demyelination and remyelination. DTI also provides data for tractography, informing about the integrity of structural connectivity between brain regions. Further MRI techniques are MR spectroscopy for metabolic profiling, T2* weighted (T2*w) MRI for monitoring infiltration of iron oxide nanoparticle-labeled peripheral immune cells, or T2* mapping to detect endogenous iron accumulation. Functional MRI (fMRI) (Fig. 1, right) is a complementary approach in the

Sergiu Groppa and Sven G. Meuth (eds.), *Translational Methods for Multiple Sclerosis Research*, Neuromethods, vol. 166, https://doi.org/10.1007/978-1-0716-1213-2_16, © Springer Science+Business Media, LLC, part of Springer Nature 2021

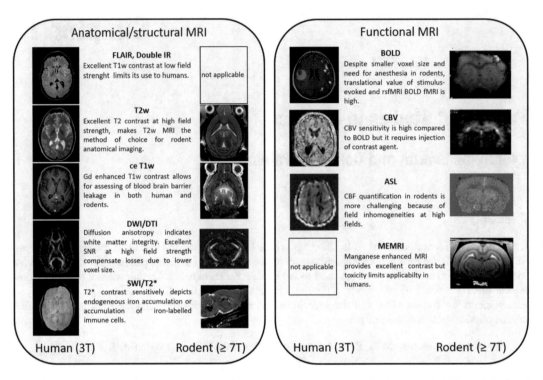

Fig. 1 Example human and rodent images acquired with established anatomical/structural (left) and functional MRI (right) methods. FLAIR-fluid attenuated inversion recovery, IR-inversion recovery, T2w-T2weighted, ce-contrast enhanced, T1w-T1weighted, DWI-diffusion weighted imaging, DTI–diffusion tensor imaging, SWI-susceptibility weighted imaging, BOLD-blood oxygenation level dependent, CBV-cerebral blood volume, ASL-arterial spin labeling, CBF-cerebral blood flow, MEMRI-manganese enhanced MRI

study of MS in humans. Among various available techniques for fMRI, blood oxygenation level dependent (BOLD) fMRI is most commonly used. Alterations in visual, motor and cognitive function have been interpreted as adaptive or compensatory mechanisms that allow normal performance despite neural damage or loss. Moreover, changes in brain function can occur not only as a consequence of neural damage or loss, but precede structural degradation, and consequently may have predictive value.

While rodent fMRI studies can provide important mechanistic insight, translation to the clinics is hampered by particular MR sequence requirements, long scan protocols or possibly toxicity issues. Rodent fMRI, generally provides a lower contrast-to-noise ratio (CNR) as compared to human studies, which is mainly due to the lower signal-to-noise ratio (SNR), resulting from the smaller image voxel volumes and the smaller size of the activated areas. Also, the anesthetics required in most animal studies influence brain's response characteristics. Therefore, a number of methodological points need to be considered carefully, including technical requirements of the experimental setting, anesthesia, animal physiology, and data analysis.

In this chapter, we will give an overview of rodent fMRI. We will outline differences compared to human fMRI and emphasize on the practical aspects of performing rodent fMRI.

2   Technical Requirements of the Experimental Setting

The fact that the rodent brain is much smaller compared to the human brain requires substantially higher spatial resolution in rodent MR protocols. Smaller voxel size, however, results in lower signal, which must be compensated by modifying the experimental setting. Higher magnetic field strengths provide higher magnetization, resulting in higher detectable signal. Higher performance gradient systems enable faster timing and thus reduced signal losses during spatial encoding for image generation. While clinical MRI systems for neuroimaging operate at 3 T (and rarely 7 T) magnetic field strength, neuroimaging in rodents is typically performed at 7 T, 9.4 T, or even higher field strength. Such high magnetic fields render T1w MRI less efficient, while T2w MRI dramatically gains contrast and becomes the method of choice for detecting MS lesions. T2*w contrast becomes even more pronounced at higher magnetic fields, rendering BOLD fMRI highly sensitive. Yet small voxel size, motion, and physiological instabilities also increase noise levels. Therefore, optimized detector settings are mandatory for successful BOLD fMRI in rodents. While in humans 32-channel array coils covering the entire brain are standard for fMRI, array coils in rodents are limited to two or four elements with only few exceptions. For most applications volume transmit-surface receive combinations of radio frequency (rf) resonators are used, which provide highest SNR, particularly in the neocortex. For deeper brain regions more complex array coils or even volume coils may be advantageous [3]. More than a factor of two in SNR can be gained by using cryogen-cooled rf resonators [4]. These, however, are inevitably subject to decreasing sensitivity for deeper brain regions.

2.1   fMRI Methods

For imaging of brain function, MRI can either be used to assess the response evoked by a specific stimulus, by a pharmacological challenge, or during rest. In stimulus-related fMRI, analyzing a stimulus-evoked response allows for investigating whether a respective neuronal pathway has been modulated or whether other brain regions eventually compensate functional loss. In rodent fMRI studies, stimuli are usually presented in a block design, to allow for averaging responses and thus increase SNR. Further, the variety of stimuli applied in small animal fMRI under anesthesia is limited to sensory stimuli (visual, olfactory, tactile, thermal, and auditory), whereas in human studies the response to cognitive tasks can be assessed. Pharmacological fMRI can be performed in rodents to probe molecular pathways, by targeting specific receptors or administering drug molecules. With resting state (rs)fMRI, the integrity of brain networks can be investigated, independently of a specific task or a pharmacological challenge. Spontaneous MRI signal fluctuations are correlated across brain regions, and resulting correlation coefficients inform about functional connectivity. Spatial resolution and brain coverage of rsfMRI outperform that of established functional neurological network analysis methods such as EEG and MEG; however, temporal resolution lags behind.

Independent of the type of stimulation or stimulation paradigm, the detected MRI signal has to ascertain physiological variations related to brain activity. For this purpose, several MRI methods exist, which are based on different physiological mechanisms and use different readouts related to specific contrast mechanisms.

2.1.1   BOLD

By far the most often employed readout in fMRI to indicate neuronal activity is the BOLD contrast, which is based on the hemodynamic response: Changes of cerebral blood flow (CBF), cerebral blood volume (CBV), and cerebral metabolic rate of oxygen consumption (CMRO2) together lead to an increased ratio of oxygenated over deoxygenated hemoglobin in the activated area. Since oxygenated hemoglobin is diamagnetic and deoxygenated hemoglobin is paramagnetic, this leads to a signal intensity increase in $T2^*$w images. The most sensitive readout of $T2^*$ contrast changes is achieved by gradient echo (GE) echo planar imaging (EPI) sequences. These, however, are prone to susceptibility artifacts, which may hamper imaging of regions like the olfactory bulb, the hypothalamus and amygdala, or generally caudal parts of the brain. Spin echo (SE) EPI sequences are much more robust regarding such artifacts, and therefore provide better image quality and better brain coverage. Sensitivity for the BOLD effect is generally lower, which results in lower amplitudes of the detected signal changes. Yet SE EPI signal more accurately reflects hemodynamic changes at the capillary level, and thus provides partly complementary information to GE EPI [5]. Recently, steady-state free precession sequences have been shown to provide higher amplitude BOLD signal changes than SE EPI [6].

2.1.2   CBV

An alternative readout for neuronal activity is CBV-related signal. In humans, this can be measured using Vascular-Space-Occupancy (VASO) MRI [7]. Blood and tissue compartments within a voxel are distinguished based on different T1 relaxation times. With a blood-nulling inversion recovery sequence, the MR signal becomes inversely proportional to CBV. For rodent fMRI at high magnetic fields, T1 differences are less pronounced, rendering VASO inefficient. Therefore, in rodent fMRI, CBV is measured by monitoring signal intensity changes in dynamic T1w or $T2^*$w imaging after intravenous injection of exogenous blood pool contrast agents, most often iron oxide nanoparticles [8]. Signal fluctuations result from spontaneous or stimulus-induced arteriole dilation, which follow increased neuronal activity, and directly represent CBV. Compared to BOLD fMRI, CNR of CBV fMRI is high but requires injection of contrast agents.

2.1.3   CBF

Measuring CBF changes provides another readout of brain activity, which is closely related to the BOLD signal. The MRI method arterial spin labeling (ASL) allows for absolute quantification of CBF. For ASL, two images are acquired, one containing signal from blood and tissue, and one with the signal from blood saturated. While ASL works well in humans [9], sensitivity, temporal resolution, and reproducibility are rather low for rodent functional studies. Issues arise from magnetic field inhomogeneities at high field strength and from the specific head and neck anatomy of rodents, impairing labeling efficiency [10]. Recently, two promising techniques have been developed to improve ASL in rodents [10]. One acquires various saturation images at different saturation angles and corrects off resonances in an elaborate post-processing, the other employs preparation scans to determine good saturation values.

2.1.4   Diffusion-Weighted fMRI

A direct readout of neuronal activity, circumventing vascular changes, has been postulated to be possible with diffusion fMRI [11]. The underlying model postulates that active neurons swell and obstruct water diffusion in the brain upon activation. To measure the related decrease in diffusion, the apparent diffusion coefficient (ADC) has to be calculated from at least two separate diffusion-weighted MRI scans. While different time courses of diffusion weighted MRI signal changes as compared to BOLD have been described for humans [12], in rodents these differences are not easy to observe [13].

2.1.5   Manganese-Enhanced MRI (MEMRI)

Using manganese-enhanced MRI (MEMRI) brain activity can be mapped in freely behaving animals [14]. Active neurons take up manganese ions (Mn^{2+}) via calcium channels. Mn^{2+} accumulates in the neurons independent of vascular activity, and signal changes in T1w MRI or T1 mapping that linearly depend on Mn^{2+} concentration [15] can be measured after prolonged exposure. Due to its cardiovascular toxicity, Mn^{2+} is usually applied by systemic, fractionated injection of a maximum tolerable dose of $MnCl_2$ over a prolonged period of time. However, blood–brain barrier crossing is limited. Continuous and slow administration of $MnCl_2$ by infusion via mini-osmotic pumps is advantageous for longitudinal studies compared to a single systemic injection of the same dose [16], avoiding acute toxic effects. MEMRI has been applied in a large number of rodent functional imaging studies and even in song birds, utilizing an intracranial application scheme to investigate brain plasticity, in particular auditory pathways in the mating season [17]. Manganese toxicity obviates its administration to humans. However, recently, Sudarshana et al. [18] applied mangafodipir (Teslascan), a chelated manganese-based contrast agent that is FDA-approved, to healthy volunteers and successfully enabled non-invasive visualization of intra- and extracranial structures that lie outside the blood–brain barrier without adverse clinical effects.

2.2 Anesthesia

A major difference to clinical examinations is that animals are usually anesthetized during fMRI, in order to immobilize them and to prevent stress due to scanner noise and mechanical restraint. However, anesthesia interferes with neural and hemodynamic activity, and with systemic and brain physiology. More previously, rodent fMRI studies were often performed under α-chloralose and urethane, because of their low impact on the respiratory and cardiovascular systems. Because of toxicity issues and slow recovery their use became less important. Isoflurane (ISO) is the most often used anesthetic in rodent MRI, because of its ease of handling and the fact that it is well tolerated by animals. However, at doses used for anatomical and structural imaging, amplitude and reproducibility of stimulus-evoked responses are rather low and rsfMRI reveal elevated global connectivity with low local specificity compared to Medetomidine (MED) sedation or awake imaging. Indeed, distinct patterns of functional connectivity are found under different anesthetic protocols [19]. To date, under consideration of their specific characteristics, ISO, MED, or a combination of MED with ISO are common choices for rodent fMRI studies [20].

2.2.1 ISO

ISO induces a sleep-like state by depression of spontaneous neuronal activity in a dose-dependent manner. ISO acts via multiple pharmacological pathways [20]: it depresses presynaptic voltage-gated sodium channels and modulates ligand-gated ion channels like GABA-A, 5-HT3, NMDA, AMPA, and nicotinic acetylcholine (ACh) receptors. At the same time ISO dilates cerebral vessels and reduces cerebral vascular resistance. In a dose-dependent manner it lowers the respiration rate and high ISO concentrations fully suppress the BOLD response. Van Alst et al. [20] showed that the low BOLD amplitude in spontaneously breathing animals resulted from a combination of a direct effect of ISO and the impact of accumulated CO_2. fMRI studies with ISO are therefore usually performed under low dose regimens (<1.2%). Anesthetic depth may then not be sufficient to tolerate scanner noise or repetitive stimulations. Ventilation and application of muscle relaxants in low-dose ISO-anesthetized animals, in order to reduce motion artifacts and maintain a stable respiration rate, may be a suitable alternative.

2.2.2 MED

MED is an α2 adrenoreceptor agonist with sedative properties [21]. Subtypes of α2 receptors in the brainstem regulate arousal and vigilance, and are responsible for their vasoconstrictive effect. Further side effects include respiratory depression and bradycardia. MED or the pharmacologically active enantiomer dexmedetomidine are usually administered by s.c. or i.v. bolus injection and concomitant continuous infusion. Doses vary between 0.04 mg/kg bolus/0.05 mg/kg/h and 0.3 mg/kg bolus and continuous infusion in rat and 0.05 mg/kg and 0.8 mg/kg in mice. Successful and stable sedation depend on the state of arousal when starting with animal preparation. Usually, induction of anesthesia and animal preparation is performed under ISO in a warm and calm environment. After MED bolus and concomitant start of continuous infusion, ISO is discontinued in a stepwise fashion. Based on the half-life of ISO and stabilization of heart and respiration rate, fMRI should not be started earlier than after 40 min waiting time. At the end of experiments, sedation can be reversed by antagonizing MED with Antisedan for longitudinal studies.

2.2.3 ISO/MED

A strategy to mitigate limited long-term instability of sedation with MED and dose dependent depression of BOLD response with ISO, is to combine anesthetics and reduce their dose by taking advantage of their additive anesthetic effects [20]. Indeed, the combination of MED and ISO provides more stable physiological conditions than continuous MED infusion alone. Stimulus-evoked BOLD responses are higher and rsfMRI derived functional connectivity pattern more closely resemble the awake state [19] as compared to ISO or MED alone.

2.2.4 Awake Imaging

To more closely reproduce the conditions of human fMRI in rodents, fMRI studies can be performed with awake animals (e.g. [22, 23]). Implementation involves specific habituation protocols, customized animal holders, and possibly efforts to obtain permission for employing awake fMRI. The goal to circumvent confounding effects of anesthesia may introduce issues like higher dropout rates, due to unstable or noncompliant animals or longer examination times to compensate loss of motion-deteriorated raw data. Repetitive training sessions or fMRI examinations may induce unwanted brain changes associated with learned helplessness and depression-like symptoms. In healthy, awake animals the spontaneous rsfMRI BOLD signal fluctuations are indeed much higher compared to anesthetized animals and functional connectivity resembles network organization pattern found in human studies. Whether awake rodent fMRI will be suitable to study brain function in MS disease models remains to be shown.

2.3 Animal Physiology/Monitoring

Impairment of baseline physiology immediately takes effect when anesthetizing an animal and regulation mechanisms may be hampered as well. In particular, body temperature and respiration rate are serious confounders during fMRI examinations and should be continuously monitored and supported. Heart rate by ECG or plethysmography and blood pressure may serve as additional physiological readouts. Regarding the latter, one should relate information gain versus additional manipulations of the animal, resulting in extra burden by surgical cannulation of a femoral artery, or additional expenditure of time for animal preparation (and anesthesia).

2.3.1 Body Temperature

Due to loss of muscular tone, impaired temperature regulation, and unfavorable surface volume ratio, anesthetized small animals reach severe hypothermia within minutes [24]. Hypothermia causes reduction of systemic physiology (heart rate, cardiac output, blood pressure, and respiratory rate) and metabolism, all potentially interfering with brain function. Not only a drop in temperature, but also unstable body temperature during fMRI exams, affect the functional response. Oxygen affinity of hemoglobin is inversely related to temperature. Body temperature changes of 2 °C induced BOLD changes of 6% [25]. Animal preparation should take place in a warm environment and continuous monitoring of rectal temperature, favorably with feedback-controlled heating during experiments, is mandatory.

2.3.2 Respiration

ISO and MED both cause respiratory depression, which is additionally aggravated when applying analgesics, in particular opioids. Anesthesia lowers respiratory minute volume by reducing the respiration rate (e.g., opioids), by reducing the tidal volume (e.g., volatile anesthetics), or both (e.g., propofol). Respiratory depression leads to an increase of blood CO_2. Hypercapnia causes cerebral vasodilation, resulting in increased baseline CBF, and decreases oxygen affinity of hemoglobin (Bohr effect). Both reduce a stimulus-evoked hemodynamic response. Mechanical ventilation with control of blood CO_2 may be advantageous to maintain a stable breathing rate and comparable blood gases between animals and different scans. Arterial blood gas analysis is the gold standard of reference. However, it only provides noncontinuous readings and the blood sampling rate is limited. Transcutaneous blood gas measurements suffer from the fact that general anesthesia depresses peripheral circulation. Capnometry for analyzing CO_2 in the expiratory air provides accurate results when the device can cope with the small tidal volume of mice and rat. Mechanical ventilation requires intubation of animals during preparation. Successful ventilation of small rodents depends on using only minimum tube lengths in order to keep airway resistance low, for example using an MRI-compatible pneumatic valve (MRI-1, CWE). The use of peripherally acting muscle relaxants (e.g., pancuronium, atracurium) prevents breathing against the ventilator and further minimizes motion artifacts during imaging. Pancuronium can be antagonized by neostigmine. Administering atropine may prevent mucus secretion and resulting tubus blockage during long examination times.

2.4 Data Analysis

Several software packages are available for analysis of sensory-evoked and rsfMRI data, such as Statistical Parametric Mapping, (www.fil.ion.ucl.ac.uk SPM), FMRIB Software Library (FSL, fsl.fmrib.ox.ac.uk), Brain Voyager (not free, www.brainvoyager.com), FreeSurfer (surfer.nmr.mgh.harvard.edu), and Analysis of Functional Neuroimages (AFNI, https://afni.nimh.nih.gov), offering diverse options. Although mostly developed for analysis of human fMRI data, all can be adapted for rodent data. We will only give a coarse overview here. Detailed methodological reviews are available, for example, by Margulies et al. [26].

2.4.1 Preprocessing

Preprocessing pipelines for fMRI data vary, but at minimum include realignment to one scan of the series to minimize motion effects, and subsequent smoothing. For group analysis, registration of single datasets to a brain template, such as an atlas, is required. While for human fMRI analysis brain templates like for example the MNI (Montreal Neurological Institute) brain are already implemented in the software packages, only a limited number of mouse and rat brain atlases is available (https://scalablebrainatlas.incf.org/; https://resource.loni.usc.edu/resources/atlases-downloads/ [27, 28]). Since brain anatomy differs between strains of mice and rats, and changes with age, generation of a customized brain template might be required.

2.4.2 Processing of Stimulus-Evoked fMRI

The most common approach for statistical analysis of fMRI data is the general linear model (GLM) [29]. It compares time courses from each voxel after stimulation with a model of the expected signal time course, using a linear regression. For this model, several basis sets of response functions are available, comprising canonical, gamma, Fourier, finite impulse response (FIR) and B-spline. SPM, for example, provides a canonical hemodynamic response function (HRF) by default which is, however, based on human datasets. The hemodynamic response of rats shows a significantly faster progression than the response of humans [30, 31]. Several studies indicate that the hemodynamic response varies with experimental conditions like anesthesia [20, 32, 33], stimulation type [34] and duration[35], and brain region [36]. Therefore, application of a species, region, and maybe even study-specific HRF may be more appropriate.

Analysis of rsfMRI data performs spatiotemporal correlations of signal fluctuations over the entire brain. Available strategies either follow model-free approaches or incorporate prior knowledge about brain architecture and networks. Independent component analysis (ICA) is one of the most popular model-free analysis options. It extracts a user-defined number of independent spatial components from fMRI data sets, which show similar time courses. Yet, completing the analysis and interpreting the results still requires a hypothesis. Separating components that represent functional meaningful signal from components that represent "unwanted" signal, like scanner noise, motion, or cardiac and respiration signal, is performed post hoc without knowledge of the "true" number of components. The highest level of data-driven analysis is provided by clustering approaches. However, even here, the "true" number of clusters is not known.

Seed-based analysis of functional connectivity, on the other hand, is hypothesis-driven, and correlates the time course of a priori selected region of interest (ROI) with the rest of the brain or with target regions. ROI definition may be specific to anatomical regions identified by registering the data to an anatomical template or to functional regions that have been detected, for example, by previous ICA in the same cohort of animals. The choice of regions, and their size, may bias results and the comparability between studies. The multi-seed region approach efficiently combines anatomical and functional features by correlating the mean time course of a seed region per brain region with the time course of every voxel in the brain, resulting in one correlation volume per brain region [37]. Graph theory can compute a number of global and local parameters from functional correlation data, such as for example average degree, strength, modularity, path length, and betweenness centrality.

Regional functional connectivity derived from time series correlations does not imply causality, which generally cannot be derived from fMRI data. One option to address directionality of influence between regions is to perform Granger causality analysis, which provides an estimate whether regional time courses of one ROI better predict future time courses of another ROI than past time courses of the latter alone. Sensitivity of this analysis, however, may be limited by the slow temporal dynamics of the BOLD signal.

3   Application of fMRI in MS Models

In MS rodent models, functional MRI has only rarely been performed. Tambalo et al. [38] acquired fMRI and morphological MRI data before disease induction (baseline) and during the relapsing and chronic phases of experimental autoimmune encephalomyelitis (EAE) in rats. Upon somatosensory stimulation of the forepaw a significant increase of the activated cortical volume, a decrease of the laterality index, and recruitment of additional areas (ipsilateral cortex, visual, secondary somatosensory regions, and thalamus) versus baseline indicated cortical reorganization/remodeling during the disease course. These findings are consistent with previously reported data from MS patients [39]. In the cuprizone mouse model of demyelination, Hübner et al. [40] performed combined rsfMRI with DTI and fiber tractography. At the peak of demyelination, they found connectivity modulations in hippocampus and in the default mode-like network. Chen et al. [41] performed MEMRI by injecting $MnCl_2$ into the visual cortex of EAE rats at the peak of disease, and acquired a series T1w images before and at several consecutive time points (1–48 h) after injection. Signal intensity changes due to the accumulation of manganese were detected in corpus callosum although diffusion tensor imaging data indicated no demyelination. Authors speculated that ion dyshomeostasis in neurons during the acute EAE attack caused an upregulation of Ca^{2+}-like accumulation and that this represented early MS pathology rather than a passive consequence of inflammatory demyelination.

4 Conclusion

The sensitivity of MRI for detecting the major pathological targets for therapeutic intervention, neuroinflammation and demyelination, so far have set the focus of neuroimaging studies to lesion detection rather than assessment of brain function. Anatomical and structural MRI findings, however, often lack correlation with disability. Different rodent models of MS emphasize specific features of the complex pathology of the human disease. fMRI studies in these disease models, particularly EAE and cuprizone, thereby may provide new insights about the interrelation between brain function and neuropathology of MS.

In order to perform robust and reproducible fMRI rodent studies, experimental protocols standardized as much as possible are necessary. Anesthetic regimen need to be tailored to species, strain, and health state of the individual animal. Monitoring and maintaining a stable physiological state is of utmost importance. Respiration rate and body temperature provide readouts for anesthetic depth and vitality, but successful functional imaging studies profit from advanced, continuous monitoring at least of blood gases by capnometry. Mechanical ventilation should be considered to prevent hypercapnia. There is a broad choice of reliable analysis strategies for stimulus-evoked and rsfMRI data, evidenced by the fact that functional network patterns are consistent across individuals, species, and scan sessions.

References

1. Bakshi R, Thompson AJ, Rocca MA et al (2008) MRI in multiple sclerosis: current status and future prospects. Lancet Neurol 7 (7):615–625. https://doi.org/10.1016/S1474-4422(08)70137-6

2. Nathoo N, Yong VW, Dunn JF (2014) Understanding disease processes in multiple sclerosis through magnetic resonance imaging studies in animal models. NeuroImage Clin 4:743–756. https://doi.org/10.1016/j.nicl.2014.04.011

3. Albers F, Wachsmuth L, van Alst TM, Faber C (2018) Multimodal functional neuroimaging by simultaneous BOLD fMRI and fiber-optic calcium recordings and optogenetic control. Mol Imaging Biol 20(2):171–182. https://doi.org/10.1007/s11307-017-1130-6

4. Baltes C, Radzwill N, Bosshard S, Marek D, Rudin M (2009) Micro MRI of the mouse brain using a novel 400 MHz cryogenic quadrature RF probe. NMR Biomed 22 (8):834–842. https://doi.org/10.1002/nbm.1396

5. Duong TQ, Yacoub E, Adriany G et al (2002) High-resolution, spin-echo BOLD, and CBF fMRI at 4 and 7 T. Magn Reson Med 48 (4):589–593. https://doi.org/10.1002/mrm.10252

6. Báez-Yánez MG, Ehses P, Mirkes C, Tsai PS, Kleinfeld D, Scheffler K (2017) The impact of vessel size, orientation and intravascular contribution on the neurovascular fingerprint of BOLD bSSFP fMRI. NeuroImage 163:13–23. https://doi.org/10.1016/j.neuroimage.2017.09.015

7. Lu H, van Zijl PCM (2012) A review of the development of vascular-space-occupancy (VASO) fMRI. NeuroImage 62(2):736–742. https://doi.org/10.1016/j.neuroimage.2012.01.013

8. Mandeville JB, Marota JJA, Kosofsky BE et al (1998) Dynamic functional imaging of relative cerebral blood volume during rat forepaw stimulation. Magn Reson Med 39(4):615–624. https://doi.org/10.1002/mrm.1910390415

9. Chuang KH, van Gelderen P, Merkle H et al (2008) Mapping resting-state functional connectivity using perfusion MRI. NeuroImage 40 (4):1595–1605. https://doi.org/10.1016/j.neuroimage.2008.01.006

10. Larkin JR, Simard MA, Khrapitchev AA et al (2018) Quantitative blood flow measurement in rat brain with multiphase arterial spin labelling magnetic resonance imaging. J Cereb Blood Flow Metab 39(8):1557–1569. https://doi.org/10.1177/0271678X18756218

11. Le BD (2007) The "wet mind": water and functional neuroimaging. Phys Med Biol 52 (7):R57–R90

12. Aso T, Urayama SI, Fukuyama H, Le Bihan D (2013) Comparison of diffusion-weighted fMRI and BOLD fMRI responses in a verbal working memory task. NeuroImage 67:25–32. https://doi.org/10.1016/j.neuroimage.2012.11.005

13. Albers F, Wachsmuth L, Schache D, Lambers H, Faber C (2019) Functional MRI readouts from BOLD and diffusion measurements differentially respond to optogenetic activation and tissue heating. Front Neurosci 13:1–16. https://doi.org/10.3389/fnins.2019.01104

14. Lin YJ, Koretsky AP (1997) Manganese ion enhances T1-weighted MRI during brain activation: an approach to direct imaging of brain function. Magn Reson Med 38(3):378–388. https://doi.org/10.1002/mrm.1910380305

15. Niehoff AC, Wachsmuth L, Schmid F, Sperling M, Faber C, Karst U (2016) Quantification of manganese enhanced magnetic resonance imaging based on spatially resolved elemental mass spectrometry. ChemistrySelect 1(2):264–266. https://doi.org/10.1002/slct.201600058

16. Eschenko O, Canals S, Simanova I, Beyerlein M, Murayama Y, Logothetis NK (2010) Mapping of functional brain activity in freely behaving rats during voluntary running using manganese-enhanced MRI: implication for longitudinal studies. NeuroImage 49 (3):2544–2555. https://doi.org/10.1016/j.neuroimage.2009.10.079

17. Van der Linden A, Van Meir V, Tindemans I, Verhoye M, Balthazart J (2004) Applications of manganese-enhanced magnetic resonance imaging (MEMRI) to image brain plasticity in song birds. NMR Biomed 17(8):602–612. https://doi.org/10.1002/nbm.936

18. Sudarshana DM, Nair G, Dwyer JT et al (2019) Manganese-enhanced MRI of the brain in healthy volunteers. Am J Neuroradiol 40 (8):1309–1316. https://doi.org/10.3174/ajnr.a6152

19. Paasonen J, Stenroos P, Salo RA, Kiviniemi V, Gröhn O (2018) Functional connectivity under six anesthesia protocols and the awake condition in rat brain. NeuroImage 172:9–20. https://doi.org/10.1016/j.neuroimage.2018.01.014

20. van Alst TM, Wachsmuth L, Datunashvili M et al (2019) Anesthesia differentially modulates

neuronal and vascular contributions to the BOLD signal. NeuroImage 195:89–103. https://doi.org/10.1016/j.neuroimage. 2019.03.057

21. Fukuda M, Vazquez AL, Zong X, Kim SG (2013) Effects of the α2-adrenergic receptor agonist dexmedetomidine on neural, vascular and BOLD fMRI responses in the somatosensory cortex. Eur J Neurosci 37(1):80–95. https://doi.org/10.1111/ejn.12024

22. Gao YR, Ma Y, Zhang Q et al (2017) Time to wake up: studying neurovascular coupling and brain-wide circuit function in the un-anesthetized animal. NeuroImage 153:382–398. https://doi.org/10.1016/j. neuroimage.2016.11.069

23. Stenroos P, Paasonen J, Salo RA et al (2018) Awake rat brain functional magnetic resonance imaging using standard radio frequency coils and a 3D printed restraint kit. Front Neurosci 12:548. https://doi.org/10.3389/fnins. 2018.00548

24. Thal SC, Plesnila N (2007) Non-invasive intraoperative monitoring of blood pressure and arterial pCO$_2$ during surgical anesthesia in mice. J Neurosci Methods 159 (2):261–267. https://doi.org/10.1016/j. jneumeth.2006.07.016

25. Vanhoutte G, Verhoye M, Van Der Linden A (2006) Changing body temperature affects the T2* signal in the rat brain and reveals hypothalamic activity. Magn Reson Med 55 (5):1006–1012. https://doi.org/10.1002/ mrm.20861

26. Margulies DS, Böttger J, Long X et al (2010) Resting developments: a review of fMRI post-processing methodologies for spontaneous brain activity. Magn Reson Mater Phys Biol Med 23(5-6):289–307. https://doi.org/10. 1007/s10334-010-0228-5

27. Schwarz AJ, Danckaert A, Reese T et al (2006) A stereotaxic MRI template set for the rat brain with tissue class distribution maps and co-registered anatomical atlas: application to pharmacological MRI. NeuroImage 32 (2):538–550. https://doi.org/10.1016/j. neuroimage.2006.04.214

28. Schweinhardt P, Fransson P, Olson L, Spenger C, Andersson JLR (2003) A template for spatial normalisation of MR images of the rat brain. J Neurosci Methods 129 (2):105–113. https://doi.org/10.1016/ S0165-0270(03)00192-4

29. Friston KJ, Frith CD, Turner R, Frackowiak RSJ (1995) Characterizing evoked hemodynamics with fMRI. NeuroImage 2 (2):157–165. https://doi.org/10.1006/ nimg.1995.1018

30. Silva AC, Koretsky AP, Duyn JH (2007) Functional MRI impulse response for BOLD and CBV contrast in rat somatosensory cortex. Magn Reson Med 57(6):1110–1118. https:// doi.org/10.1002/mrm.21246

31. De Zwart JA, Silva AC, Van Gelderen P et al (2005) Temporal dynamics of the BOLD fMRI impulse response. NeuroImage 24 (3):667–677. https://doi.org/10.1016/j. neuroimage.2004.09.013

32. Masamoto K, Kanno I (2012) Anesthesia and the quantitative evaluation of neurovascular coupling. J Cereb Blood Flow Metab 32 (7):1233–1247. https://doi.org/10.1038/ jcbfm.2012.50

33. Schlegel F, Schroeter A, Rudin M (2015) The hemodynamic response to somatosensory stimulation in mice depends on the anesthetic used: implications on analysis of mouse fMRI data. NeuroImage 116:40–49. https://doi. org/10.1016/j.neuroimage.2015.05.013

34. Albers F, Schmid F, Wachsmuth L, Faber C (2018) Line scanning fMRI reveals earlier onset of optogenetically evoked BOLD response in rat somatosensory cortex as compared to sensory stimulation. NeuroImage 164:144–154. https://doi.org/10.1016/j. neuroimage.2016.12.059

35. Vazquez AL, Noll DC (1998) Nonlinear aspects of the BOLD response in functional MRI. NeuroImage 7(2):108–118. https:// doi.org/10.1006/nimg.1997.0316

36. Lambers H, Segeroth M, Albers F, Wachsmuth L, van Alst TM, Faber C (2020) A cortical rat hemodynamic response function for improved detection of BOLD activation under common experimental conditions. NeuroImage 208:116446. https://doi.org/10. 1016/j.neuroimage.2019.116446

37. Kreitz S, Alonso B de C, Uder M, Hess A (2018) A new analysis of resting state connectivity and graph theory reveals distinctive short-term modulations due to whisker stimulation in rats. Front Neurosci 12:1–19. https:// doi.org/10.3389/fnins.2018.00334

38. Tambalo S, Peruzzotti-Jametti L, Rigolio R et al (2015) Functional magnetic resonance imaging of rats with experimental autoimmune encephalomyelitis reveals brain cortex remodeling. J Neurosci 35(27):10088–10100. https://doi.org/10.1523/JNEUROSCI. 0540-15.2015

39. Reddy H (2002) Functional brain reorganization for hand movement in patients with multiple sclerosis: defining distinct effects of injury and disability. Brain 125(12):2646–2657. https://doi.org/10.1093/brain/awf283

40. Hübner NS, Mechling AE, Lee H-L et al (2017) The connectomics of brain demyelination: functional and structural patterns in the cuprizone mouse model. NeuroImage 146:1–18. https://doi.org/10.1016/J.NEUROIMAGE.2016.11.008

41. Chen CCV, Zechariah A, Hsu YH, Chen HW, Yang LC, Chang C (2008) Neuroaxonal ion dyshomeostasis of the normal-appearing corpus callosum in experimental autoimmune encephalomyelitis. Exp Neurol 210 (2):322–330. https://doi.org/10.1016/j.expneurol.2007.11.008

Chapter 17

Noninvasive Electrophysiology

Matthias Grothe

Abstract

Since the original development of noninvasive electrophysiology its role has evolved. In this chapter, we outline the utilization of noninvasive electrophysiology, especially evoked potentials, in research using animal models and for human diagnostic, prognostic, and therapeutic purposes in multiple sclerosis—emphasizing the crucial roles of EPs to date.

Key words Noninvasive electrophysiology, Evoked potentials, Diagnostic, Prognosis, Therapy

1 Introduction

Noninvasive electrophysiology enables in vivo exploration of functional systems. First reports described evoked responses to electrical stimulation of peripheral nerves in the 1940s [1]. Since then, different methods have been developed to investigate the visual (visual evoked potential; VEP), auditory (brainstem auditory evoked potential; BAEP), somatosensory (somatosensory evoked potential; SEP), and motor (motor evoked potential; MEP) systems [2]. Evoked potentials (EPs) are recorded as electrophysiological responses to various types of stimuli and have a wide range of applications in neurological research and medicine [2, 3]. In general, measurements include EP amplitude, morphology (presence or absence of components), latency, interpeak latency, and dispersion, and comparisons can be made between the body sides and with reference values. About 50 years ago, the first EP studies were performed in multiple sclerosis (MS) providing the first descriptions of abnormal and delayed SEP [4], delayed VEP in optic neuritis (ON) [5], and later the dispersion and delays in BAEP [6] and MEP [7].

To date, studies of MS have mainly focused on diagnostic and therapeutic purposes to demonstrate subclinical lesions and monitor disease progression. However, these methods—especially in animal models—have also provided insight into the underlying

Sergiu Groppa and Sven G. Meuth (eds.), *Translational Methods for Multiple Sclerosis Research*, Neuromethods, vol. 166, https://doi.org/10.1007/978-1-0716-1213-2_17, © Springer Science+Business Media, LLC, part of Springer Nature 2021

pathophysiological processes of the disease [3, 8]. EPs are widely used due to their many advantages; they are noninvasive, cheap, free of severe side effects and they provide information regarding the status of specific functional systems. Additionally, compared to behavioral tests, EP measurements are more sensitive to structural damage as proven in humans and in animal studies [9, 10].

2   Noninvasive Electrophysiology in Animal Models

Over many years, neurophysiological investigations with EPs have been performed in animal models of MS [11, 12]. There is ongoing debate regarding the similarities and differences between various animal models of MS and MS in humans [13]. Neurophysiological investigation approaches help clarify the differences and similarities of the underlying pathological processes, therefore influencing further investigations. Experimental autoimmune/allergic encephalomyelitis (EAE) is the most commonly studied animal model, but noninvasive electrophysiology studies have also been performed in virus-induced models, such as the Theiler's murine encephalomyelitis virus (TMEV) model and toxin-induced models of demyelination, such as the cuprizone or lysolecithin model [11, 12, 14].

In one of the first neurophysiological studies in EAE, Bilbool and colleagues investigated two different guinea pig EAE models with bovine white matter extract or myelin basic protein as the EAE-inducing agent [15]. Despite histopathological differences between the models no difference in EPs was observed, with both models exhibiting abnormal wave forms and prolonged latencies in VEP preceding the development of clinical signs [15]. Neurophysiological alterations in the absence of demyelinated plaques were also reported in a study of acute and chronic EAE in guinea pigs, which exhibited alterations of VEP, SEP, and BEAP at 2 weeks after sensitization [16]. Research over the following years greatly increased our knowledge about animal models, due to immunological, imaging, and histopathological advancements [17–19]. There are now many different EAE models available, enabling a wide range of pathophysiological insights [11, 20, 21].

In a chronic EAE model, Amadio et al. used transcranial electrical stimulation to investigate MEP latencies and amplitudes [22]. The early MEP latencies increased prior to the development of clinical impairments, then plateaued for several weeks, followed by a decrease of amplitude until complete EP loss. These animals exhibited early demyelination, which peaked simultaneously with the maximum delay in latencies, followed by a slow increase of axonal damage until the end of the study—which explains the neurophysiological results [22]. In a more recent study of VEP in rats the findings indicated that early disturbances of latencies are associated with inflammation, while the later delays are related to inflammation, axonal loss and demyelination [23]. On the other hand, in a rat EAE model, optic neuritis-associated VEP loss was associated with apoptosis of retinal ganglion cells [24]. Furthermore, in both non-immunized and EAE mice, MEP latency was found to be associated with genes on different chromosomal loci, suggesting a genetic predisposition for susceptibility based on myelin structure and composition [25]. In addition to the classic EAE models, there are also transgenic mice that spontaneously manifest EAE (sEAE) [26]—for example, shiverer mice, which have little or no CNS myelin and exhibit prolonged VEP latencies [27]. Additionally, there is a transgenic mouse model of oligodendrocyte ablation and demyelination, which exhibits early increases of AEP and SEP latencies, followed by spontaneous recovery in a later phase [28].

In contrast to the EAE models, in the TMEV model, axonal damage precedes demyelination [29]. Accordingly, neurophysiological investigations of TMEV mice have revealed early amplitude reduction and steadily increasing SEP and MEP latencies up to months after infection [30, 31]. On the other hand, investigations of toxic models, such as the cuprizone and lysolecithin models, have provided great insight regarding demyelination and remyelination [11, 21, 32]. Mozafari et al. investigated VEP alterations in a lysolecithin rat model and found that latencies and amplitudes deteriorated until day 14 after injection, followed by recovery [33]. Moreover, the recovery was accompanied by high expression of MBP mRNA suggesting myelin repair processes [33].

Another more recent approach to induce demyelination involves feeding irradiated food to healthy adult cats, which results in non-inflammatory demyelination, mainly of the spinal cord, optic nerves and, to a more less extent, the brain [34]. One advantage of this animal model is its reversibility; returning to a normal diet early in the disease process enables functional recovery through extensive remyelination [34]. In this model, demyelination is accompanied by prolongation of VEP latencies, which is somewhat reversible upon return to normal diet, although VEP latencies remain prolonged compared to controls [35].

In addition to studies of underlying pathophysiology in animal models, non-invasive electrophysiology has also been used to monitor pharmacological approaches. EP has been used to help demonstrate the therapeutic effects of several approved drugs, including methylprednisolone [36], glatiramer acetate [37], fingolimod [38] and teriflunomide [39]. In each of these studies, drug administration protected against induced EP changes (amplitude reduction and delayed latencies) and enhanced recovery to some extent [38, 39]. Noninvasive neurophysiology has also been useful in ongoing research examining the pharmacological promotion of functional recovery, for example with anti-LINGO [40].

Overall, noninvasive electrophysiology methods have been widely used to study animal models of MS for several years, enabling in vivo investigations of the underlying pathophysiology, differentiation between MS animal models and improved translation to human MS research, as well as the development of interventional techniques. More recent approaches, involving the combination of EPs and diffusion tensor imaging (DTI) in EAE [41], have enabled further utilization with translational benefits for both animal and human research.

3   Noninvasive Electrophysiology in the Diagnosis of Multiple Sclerosis

The diagnostic criteria for MS reflect ongoing adjustments to the current available clinical, paraclinical and imaging methods, with the last revision in 2017 [42]. The early Poser criteria from 1983 included EPs to demonstrate paraclinical evidence of a structural lesion if two attacks occurred in the absence of evidence of two separate clinical lesions [43]. In the next revision (the McDonald criteria of 2001) only VEPs remained part of the recommended diagnostic algorithm [44]. The 2010 revision no longer included EPs at all [45]. This change was prompted by the broad use of neuroimaging (particularly MRI) in MS diagnosis, based on its high sensitivity and specificity for diagnosing MS [46]. Nevertheless, the inclusion of EP in the diagnostic algorithm provides important information.

VEPs are the most sensitive EPs, with pathological findings in up to 90% of MS patients [47–49]. Additionally, some studies report pathological VEP in 40–60% of patients, even without history of ON [49, 50]. More recently, multifocal VEPs have been developed, which allow simultaneous assessment of different sectors of the visual field, with superior sensitivity and specificity compared to standard or full-field VEP [51, 52]. Multifocal VEP may confer advantages in detecting smaller optic nerve lesions, especially retrochiasmal lesions; however, the validation of this method is hindered by lack of standardization between the available systems and small sample sizes in the available studies, at least at this stage [52].

Optical coherence tomography (OCT) is a low-coherence interferometric method for visualizing axonal and neuronal retinal fiber loss with very high spatial resolution, which is widely used for visual system assessment in MS [53, 54]. Both OCT and VEP methods have comparable sensitivity for ON diagnosis [48, 49], and the combination of both methods seems to further improve the sensitivity for ON detection [48, 55]; however, further research is needed to evaluate this promising benefit.

After ON, VEP latencies of the affected eye show rapid recovery between 3 and 6 months after onset, while the electrophysiological changes can last up to 24 months [56, 57]. Despite the rapid recovery, abnormal VEP persists in 75% of patients with recovered visual accuracy [58]. Additionally, significant prolongation of the VEP latency can be found in the asymptomatic eye over time [57] even in MS patients exhibiting prolonged latencies without clinical ON [59].

VEP can also be used for differential diagnostic considerations. Applying a multimodal approach for the diagnosis of MS mimics, normal VEP showed a negative predictive value of 92.5%. Even in multivariate analysis, normal VEP was an independent predictor of an alternative diagnosis [60]. Furthermore, a proposed OCT-VEP index exhibits a specificity of 89% for discriminating ON between MS and NMOSD [55].

Standard SEP techniques can only be used to assess the dorsal column-lemniscal sensory system [61]. Moreover, SEP abnormalities may be due to spinal, cerebral or even peripheral sensory lesions [2], such that SEP exhibits a very low sensitivity. Older studies have also suggested that serial SEP measurements are not valid for the assessment of clinical progression [62], but are sensitive for increasing body temperature among MS patients but not healthy controls [63].

SEP abnormalities can be detected in up to 50% of MS patients, even during their first clinical relapse [64] and correlate with the Expanded Disability Status Scale (EDSS) [65] and its sensory functional system [66]. Furthermore, SEP alterations in MS patients are more common in the presence of pyramidal tract lesions [62]. Interestingly, among MS patients lacking SEP abnormalities in the lower limb, less than 10% will have abnormal SEP in the upper limbs [3]. Moreover, in a group of 161 patients with progressive MS, repeated intrathecal triamcinolone application reduced latencies of SEP, which was accompanied by significant extension of walking distance [67].

Limited evidence supports the use of SEP for differential-diagnostic considerations. In a small study including 12 aquaporin-4 antibody-positive patients and 60 MS patients, the MS patients more commonly showed absent or delayed SEP, but this finding was not statistically significant [68]. In a cohort including 18 NMOSD patients and 28 patients with MS, upper extremity SEP absence was significantly more common in MS, and normal upper extremity SEPs were significantly more common in NMOSD [69]. Another study reported a high rate of SEP alterations (nearly 50%) in patients with myelin oligodendrocyte glycoprotein antibodies, even among patients without clinical or MRI evidence of sensory impairment, supporting its possible diagnostic application [70].

Brainstem auditory evoked potentials have been recommended for MS diagnostics since 1990, when Sand et al. reported abnormal BAEPs in 42% of MS patients with normal VEPs and SEPs and in 38% of MS patients with normal cerebrospinal fluid [71]. However, the diagnostic use of BAEP has not been as widespread as the use of other EPs. In a study of 86 MS patients, BAEP interpeak latencies discriminated between 26 severely fatigued MS patients and 40 controls [72]. Moreover, MS patients with additional epilepsy exhibit pathological BAEP alterations compared to MS patients without epilepsy and MS patients taking antiepileptic drugs for indications other than epilepsy [73]. Since pathological BAEP has never been detected in NMOSD patients, it has been suggested that it might be clinically useful for distinguishing between NMOSD and MS [69].

Compared to other EPs, MEPs are different in that they are induced by transcranial magnetic stimulation (TMS) exciting efferent cortical motor neurons. The triggered action potentials spread along the corticospinal pathways, resulting in a MEP in a target muscle that can be recorded and analyzed. For specific assessment of the central motor pathway, the latency from the cervical spine (with the coil centered at C6) or the lumbar spine (with the coil centered at L4) must be subtracted from the latency of the motor cortex MEP, yielding the central motor conduction time (CMCT) of the upper limb or lower limb, respectively. A prolonged CMCT has a reliability of 0.83 for MS diagnosis, supporting its utilization in clinical practice [74].

In addition to CMCT and information regarding MEP latency and MEP amplitude, several other TMS protocols with single- or paired-pulse stimulation have also been investigated in MS [75, 76]. As recently reviewed [76], broad literature discusses TMS and different clinical variables, such as disease severity [77], motor skills [78] and fatigue [79], suggesting that higher clinical impairments are associated with alterations of single-pulse TMS amplitude or latency. Data also indicate that paired-pulse protocols, such as short intracortical inhibition (SICI) or intracortical facilitation (ICF), may help in differentiating relapsing remitting and secondary progressive disease courses [77], exhibit alterations during acute relapse [80] or with fatigue [81] and are modulated by immunomodulatory treatment [82] or high-dose steroids [83]. These findings are promising but must be validated in larger cohorts. In an interesting translational approach, cerebrospinal fluid obtained from MS patients during acute relapse (proven by gadolinium enhancement on MRI) increased the spontaneous glutamate-mediated excitatory postsynaptic current frequency in mouse corticostriatal brain slices, which could be blocked using glutamate receptor antagonists [84]. Moreover, acute inflammatory cytokine levels in these relapsing MS patients were associated with the glutamatergic paired-pulse TMS parameter ICF [84]. This study is an excellent example of a translational approach between animal data and non-invasive electrophysiology, which provides useful tools and knowledge to be used in human research.

Few studies have investigated the utilization of TMS for differential diagnostic considerations. In a study comparing MS, stroke, hereditary spastic, and psychogenic patients with healthy controls, MS patients showed a prolonged MEP duration during rest and shortened MEP duration during voluntary contraction [85]. Additionally, CMCT is more commonly absent in NMOSD and more commonly delayed in MS, and MEP recruitment curves show a lower slope in NMOSD, suggesting that MEP may be valuable for clinical diagnostics [69, 86].

4 Noninvasive Electrophysiology in the Prognosis of Multiple Sclerosis

There are many approaches using non-invasive electrophysiology as a marker for MS development and progression. Such studies use either a single EP in association with a clinical outcome parameter or an EP score that summarizes multiple EPs [50]. In principle, more EP alterations will be associated with worse prognosis. The predictive value is influenced by acute relapse, suggesting that such investigations should be performed during a relapse-free interval [87].

Even early studies demonstrated the prognostic value of EPs in MS patients. In 1988, Hume and Waxman retrospectively investigated a cohort of 277 neurological patients, analyzing the clinical course over 30 months and its associations with abnormal VEP, SSEP, and BAEP [88]. Among patients suspected of having MS, those exhibiting a pathological EP had a 71% likelihood of clinical progression, compared to a 12% likelihood among patients with normal EPs [88]. This predictive value of EPs for disease progression was also evident for VEP, SEP, and BAEP among CIS patients [89] and for VEP among RIS patients [90]. To date, the MAGNIMS recommendations include pathological VEP as a sign of increased risk for developing MS in RIS [91].

Several studies have investigated the value of EPs for predicting long-term disability with longitudinal data for up to 20 years of progression [92]. As recently reviewed, data from nearly 1000 patients reveal a strong relationship between baseline EPs and future disability, as measured by the EDSS [50]. Moreover, other investigations show that baseline EPs are associated with other clinical scores, such as the timed 25-foot walk test and the Multiple Sclerosis Walking Scale [65], or with the probability of future falls [93].

Various approaches can be used to combine the measurements of different EPs into an EP score. The global EP score, originally developed by Leocani et al. [47] and modified by Invernizzi et al. [94], constitutes a simple way to summarize 12 EPs: VEP, BAEP, SEP upper limb, SEP lower limb, MEP upper limb, and MEP lower limb, each for the right and left sides. Each EP is scored from 0 (normal EP) to 3 (absence of a major component), generating a global score that ranges from 0 to 36 (Table 1).

A baseline global EP score of ≤4 in RRMS had a positive predictive value of 0.8 for a stable clinical course (i.e., without EDSS progression) for up to 15 years after disease onset [95]. Among MS patients with a baseline global EP score of ≤4 only 9% reached an EDSS of >4.0 after 15 years, and 0% reached an EDSS of >6. In contrast, among MS patients with a baseline global EP score of >4 45% reached an EDSS of >4.0 and 17% reached an EDSS of >6.0 after 15 years [95]. With regards to the individual EP modality and its relation to long-term disability, the best correlation was identified for MEP ($r = 0.6$), followed by SEP ($r = 0.53$), BAEP ($r = 0.43$) and VEP ($r = 0.28$), with VEP not reaching statistical significance [95]. In another approach, VEP and MEP of the upper limb are used to calculate a z-transformed standardized EP value, which is used with different regression models to predict EDSS after 20 years [92]. Applying this method with a model including age, therapy status and standardized EPs, Schlaeger et al. explained 58% of variability in the EDSS at year 20 [92].

Few studies have focused on the prognostic outcome of relapses in MS. In one study, paired associative stimulation (PAS)—a neurophysiological method of assessing synaptic long-term potentiation (LTP) in vivo—was used on 22 MS patients during relapse and after recovery [96]. Compared to the normal PAS-induced LTP observed in patients with full recover, reduced PAS-induced LTP at 12 weeks after relapse was associated with worse recovery, suggesting that this neurophysiological method has prognostic value for functional recovery after relapse [96]. A recent study also demonstrated that recovery after motor relapse in MS was associated with interhemispheric TMS measurements with an increased duration of the ipsilateral silent period predicting reduced dexterity motor recovery [97]. Additional data are needed in this field to support a better understanding of the prognostic use of neurophysiological measurements in MS relapse.

5   Noninvasive Electrophysiology in Assessing Pharmacological Effects in Multiple Sclerosis

Noninvasive electrophysiology can also serve as an objective marker of pharmacological treatment success. However, interventions with measurements like EPs only exhibit small effect sizes. Therefore, a large sample must be examined to produce significant results which is uncommon in the existing literature.

Initial studies have revealed that latencies in EPs increase during the treatment period, regardless of the specific treatment. In a

Table 1
Global EP score according to Leocani et al. [47]

Evoked potential	0, Normal		1, increased latency		2, increased latency and morphological abnormality in major component of EP		3, absence of major component	
Side of EP	Right	Left	Right	Left	Right	Left	Right	Left
VEP								
BAEP								
SEP-UL								
SEP-LL								
MEP-UL								
MEP-LL								
Sum score								

Scores range from 0 to 36, with higher score indicating more severe EP alterations
Predictive value has been assessed up to 15 years [95]

large sample of MS patients treated with azathioprine or placebo, the results revealed neurophysiological and clinical deterioration in all groups [98]. On the other hand, an early and small study of 7 MS patients treated with 4-aminopyridine demonstrated that improved VEP (latency or amplitude) was associated with clinical benefit (visual accuracy) [99]. Evidence also shows improved electrophysiological measurements in MS following treatment with several approved immunomodulatory drugs—for example, interferon beta treatment exerts a positive effect on MEP [100], and natalizumab and fingolimod improve VEP and SEP [101, 102]. Additionally, in secondary progressive multiple sclerosis, autologous mesenchymal stem cell treatment resulted in improved VEP latencies [103]. Prospective studies have also revealed VEP improvements following treatment with drugs that are currently in development for MS, such as biotin [104], simvastatin [105], opicinumab [106], and clemastine [107]; however, data regarding other neurophysiological measurements are lacking. Although remyelination or neuroprotection is regarded as the mechanism underlying these EP improvements, further studies are needed to verify these results.

Only one study to date has investigated treatment during acute relapse. A three-day course of therapy with 1 g methylprednisolone per day resulted in a reduction of SICI and an increase of ICF [83]. The authors concluded that steroid treatment altered the balance between GABAergic inhibition and glutamatergic facilitation with a net facilitative effect [83].

6   Conclusion

For several decades, noninvasive electrophysiology has been used in the field of multiple sclerosis research, enabling in vivo insights into pathophysiology, and providing an outstanding method for translating animal data into human research. Noninvasive electrophysiology techniques have been used to investigate animal models of MS (such as EAE), and the findings have been embedded in current knowledge and directly used for further advancements. These methods are applied in well-known models, and also used to help better understand new animal models, such as demyelination induced by a diet of irradiated food [34]. EPs are also widely used in MS diagnostics for detecting subclinical lesions [2] and for differential diagnostic considerations [55]. Moreover, EPs will be valuable for future research, especially when applying more advanced methods, like paired-pulse TMS [76], and in combination with other techniques, such as MRI [108]. Finally, neurophysiological information regarding functional system integrity is of prognostic relevance, for investigating MS development and progression [95], and is sensitive to drug interventions, enabling the monitoring of structural damage and repair which will continue to be useful in future clinical trials [50].

References

1. Dawson GD (1947) Cerebral responses to electrical stimulation of peripheral nerve in man. J Neurol Neurosurg Psychiatry 10:134–140

2. Lascano AM, Lalive PH, Hardmeier M, Fuhr P, Seeck M (2017) Clinical evoked potentials in neurology: a review of techniques and indications. J Neurol Neurosurg Psychiatry 88:688–696

3. Walsh P, Kane N, Butler S (2005) The clinical role of evoked potentials. J Neurol Neurosurg Psychiatry 76(Suppl 2):ii16–ii22

4. Halliday AM, Wakefield GS (1963) Cerebral evoked potentials in patients with dissociated sensory loss. J Neurol Neurosurg Psychiatry 26:211–219

5. Halliday AM, McDonald WI, Mushin J (1972) Delayed visual evoked response in optic neuritis. Lancet 1:982–985

6. Lacquaniti F, Benna P, Gilli M, Troni W, Bergamasco B (1979) Brain stem auditory evoked potentials and blink reflexes in quiescent multiple sclerosis. Electroencephalogr Clin Neurophysiol 47:607–610

7. Cowan JM, Rothwell JC, Dick JP, Thompson PD, Day BL, Marsden CD (1984) Abnormalities in central motor pathway conduction in multiple sclerosis. Lancet 2:304–307

8. Fuhr P, Kappos L (2001) Evoked potentials for evaluation of multiple sclerosis. Clin Neurophysiol 112:2185–2189

9. Ali AH, Agrawal G, Walczak P, Maybhate A, Bulte JW, Kerr DA (2010) Evoked potential and behavioral outcomes for experimental autoimmune encephalomyelitis in Lewis rats. Neurol Sci 31:595–601

10. Gronseth GS, Ashman EJ (2000) Practice parameter: the usefulness of evoked potentials in identifying clinically silent lesions in patients with suspected multiple sclerosis (an evidence-based review): report of the quality standards Subcommittee of the American Academy of Neurology. Neurology 54:1720–1725

11. Procaccini C, De Rosa V, Pucino V, Formisano L, Matarese G (2015) Animal models of multiple sclerosis. Eur J Pharmacol 759:182–191

12. Kipp M, Nyamoya S, Hochstrasser T, Amor S (2017) Multiple sclerosis animal models: a

clinical and histopathological perspective. Brain Pathol 27:123–137

13. Milo R, Korczyn AD, Manouchehri N, Stuve O (2020) The temporal and causal relationship between inflammation and neurodegeneration in multiple sclerosis. Mult Scler 26 (8):876–886. https://doi.org/10.1177/1352458519886943

14. Lassmann H, Bradl M (2017) Multiple sclerosis: experimental models and reality. Acta Neuropathol 133:223–244

15. Bilbool N, Kaitz M, Feinsod M, Soffer D, Abramsky O (1983) Visual evoked potentials in experimental allergic encephalomyelitis. J Neurol Sci 60:105–115

16. Gambi D, Fulgente T, Melchionda D, Onofrj M (1996) Evoked potential (EP) alterations in experimental allergic encephalomyelitis (EAE): early delays and latency reductions without plaques. Ital J Neurol Sci 17:23–33

17. Seewann A, Kooi EJ, Roosendaal SD, Barkhof F, van der Valk P, Geurts JJ (2009) Translating pathology in multiple sclerosis: the combination of postmortem imaging, histopathology and clinical findings. Acta Neurol Scand 119:349–355

18. Filippi M, Rocca MA, Barkhof F, Bruck W, Chen JT, Comi G, DeLuca G, De Stefano N, Erickson BJ, Evangelou N, Fazekas F, Geurts JJ, Lucchinetti C, Miller DH, Pelletier D, Popescu BF, Lassmann H, M.R.I.f.i.M.S. w (2012) Attendees of the Correlation between Pathological, Association between pathological and MRI findings in multiple sclerosis. Lancet Neurol 11:349–360

19. Dendrou CA, Fugger L, Friese MA (2015) Immunopathology of multiple sclerosis. Nat Rev Immunol 15:545–558

20. Mix E, Meyer-Rienecker H, Hartung HP, Zettl UK (2010) Animal models of multiple sclerosis--potentials and limitations. Prog Neurobiol 92:386–404

21. Ransohoff RM (2012) Animal models of multiple sclerosis: the good, the bad and the bottom line. Nat Neurosci 15:1074–1077

22. Amadio S, Pluchino S, Brini E, Morana P, Guerriero R, Martinelli Boneschi F, Comi G, Zaratin P, Muzio V, del Carro U (2006) Motor evoked potentials in a mouse model of chronic multiple sclerosis. Muscle Nerve 33:265–273

23. Castoldi V, Marenna S, d'Isa R, Huang SC, De Battista D, Chirizzi C, Chaabane L, Kumar D, Boschert U, Comi G, Leocani L (2020) Non-invasive visual evoked potentials to assess optic nerve involvement in the dark agouti rat model of experimental autoimmune encephalomyelitis induced by myelin oligodendrocyte glycoprotein. Brain Pathol 30 (1):137–150. https://doi.org/10.1111/bpa.12762(2019)

24. Meyer R, Weissert R, Diem R, Storch MK, de Graaf KL, Kramer B, Bahr M (2001) Acute neuronal apoptosis in a rat model of multiple sclerosis. J Neurosci 21:6214–6220

25. Mazon Pelaez I, Vogler S, Strauss U, Wernhoff P, Pahnke J, Brockmann G, Moch H, Thiesen HJ, Rolfs A, Ibrahim SM (2005) Identification of quantitative trait loci controlling cortical motor evoked potentials in experimental autoimmune encephalomyelitis: correlation with incidence, onset and severity of disease. Hum Mol Genet 14:1977–1989

26. Gupta AA, Ding D, Lee RK, Levy RB, Bhattacharya SK (2012) Spontaneous ocular and neurologic deficits in transgenic mouse models of multiple sclerosis and noninvasive investigative modalities: a review. Invest Ophthalmol Vis Sci 53:712–724

27. Martin M, Hiltner TD, Wood JC, Fraser SE, Jacobs RE, Readhead C (2006) Myelin deficiencies visualized in vivo: visually evoked potentials and T2-weighted magnetic resonance images of shiverer mutant and wild-type mice. J Neurosci Res 84:1716–1726

28. Farley BJ, Morozova E, Dion J, Wang B, Harvey BD, Gianni D, Wipke B, Cadavid D, Wittmann M, Hajos M (2019) Evoked potentials as a translatable biomarker to track functional remyelination. Mol Cell Neurosci 99:103393

29. Tsunoda I, Kuang LQ, Libbey JE, Fujinami RS (2003) Axonal injury heralds virus-induced demyelination. Am J Pathol 162:1259–1269

30. Iuliano BA, Schmelzer JD, Thiemann RL, Low PA, Rodriguez M (1994) Motor and somatosensory evoked potentials in mice infected with Theiler's murine encephalomyelitis virus. J Neurol Sci 123:186–194

31. McGavern DB, Murray PD, Rivera-Quinones C, Schmelzer JD, Low PA, Rodriguez M (2000) Axonal loss results in spinal cord atrophy, electrophysiological abnormalities and neurological deficits following demyelination in a chronic inflammatory model of multiple sclerosis. Brain 123(Pt 3):519–531

32. Skripuletz T, Gudi V, Hackstette D, Stangel M (2011) De- and remyelination in the CNS white and grey matter induced by cuprizone: the old, the new, and the unexpected. Histol Histopathol 26:1585–1597

33. Mozafari S, Sherafat MA, Javan M, Mirnajafi-Zadeh J, Tiraihi T (2010) Visual evoked potentials and MBP gene expression imply endogenous myelin repair in adult rat optic nerve and chiasm following local lysolecithin induced demyelination. Brain Res 1351:50–56

34. Duncan ID, Brower A, Kondo Y, Curlee JF Jr, Schultz RD (2009) Extensive remyelination of the CNS leads to functional recovery. Proc Natl Acad Sci U S A 106:6832–6836

35. Heidari M, Radcliff AB, McLellan GJ, Ver Hoeve JN, Chan K, Kiland JA, Keuler NS, August BK, Sebo D, Field AS, Duncan ID (2019) Evoked potentials as a biomarker of remyelination. Proc Natl Acad Sci U S A 116 (52):27074–27083. https://doi.org/10.1073/pnas.1906358116(2019)

36. Diem R, Sattler MB, Merkler D, Demmer I, Maier K, Stadelmann C, Ehrenreich H, Bahr M (2005) Combined therapy with methylprednisolone and erythropoietin in a model of multiple sclerosis. Brain 128:375–385

37. Maier K, Kuhnert AV, Taheri N, Sattler MB, Storch MK, Williams SK, Bahr M, Diem R (2006) Effects of glatiramer acetate and interferon-beta on neurodegeneration in a model of multiple sclerosis: a comparative study. Am J Pathol 169:1353–1364

38. Balatoni B, Storch MK, Swoboda EM, Schonborn V, Koziel A, Lambrou GN, Hiestand PC, Weissert R, Foster CA (2007) FTY720 sustains and restores neuronal function in the DA rat model of MOG-induced experimental autoimmune encephalomyelitis. Brain Res Bull 74:307–316

39. Merrill JE, Hanak S, Pu SF, Liang J, Dang C, Iglesias-Bregna D, Harvey B, Zhu B, McMonagle-Strucko K (2009) Teriflunomide reduces behavioral, electrophysiological, and histopathological deficits in the dark agouti rat model of experimental autoimmune encephalomyelitis. J Neurol 256:89–103

40. Zhang Y, Zhang YP, Pepinsky B, Huang G, Shields LB, Shields CB, Mi S (2015) Inhibition of LINGO-1 promotes functional recovery after experimental spinal cord demyelination. Exp Neurol 266:68–73

41. Nishioka C, Liang HF, Chung CF, Sun SW (2017) Disease stage-dependent relationship between diffusion tensor imaging and electrophysiology of the visual system in a murine model of multiple sclerosis. Neuroradiology 59:1241–1250

42. Thompson AJ, Banwell BL, Barkhof F, Carroll WM, Coetzee T, Comi G, Correale J, Fazekas F, Filippi M, Freedman MS, Fujihara K, Galetta SL, Hartung HP, Kappos L, Lublin FD, Marrie RA, Miller AE, Miller DH, Montalban X, Mowry EM, Sorensen PS, Tintore M, Traboulsee AL, Trojano M, Uitdehaag BMJ, Vukusic S, Waubant E, Weinshenker BG, Reingold SC, Cohen JA (2018) Diagnosis of multiple sclerosis 2017: revisions of the McDonald criteria. Lancet Neurol 17:162–173

43. Poser CM, Paty DW, Scheinberg L, McDonald WI, Davis FA, Ebers GC, Johnson KP, Sibley WA, Silberberg DH, Tourtellotte WW (1983) New diagnostic criteria for multiple sclerosis: guidelines for research protocols. Ann Neurol 13:227–231

44. McDonald WI, Compston A, Edan G, Goodkin D, Hartung HP, Lublin FD, McFarland HF, Paty DW, Polman CH, Reingold SC, Sandberg-Wollheim M, Sibley W, Thompson A, van den Noort S, Weinshenker BY, Wolinsky JS (2001) Recommended diagnostic criteria for multiple sclerosis: guidelines from the international panel on the diagnosis of multiple sclerosis. Ann Neurol 50:121–127

45. Polman CH, Reingold SC, Banwell B, Clanet M, Cohen JA, Filippi M, Fujihara K, Havrdova E, Hutchinson M, Kappos L, Lublin FD, Montalban X, O'Connor P, Sandberg-Wollheim M, Thompson AJ, Waubant E, Weinshenker B, Wolinsky JS (2011) Diagnostic criteria for multiple sclerosis: 2010 revisions to the McDonald criteria. Ann Neurol 69:292–302

46. Traboulsee AL, Li DK (2006) The role of MRI in the diagnosis of multiple sclerosis. Adv Neurol 98:125–146

47. Leocani L, Rovaris M, Boneschi FM, Medaglini S, Rossi P, Martinelli V, Amadio S, Comi G (2006) Multimodal evoked potentials to assess the evolution of multiple sclerosis: a longitudinal study. J Neurol Neurosurg Psychiatry 77:1030–1035

48. Di Maggio G, Santangelo R, Guerrieri S, Bianco M, Ferrari L, Medaglini S, Rodegher M, Colombo B, Moiola L, Chieffo R, Del Carro U, Martinelli V, Comi G, Leocani L (2014) Optical coherence tomography and visual evoked potentials: which is more sensitive in multiple sclerosis? Mult Scler 20:1342–1347

49. Behbehani R, Ahmed S, Al-Hashel J, Rousseff RT, Alroughani R (2017) Sensitivity of visual evoked potentials and spectral domain optical coherence tomography in early relapsing remitting multiple sclerosis. Mult Scler Relat Disord 12:15–19

50. Hardmeier M, Leocani L, Fuhr P (2017) A new role for evoked potentials in MS?

Repurposing evoked potentials as biomarkers for clinical trials in MS. Mult Scler 23:1309–1319

51. Klistorner A, Fraser C, Garrick R, Graham S, Arvind H (2008) Correlation between full-field and multifocal VEPs in optic neuritis. Doc Ophthalmol 116:19–27

52. Pihl-Jensen G, Schmidt MF, Frederiksen JL (2017) Multifocal visual evoked potentials in optic neuritis and multiple sclerosis: a review. Clin Neurophysiol 128:1234–1245

53. Thomas D, Duguid G (2004) Optical coherence tomography--a review of the principles and contemporary uses in retinal investigation. Eye (Lond) 18:561–570

54. Nolan-Kenney RC, Liu M, Akhand O, Calabresi PA, Paul F, Petzold A, Balk L, Brandt AU, Martinez-Lapiscina EH, Saidha S, Villoslada P, Al-Hassan AA, Behbehani R, Frohman EM, Frohman T, Havla J, Hemmer B, Jiang H, Knier B, Korn T, Leocani L, Papadopoulou A, Pisa M, Zimmermann H, Galetta SL, Balcer LJ, International Multiple Sclerosis Visual System Consortium (2019) Optimal intereye difference thresholds by optical coherence tomography in multiple sclerosis: An international study. Ann Neurol 85:618–629

55. Vabanesi M, Pisa M, Guerrieri S, Moiola L, Radaelli M, Medaglini S, Martinelli V, Comi G, Leocani L (2019) In vivo structural and functional assessment of optic nerve damage in neuromyelitis optica spectrum disorders and multiple sclerosis. Sci Rep 9:10371

56. Alshowaeir D, Yannikas C, Garrick R, Van Der Walt A, Graham SL, Fraser C, Klistorner A (2015) Multifocal VEP assessment of optic neuritis evolution. Clin Neurophysiol 126:1617–1623

57. Brusa A, Jones SJ, Plant GT (2001) Long-term remyelination after optic neuritis: a 2-year visual evoked potential and psychophysical serial study. Brain 124:468–479

58. Naismith RT, Tutlam NT, Xu J, Shepherd JB, Klawiter EC, Song SK, Cross AH (2009) Optical coherence tomography is less sensitive than visual evoked potentials in optic neuritis. Neurology 73:46–52

59. Niklas A, Sebraoui H, Hess E, Wagner A, Then Bergh F (2009) Outcome measures for trials of remyelinating agents in multiple sclerosis: retrospective longitudinal analysis of visual evoked potential latency. Mult Scler 15:68–74

60. Calabrese M, Gasperini C, Tortorella C, Schiavi G, Frisullo G, Ragonese P, Fantozzi R, Prosperini L, Annovazzi P, Cordioli C, Di Filippo M, Ferraro D, Gajofatto A, Malucchi S, Lo Fermo S, De Luca G, Stromillo ML, Cocco E, Gallo A, Paolicelli D, Lanzillo R, Tomassini V, Pesci I, Rodegher ME, Solaro C, R. group (2019) "Better explanations" in multiple sclerosis diagnostic workup: A 3-year longitudinal study. Neurology 92:e2527–e2537

61. Cruccu G, Aminoff MJ, Curio G, Guerit JM, Kakigi R, Mauguiere F, Rossini PM, Treede RD, Garcia-Larrea L (2008) Recommendations for the clinical use of somatosensory-evoked potentials. Clin Neurophysiol 119:1705–1719

62. Davis SL, Aminoff MJ, Panitch HS (1985) Clinical correlations of serial somatosensory evoked potentials in multiple sclerosis. Neurology 35:359–365

63. Matthews WB, Read DJ, Pountney E (1979) Effect of raising body temperature on visual and somatosensory evoked potentials in patients with multiple sclerosis. J Neurol Neurosurg Psychiatry 42:250–255

64. Rot U, Mesec A (2006) Clinical, MRI, CSF and electrophysiological findings in different stages of multiple sclerosis. Clin Neurol Neurosurg 108:271–274

65. Kiylioglu N, Parlaz AU, Akyildiz UO, Tataroglu C (2015) Evoked potentials and disability in multiple sclerosis: a different perspective to a neglected method. Clin Neurol Neurosurg 133:11–17

66. Jung P, Beyerle A, Ziemann U (2008) Multimodal evoked potentials measure and predict disability progression in early relapsing-remitting multiple sclerosis. Mult Scler 14:553–556

67. Hellwig K, Stein FJ, Przuntek H, Muller T (2004) Efficacy of repeated intrathecal triamcinolone acetonide application in progressive multiple sclerosis patients with spinal symptoms. BMC Neurol 4:18

68. Watanabe A, Matsushita T, Doi H, Matsuoka T, Shigeto H, Isobe N, Kawano Y, Tobimatsu S, Kira J (2009) Multimodality-evoked potential study of anti-aquaporin-4 antibody-positive and -negative multiple sclerosis patients. J Neurol Sci 281:34–40

69. Ohnari K, Okada K, Takahashi T, Mafune K, Adachi H (2016) Evoked potentials are useful for diagnosis of neuromyelitis optica spectrum disorder. J Neurol Sci 364:97–101

70. Jarius S, Ruprecht K, Kleiter I, Borisow N, Asgari N, Pitarokoili K, Pache F, Stich O, Beume LA, Hummert MW, Ringelstein M, Trebst C, Winkelmann A, Schwarz A, Buttmann M, Zimmermann H, Kuchling J,

Franciotta D, Capobianco M, Siebert E, Lukas C, Korporal-Kuhnke M, Haas J, Fechner K, Brandt AU, Schanda K, Aktas O, Paul F, Reindl M, Wildemann B, in cooperation with the Neuromyelitis Optica Study Group (NEMOS) (2016) MOG-IgG in NMO and related disorders: a multicenter study of 50 patients. Part 2: Epidemiology, clinical presentation, radiological and laboratory features, treatment responses, and long-term outcome. J Neuroinflammation 13:280

71. Sand T, Sjaastad O, Romslo I, Sulg I (1990) Brain-stem auditory evoked potentials in multiple sclerosis: the relation to VEP, SEP and CSF immunoglobulins. J Neurol 237:376–378

72. Pokryszko-Dragan A, Bilinska M, Gruszka E, Kusinska E, Podemski R (2015) Assessment of visual and auditory evoked potentials in multiple sclerosis patients with and without fatigue. Neurol Sci 36:235–242

73. Papathanasiou ES, Pantzaris M, Myrianthopoulou P, Kkolou E, Papacostas SS (2010) Brainstem lesions may be important in the development of epilepsy in multiple sclerosis patients: an evoked potential study. Clin Neurophysiol 121:2104–2110

74. Ravnborg M, Liguori R, Christiansen P, Larsson H, Sorensen PS (1992) The diagnostic reliability of magnetically evoked motor potentials in multiple sclerosis. Neurology 42:1296–1301

75. Simpson M, Macdonell R (2015) The use of transcranial magnetic stimulation in diagnosis, prognostication and treatment evaluation in multiple sclerosis. Mult Scler Relat Disord 4:430–436

76. Snow NJ, Wadden KP, Chaves AR, Ploughman M (2019) Transcranial magnetic stimulation as a potential biomarker in multiple sclerosis: a systematic review with recommendations for future research. Neural Plast 2019:6430596

77. Vucic S, Burke T, Lenton K, Ramanathan S, Gomes L, Yannikas C, Kiernan MC (2012) Cortical dysfunction underlies disability in multiple sclerosis. Mult Scler 18:425–432

78. Nantes JC, Zhong J, Holmes SA, Narayanan S, Lapierre Y, Koski L (2016) Cortical damage and disability in multiple sclerosis: relation to Intracortical inhibition and facilitation. Brain Stimul 9(4):566–573. https://doi.org/10.1016/j.brs.2016.01.003 (2016)

79. Yusuf A, Koski L (2013) A qualitative review of the neurophysiological underpinnings of fatigue in multiple sclerosis. J Neurol Sci 330:4–9

80. Caramia MD, Palmieri MG, Desiato MT, Boffa L, Galizia P, Rossini PM, Centonze D, Bernardi G (2004) Brain excitability changes in the relapsing and remitting phases of multiple sclerosis: a study with transcranial magnetic stimulation. Clin Neurophysiol 115:956–965

81. Liepert J, Mingers D, Heesen C, Baumer T, Weiller C (2005) Motor cortex excitability and fatigue in multiple sclerosis: a transcranial magnetic stimulation study. Mult Scler 11:316–321

82. Landi D, Vollaro S, Pellegrino G, Mulas D, Ghazaryan A, Falato E, Pasqualetti P, Rossini PM, Filippi MM (2015) Oral fingolimod reduces glutamate-mediated intracortical excitability in relapsing-remitting multiple sclerosis. Clin Neurophysiol 126:165–169

83. Ayache SS, Creange A, Farhat WH, Zouari HG, Mylius V, Ahdab R, Abdellaoui M, Lefaucheur JP (2014) Relapses in multiple sclerosis: effects of high-dose steroids on cortical excitability. Eur J Neurol 21:630–636

84. Rossi S, Furlan R, De Chiara V, Motta C, Studer V, Mori F, Musella A, Bergami A, Muzio L, Bernardi G, Battistini L, Martino G, Centonze D (2012) Interleukin-1beta causes synaptic hyperexcitability in multiple sclerosis. Ann Neurol 71:76–83

85. Brum M, Cabib C, Valls-Sole J (2015) Clinical value of the assessment of changes in MEP duration with voluntary contraction. Front Neurosci 9:505

86. Manogaran P, Vavasour I, Borich M, Kolind SH, Lange AP, Rauscher A, Boyd L, Li DK, Traboulsee A (2016) Corticospinal tract integrity measured using transcranial magnetic stimulation and magnetic resonance imaging in neuromyelitis optica and multiple sclerosis. Mult Scler 22:43–50

87. Schlaeger R, D'Souza M, Schindler C, Grize L, Kappos L, Fuhr P (2014) Prediction of MS disability by multimodal evoked potentials: investigation during relapse or in the relapse-free interval? Clin Neurophysiol 125:1889–1892

88. Hume AL, Waxman SG (1988) Evoked potentials in suspected multiple sclerosis: diagnostic value and prediction of clinical course. J Neurol Sci 83:191–210

89. Pelayo R, Montalban X, Minoves T, Moncho D, Rio J, Nos C, Tur C, Castillo J, Horga A, Comabella M, Perkal H, Rovira A, Tintore M (2010) Do multimodal evoked potentials add information to MRI in clinically isolated syndromes? Mult Scler 16:55–61

90. Lebrun C, Bensa C, Debouverie M, Wiertlevski S, Brassat D, de Seze J, Rumbach L, Pelletier J, Labauge P, Brochet B, Tourbah A, Clavelou P, Club Francophone de la Sclérose en Plaques (2009) Association between clinical conversion to multiple sclerosis in radiologically isolated syndrome and magnetic resonance imaging, cerebrospinal fluid, and visual evoked potential: follow-up of 70 patients. Arch Neurol 66:841–846

91. De Stefano N, Giorgio A, Tintore M, Pia Amato M, Kappos L, Palace J, Yousry T, Rocca MA, Ciccarelli O, Enzinger C, Frederiksen J, Filippi M, Vrenken H, Rovira A, M.s. group (2018) Radiologically isolated syndrome or subclinical multiple sclerosis: MAGNIMS consensus recommendations. Mult Scler 24:214–221

92. Schlaeger R, Schindler C, Grize L, Dellas S, Radue EW, Kappos L, Fuhr P (2014) Combined visual and motor evoked potentials predict multiple sclerosis disability after 20 years. Mult Scler 20:1348–1354

93. Chinnadurai SA, Gandhirajan D, Srinivasan AV, Kesavamurthy B, Ranganathan LN, Pamidimukkala V (2018) Predicting falls in multiple sclerosis: do electrophysiological measures have a better predictive accuracy compared to clinical measures? Mult Scler Relat Disord 20:199–203

94. Invernizzi P, Bertolasi L, Bianchi MR, Turatti M, Gajofatto A, Benedetti MD (2011) Prognostic value of multimodal evoked potentials in multiple sclerosis: the EP score. J Neurol 258:1933–1939

95. London F, El Sankari S, van Pesch V (2017) Early disturbances in multimodal evoked potentials as a prognostic factor for long-term disability in relapsing-remitting multiple sclerosis patients. Clin Neurophysiol 128:561–569

96. Mori F, Kusayanagi H, Nicoletti CG, Weiss S, Marciani MG, Centonze D (2014) Cortical plasticity predicts recovery from relapse in multiple sclerosis. Mult Scler 20:451–457

97. Chieffo R, Straffi L, Inuggi A, Coppi E, Moiola L, Martinelli V, Comi G, Leocani L (2019) Changes in cortical motor outputs after a motor relapse of multiple sclerosis. Mult Scler J Exp Transl Clin 5:2055217319866480

98. Nuwer MR, Packwood JW, Myers LW, Ellison GW (1987) Evoked potentials predict the clinical changes in a multiple sclerosis drug study. Neurology 37:1754–1761

99. Davis FA, Stefoski D, Rush J (1990) Orally administered 4-aminopyridine improves clinical signs in multiple sclerosis. Ann Neurol 27:186–192

100. Feuillet L, Pelletier J, Suchet L, Rico A, Ali Cherif A, Pouget J, Attarian S (2007) Prospective clinical and electrophysiological follow-up on a multiple sclerosis population treated with interferon beta-1 a: a pilot study. Mult Scler 13:348–356

101. Meuth SG, Bittner S, Seiler C, Gobel K, Wiendl H (2011) Natalizumab restores evoked potential abnormalities in patients with relapsing-remitting multiple sclerosis. Mult Scler 17:198–203

102. Iodice R, Carotenuto A, Dubbioso R, Cerillo I, Santoro L, Manganelli F (2016) Multimodal evoked potentials follow up in multiple sclerosis patients under fingolimod therapy. J Neurol Sci 365:143–146

103. Connick P, Kolappan M, Crawley C, Webber DJ, Patani R, Michell AW, Du MQ, Luan SL, Altmann DR, Thompson AJ, Compston A, Scott MA, Miller DH, Chandran S (2012) Autologous mesenchymal stem cells for the treatment of secondary progressive multiple sclerosis: an open-label phase 2a proof-of-concept study. Lancet Neurol 11:150–156

104. Sedel F, Papeix C, Bellanger A, Touitou V, Lebrun-Frenay C, Galanaud D, Gout O, Lyon-Caen O, Tourbah A (2015) High doses of biotin in chronic progressive multiple sclerosis: a pilot study. Mult Scler Relat Disord 4:159–169

105. Tsakiri A, Kallenbach K, Fuglo D, Wanscher B, Larsson H, Frederiksen J (2012) Simvastatin improves final visual outcome in acute optic neuritis: a randomized study. Mult Scler 18:72–81

106. Klistorner A, Chai Y, Leocani L, Albrecht P, Aktas O, Butzkueven H, Ziemssen T, Ziemssen F, Frederiksen J, Xu L, Cadavid D, Investigators RM-V (2018) Assessment of Opicinumab in acute optic neuritis using multifocal visual evoked potential. CNS Drugs 32:1159–1171

107. Green AJ, Gelfand JM, Cree BA, Bevan C, Boscardin WJ, Mei F, Inman J, Arnow S, Devereux M, Abounasr A, Nobuta H, Zhu A, Friessen M, Gerona R, von Budingen HC, Henry RG, Hauser SL, Chan JR (2017) Clemastine fumarate as a remyelinating therapy for multiple sclerosis (ReBUILD): a randomised, controlled, double-blind, crossover trial. Lancet 390:2481–2489

108. Leocani L, Rocca MA, Comi G (2016) MRI and neurophysiological measures to predict course, disability and treatment response in multiple sclerosis. Curr Opin Neurol 29:243–253

Part V

Translational View on Therapeutic Strategies

Chapter 18

Targeting Acute Inflammation

Felix Luessi

Abstract

Understanding the immune response in the central nervous system (CNS) is essential for the development of new therapeutic concepts in multiple sclerosis (MS), which differs considerably from other autoimmune diseases. Particular properties of inflammatory processes in the CNS, which is often referred to as an immune privileged site, imply distinct features of MS in terms of disease initiation, perpetuation, and therapeutic accessibility. Furthermore, the CNS is a stress-sensitive organ with low-capacity for self-renewal and is highly prone to bystander damage caused by CNS inflammation.

In this chapter, we review emerging data on the contributions of both autoimmune inflammation and progressive neurodegeneration underlying the pathophysiology of multiple sclerosis (MS) in patients and experimental autoimmune encephalomyelitis (EAE) in rodents. Subsequently, we report on efforts to translate the recent identification of molecules responsible for the acute inflammatory processes in MS into clinical practice with the aim of developing selective regimens.

Key words Multiple sclerosis, Acute inflammation, Drug targets

1 Introduction

In multiple sclerosis (MS), the interplay between environmental factors and susceptibility genes triggers the infiltration of circulating myelin-specific autoreactive lymphocytes into the CNS, leading to inflammation, demyelination, and neuronal injury. Relapses are considered to be the clinical manifestation of acute inflammatory demyelination in the CNS, and disability progression is thought to reflect chronic demyelination, gliosis, and axonal loss as well as neuronal injury. Data obtained over the years point to a complex interplay between environment (e.g., the near-absolute requirement of Epstein–Barr virus exposure), immunogenetics (strong associations with a large number of immune genes), and an autoimmune inflammatory response resulting in demyelination and neurodegenerative processes. Viewing MS as both inflammatory and neurodegenerative disease has major implications for therapy, with CNS protection and repair being needed in addition to controlling inflammation [1]. Up to now treatment options

Sergiu Groppa and Sven G. Meuth (eds.), *Translational Methods for Multiple Sclerosis Research*, Neuromethods, vol. 166, https://doi.org/10.1007/978-1-0716-1213-2_18, © Springer Science+Business Media, LLC, part of Springer Nature 2021

established do not sufficiently prevent the accumulation of tissue damage and clinical disability in patients with MS. Over the past decade intensive research efforts have been expended to develop therapies for MS which may empower patients with a level of independence not presently possible. In this chapter we describe a selection of recently identified molecules responsible for the acute inflammatory processes in MS and its animal model, experimental autoimmune encephalitis (EAE), and review their potential as drug targets.

2   Recent Insights into the Pathogenesis of Multiple Sclerosis

Whereas MS was traditionally considered to be an inflammatory demyelinating disease of the central nervous system, which leaves the axons largely intact at least at onset of the disease [2], research in recent years has shown that neuronal damage processes also play an important role already early in the pathogenesis of MS. State of the art histopathological analyses of brain tissue and neuroimaging studies demonstrated significant damage to neuronal structures with axonal loss and neurodegeneration which already occurs in early disease stage and most likely leads to irreversible neurological impairment [3, 4]. Axonal pathology is particularly pronounced in active and chronic active MS lesions throughout the disease course and is closely associated with the presence of immune cells [5–7]. In addition to axonal damage, either immediate or subsequent to acute inflammatory infiltration, neurodegeneration continues in the progressive stage of the disease [1]. Quantitative morphological studies also detected neuronal damage within the normal appearing white and gray matter, devoid of obvious demyelinating lesions [8–10]. These observations have led to the hypothesis that the destruction of myelin and neurons might at least partially represent independent processes.

2.1   Mechanisms of Neuronal Injury

Improving our understanding of the mechanisms underlying neuronal injury in MS is a major challenge in experimental neuroimmunology. The underlying disease pathophysiology is complex and involves the key features of the disease, which are inflammation, demyelination, astrogliosis, and neuronal pathology. The potential causes of acute and chronic neuronal and axonal injury are bystander damage by proinflammatory neurotoxic substances, direct damage processes, which involve cell contact-dependent mechanisms, and demyelination dependent metabolic disturbances in the denuded axons.

2.1.1   Immune Cell-Mediated Neuronal/ Axonal Injury

The inflammatory infiltrates of active and chronic active MS lesions consist predominantly of CD4+ T cells, CD8+ T cells, and activated microglia/macrophages [5, 11]. Because of the correlation between the degree of inflammation and neuronal injury [12], exposure to the inflammatory milieu has been proposed as trigger of neuronal injury [13]. However, also direct cell-mediated mechanisms have been postulated as cause of neuronal pathology [14].

Endogenous microglia cells in the CNS are dynamic surveillants of brain parenchyma integrity and rapidly react to potential threats by encapsulation of dangerous foci, removal of apoptotic cells, and assistance with tissue regeneration in toxin-induced demyelination [15, 16]. In the context of non-autoimmune, pathogen-associated inflammation, microglial cells protect the neuronal compartment [17]. Contrarily in MS, microglia and macrophages are shifted towards a strongly proinflammatory phenotype and may potentiate neuronal damage by releasing proinflammatory cytokines (i.e., tumor necrosis factor (TNF)-α, interleukin (IL)-1β, and IL-6) and proinflammatory molecules like nitric oxide (NO), proteolytic enzymes, and free radicals [18–20]. In the MS animal model, experimental autoimmune encephalomyelitis (EAE), paralysis of microglia in vivo resulted in substantial amelioration of the clinical signs and in strong reduction of CNS inflammation, demonstrating their active involvement in damage processes [21]. In the light of this, it is questionable whether microglia and monocyte-derived macrophages, the very last downstream effector cells in the immune reaction, actually have the capacity to influence their fate. It is more likely that the adaptive immune system orchestrates the attack against CNS cells and drives microglia and macrophages to attack neurons and oligodendrocytes.

Clonally expanded CD8+ T cells have been shown within MS lesions as well as in the cerebrospinal fluid (CSF) of MS patients [22, 23]. However, the significance of these CD8+ T cells in MS pathogenesis is controversial since there is evidence for a suppressor function that inhibits pathogenic autoreactive CD4+ T cells [24–26] and evidence for a tissue-damaging role because a significant correlation between the extent of axonal damage and the number of CD8+ T cells has been reported [6, 12]. In accordance with the latter observation, major histocompatibility complex (MHC) class I-restricted CD8+ T cells were found to induce neuronal cell death in certain immunological constellations in cultured neurons and hippocampal brain slices [27, 28]. In addition, transection of MHC class I-expressing neurites by CD8+ T cells has been described [29], a process that might also contribute to pathology in human disease. The contribution of CD4+ T cells to neuronal injury is a matter of debate. The genetic risk of MS and EAE is, to a substantial degree, conferred by MHC class II alleles and other genes involved in the differentiation and activity of T helper cell phenotypes [30]. An affinity between invading activated CD4+ T cells and neurons had not been considered to date as neurons do not express MHC class

II molecules. However, due to recent advances in deep-tissue imaging using two-photon microscopy, interactions between neurons and immune cells can be investigated in vivo and in organotypic microenvironments. This revealed that encephalitogenic CD4 + T cells possess marked migratory capacities within the CNS parenchyma and directly interact with the soma and processes of neurons, partially leading to cell death [31]. Amongst others the death ligand TNF-related apoptosis-inducing ligand (TRAIL) as a T cell-associated effector molecule contributes to the induction of neuronal apoptosis. It has been shown that TRAIL expressed by CD4+ T cells induces death of neurons in the inflamed brain and promotes EAE [32, 33]. Using in vivo live imaging in EAE, a recent study has confirmed a direct contact between CD4+ T cells, in particular T helper 17 (Th17) cells, and neurons, resulting in neuronal dysfunction and subsequently cell death [34]. The antigen-independent Th17-mediated neuronal injury was LFA-1-dependent and potentially reversible and thus accessible by therapeutic approaches. These findings suggest that once they reach the CNS, CD4+ T cells are directly involved in local neuronal damage processes [35].

2.1.2 Axonal/Neuronal Degeneration as a Consequence of Inflammatory Demyelination

Although irreversible neurological disability in MS patients results from axonal degeneration [36, 37], knowledge of the mechanisms by which demyelinated axons degenerate is far from complete. The "virtual hypoxia hypothesis" postulates that demyelination increases the energy demand in denuded axons [13]. To safeguard nerve conduction since the voltage-gated Na^+ channels that are usually concentrated in axons that have incomplete myelination, larger numbers of Na^+ channels are needed to compensate for loss of saltatory axon potential propagation [38, 39]. However, higher numbers of Na^+ channels necessitate an increased energy supply to restore trans-axolemmal Na^+ and K^+ gradients. In addition, an impaired axoplasmatic ATP production in chronically demyelinated axons due to mitochondrial dysfunction has been described [40]. The function of mitochondrial respiratory chain complex I and III was reduced by 40–50% in mitochondrial-enriched preparations from the motor cortex of MS patients [41]. Furthermore, defects of mitochondrial respiratory chain complex IV have been reported [42, 43] and have been associated with hypoxia-like tissue injury [44] and reduced brain NAA concentration [45]. The combination of increased energy requirements and compromised ATP production as a result of demyelination leads to a vicious circle by the loss of Na^+/K^+ ATPase [46], which contributes to an increased intracellular Na^+. Consequently, Ca^{2+} is released from intracellular stores [47] and the direction of the Na^+/Ca^{2+}-exchanger is reversed, resulting in additional extracellular Ca^{2+} influx [48]. That, in turn, leads to Ca^{2+}-mediated degenerative responses such as cytoskeleton disruption and cell death [49, 50].

Aside from the summarized dramatic ion and energy imbalances following demyelination, the lack of structural as well as trophic support to axons provided by myelin and oligodendrocytes also contributes to neuronal damage [51, 52]. In vitro evidence suggests that oligodendrocytes produce trophic factors such as insulin-like growth factor-type 1 and neuregulin that promote normal axon function and survival [53, 54]. Moreover, mice lacking structural components of compact myelin such as proteolipid protein (PLP) demonstrated a late onset, slowly progressing axonopathy [55]. However, despite an indication of primary pathological changes of oligodendrocytes in some patients suffering from relapsing remitting multiple sclerosis (RRMS) [56], direct evidence for oligodendrocyte dysfunction independent of and prior to inflammation in classic MS is missing.

3   Potential Therapeutic Strategies Targeting Acute Inflammation

The limitations of established therapies for MS are well-known and include treatment adherence and convenience issues, partial efficacy, and, in rare cases, a risk of potentially life-threatening adverse events. To overcome these limitations research efforts have turned toward identifying drug targets for novel medications that have greater efficacy via anti-inflammatory and neuroprotective effects which will be discussed below.

3.1 β-HMG-CoA Reductase

Blockade of the β-3-hydroxy-3-methylglutaryl coenzyme A (HMG-CoA) reductase by statins, primarily used as effective cholesterol-lowering agents, results in interference of T cell cycle progression and plays a beneficial role in chronic neuroinflammation [57]. Increasing evidence suggest that the pharmacological effects of statins relate not only to cholesterol-lowering but also to anti-inflammatory effects associated with reduced serum concentrations of C-reactive protein and a lowered incidence of acute rejection after heart transplantation [58, 59]. Concerning the underlying mechanisms, Kwak et al. demonstrated that statins inhibit the IFN-γ-induced expression of MHC class II on most antigen-presenting cells by suppression of the inducible promotor IV of the MHC class II transactivator CIITA [60]. Statins have been reported to reduce clinical manifestation in the inflammatory Lewis rat EAE model [61] and to prevent or reverse chronic and relapsing paralysis in a PLP-induced EAE model in the SJL mouse strain [62]. In addition statins decrease T cell proliferation mediated by direct T cell receptor engagement independently of MHC class II and LFA-1 [31]. The underlying mechanism for the inhibition of T cell response in EAE is the interference with cell cycle regulation represented by downregulation of CDK-4 and upregulation of p27^{kip1}. Anergy induction by statins was dependent on HMG-CoA reductase, required IL-10 signaling, and was associated with phosphorylation of Erk1 [57]. Taken together, the plethora of these findings demonstrates the pronounced immunomodulatory effects of statins and suggests their therapeutic potential. This perspective has been confirmed in an MRI-based study in patients with RRMS, treatment with atorvastatin, alone or in combination with IFN-β, led to a substantial reduction in the number and volume of contrast enhancing lesions (CEL) [63]. Moreover, a clinical study in RRMS suggested that adding statins to IFN-β may reduce the relapse rate compared to IFN-β alone [64].

Considering that statins inhibit the synthesis of cholesterol and the organ richest in sterol is the brain, several studies have been performed with statins on various target cells in the CNS. However, reports investigating the direct impact of statins on neurons showed a rather complex trend: previous studies using high micromolar concentrations of statins demonstrated an induction of apoptosis in transformed neuronal cell lines [65], while recent studies show that statins actually protect from glutamate-mediated excitotoxicity when given at low concentrations to neuronal cultures [66]. Thus, neurotoxic and neuroprotective effects of statins are markedly dose-dependent. Clinical and imaging studies on statins in neuroinflammatory conditions will determine the outcome of therapy on neuronal fate.

3.2 NF-κB

NF-κB is a ubiquitous transcription factor that transmits a large number of extracellular signals involved in both innate and adaptive immune responses from the cell surface to the nucleus. In mammals, NF-κB comprises a family of five protein subunits, p50, p52, p65, c-Rel, and Rel-B, which share an N-terminal 300-amino acid Rel homology domain that allows them to dimerize, translocate to the nucleus, and bind to specific DNA sequences known as κB sites [67]—thereby controlling gene expression in inflammation, immunity, cell proliferation, and apoptosis [68]. Two major NF-κB activating signal transduction pathways have been described, the classical and alternative pathways [69]. NF-κB is an important mediator in activation of T cells and their expression of proinflammatory cytokines, antibody production by B cells, cytokine production by dendritic cells and macrophages, and in the regulation of the susceptibility of these cells to apoptosis [68]. Thus, NF-κB is a key element in the regulation of immune cells involved in the pathogenesis of MS.

Studies which have addressed the role of NF-κB activation in EAE found that p50 and p65, which form heterodimers involved in the classical NF-κB activation pathway, are the prototypic inducible NF-κB subunits in the CNS during EAE [70, 71]. This is supported by findings from van Loo et al. [72] who showed that CNS-restricted ablation of two major classical NF-κB pathway activators, namely NEMO and IKKβ, ameliorated EAE pathology. These findings suggest that classical NF-κB activation in CNS cells has mainly pathogenic effects.

Several studies have investigated the localization of NF-κB in MS brain tissue. In active MS plaques, p50, p65, c-Rel, and IκB subunits were found in the nuclei of infiltrating macrophages [73], and p65 was also present in the nuclei of hypertrophic astrocytes as well as some oligodendrocytes [74]. Perivascular infiltrating lymphocytes in active plaques were reported to have c-Rel in their nuclei [73]. Furthermore, gene-microarray analysis of brain tissue demonstrated that genes associated with NF-κB are upregulated in tissue from patients with MS compared to controls [75].

Inhibition of NF-κB is therefore a potential therapeutic target for RRMS, where inflammatory CNS infiltration correlates closely to relapses. Sites to target include signal transduction that activates IκB kinase (IKK) complex, IκB degradation, NF-κB translocation, and inhibition of NF-κB binding to DNA. Further down the NF-κB pathway, molecules that inhibit the activation of IKK complex were strongly protective against the induction of EAE in mice [76]. Utilizing agents that prevent IκB degradation is also an effective strategy to inhibit NF-κB activation in EAE. Aktas et al. found that EGCG, which blocks catalytic activities of the 20S/26S proteasome complex, resulted in the accumulation of IκBα and protected from relapses in EAE [77]. In another study, thymoquinone has been reported to inhibit NF-κB activation specifically by suppressing the direct binding of nuclear p65 to DNA [78], and to ameliorate signs of EAE by preventing perivascular cuffing and infiltration of mononuclear cells into the brain [79]. However, one caveat in studies on the effects of NF-κB inhibitors in EAE is that none of them has addressed possible changes in measures such as oligodendrocyte or neuronal cell apoptosis. Therapeutic efficacy must be balanced against the toxicity associated with inhibition of a ubiquitous cellular pathway. Interference with NFkB pathways is discussed as one mechanism of action for new therapies, which has been investigated in the clinical evaluation of laquinimod.

3.3 Sphingosine 1-Phosphate Receptors

Sphingosine 1-phosphate (S1P), a metabolite of sphingolipids, has essential signaling functions in the immune system and the CNS, mediated via G protein-coupled receptor subtypes, S1P receptors (S1PRs)$_{1-5}$. Thereby, S1P regulates diverse cellular responses, including proliferation, differentiation, survival, motility, chemoattraction, and cell–cell adherence [80]. Resting T cells and B cells express S1PR$_1$ and lower levels of S1PR$_4$ and S1PR$_3$ [81]. S1P-S1PR$_1$ interaction plays a key role in lymphocyte trafficking. All S1PRs are expressed in the CNS. Oligodendrocytes predominantly express S1PR$_1$ and S1PR$_5$ which have been shown to be involved in process dynamics, oligodendrocyte migration and differentiation, survival, cross talk with neurotrophins, and remyelination

[82]. Astrocytes primarily express $S1PR_1$ and $S1PR_3$ that enhance astrocyte proliferation, migration, and secretion of glial cell-derived neurotrophic factor upon S1P stimulation [83]. In neurons that can express all five S1PRs, S1P signaling is crucial for neural development and has roles in differentiation, excitability, process extension, and calcium signaling [84]. Taken together, S1PRs are attractive drug targets for immunomodulatory and possibly neuromodulatory therapies.

Fingolimod (FTY720), a S1PR agonist, is reversibly phosphorylated to Fingolimod-P, the active moiety, by sphingosine kinase (SphK) 2 [85]. Fingolimod-P binds with high affinity to $S1PR_1$ and $S1PR_{3-5}$. Binding to $S1PR_1$, initially causes agonist effects, which are followed by aberrant receptor phosphorylation, long-lasting internalization, ubiquitination, and proteasomal receptor degradation, leading to functional antagonism [86]. A large number of studies have reported that Fingolimod is effective in EAE [87]. In two phase 3 clinical trials for RRMS, Fingolimod ameliorated the relapse rate, the risk of disability progression and CNS lesions measured by MRI [88, 89]. Because of these encouraging results, fingolimod was approved by the European Medicines Agency in March 2011 as first oral therapy for RRMS.

Because Fingolimod-mediated immune regulation mainly depends on its effects on $S1PR_1$ and binding to other receptors in part causes side effects, development of next-generation S1PR modulators by targeting S1PR selectively, is underway. Recently, siponimod has been approved as the first next-generation S1PR modulator for patients with active SPMS. Siponimod has an oral formulation with blood–brain barrier penetration [90]. The agent is designed for increased S1PR specificity which may limit toxicity. In a phase 3 randomized clinical trial in SPMS patients, siponimod significantly reduced the risk of disease progression, including impact on physical disability and cognitive decline [91].

Other strategies to target the ShpK-S1P-S1PR axis include SphK inhibition and the use of a specific monoclonal antibody to S1P. However, the effects of these therapeutic strategies on EAE have not been reported.

3.4 PPARγ

Peroxisome proliferator-activated receptors (PPARs) are nuclear receptor transcription factors that play a key role in the regulation of development, metabolism and inflammation [92]. Three PPAR isotypes, α, δ, and γ, have distinct actions on cellular physiology and show specificity in their ligand-binding properties. Upon activation with specific ligands, PPARs heterodimerize with retinoic acid receptors (RXRs), bind to the response elements and induce the expression of target genes. PPARγ and PPARα ligands have been demonstrated to exert anti-inflammatory effects, including the suppression of TNF-α and IL-1β, inducible nitric oxide synthase (iNOS), and cyclooxygenase type 2 gene expression as well as the inhibition of chemokine expression [93]. In addition, PPARγ agonists inhibit proliferation, block cytokine production, and induce apoptosis in T cells [94]. The potent anti-inflammatory and anti-proliferative effects of PPARγ agonism render PPARγ a potential drug target in MS.

Data from animal models show that PPARγ deficient heterozygous mice develop an exacerbated EAE with an augmented Th1 response, which suggests a physiological role for PPARγ in the regulation of CNS inflammation [95]. The effects of PPARγ agonists in modulating EAE were first investigated by Niino et al. who demonstrated that the thiazolidinedione troglitazone, originally designed as antidiabetic drug, inhibited the development of EAE, but did not alter T cell proliferation or T cell production of IFN-γ in vitro [96]. Subsequently, the thiazolidinedione pioglitazone was shown to inhibit monophasic EAE [93]. Although pioglitazone had no effect on the initial phase of relapsing-remitting EAE, disease severity was reduced upon subsequent relapses. In another study, the natural PPARγ agonist 15d-PGJ$_2$, a cyclopentone prostaglandin, inhibited the development of EAE when administered prior to or following onset of disease [97]. Moreover, 15d-PGJ$_2$ inhibited the proliferation of T cells and inhibited IFN-γ and IL-4 production of these cells in vitro. Natarajan et al. reported that PPARγ agonists 15d-PGJ$_2$ and ciglitazone decreased IL-12 expression and differentiation of Th$_1$ cells which was associated with decreased severity of active and passive EAE [98]. Interestingly, the combination of the PPARγ agonist 15d-PGJ$_2$ and the RXR agonist 9-cis retinoic acid exerts additive anti-inflammatory effects on EAE [99]. A recent study reported that PPARγ agonists promoted oligodendrocyte progenitor cell differentiation and enhanced their antioxidant defenses in vitro, suggesting that in addition to their known anti-inflammatory effects, PPARγ agonists may promote recovery from demyelination by direct effects on oligodendrocytes [100].

Schmidt et al. investigated the effects PPARγ agonists on the function of peripheral blood mononuclear cells (PBMCs) from MS patients and healthy donors [101]. In this study, PPARγ agonists decreased T cell proliferation and production of the cytokines TNF-α and IFN-γ by PBMCs. A recent study investigated the effects of pioglitazone on diffusion tensor imaging indices in MS patients [102]. The findings of this trial suggest that pioglitazone may reduce new lesion development and may improve anisotropy and mean diffusivity at the site of degenerating tissue. However, reports on the clinical use of PPARγ agonists in MS are presently fragmentary [103–105]. Thus, systemic clinical trials are needed to determine the safety and efficacy of PPARγ agonists for the treatment of MS.

3.5 Matrix Metalloproteinases

Matrix metalloproteinases (MMPs) constitute a family of proteolytic enzymes, which are involved in many physiologic functions including cell survival and death, and signal transduction [106]. Excessive upregulation of MMPs has been found in various human neurologic diseases including MS [107]. Besides other activities, minocycline exhibits a dose-dependent effect in reducing MMPs activity. In addition, minocycline decreases clinical severity, inflammation and neuropathology in EAE [108]. On the basis of these experimental findings application of minocycline has been successfully tested in 10 patients with RRMS with regard to clinical and MRI findings [109, 110]. Furthermore, minocycline combined with glatiramer acetate reduces the number of CEL and new T2 lesions compared to glatiramer acetate alone. However, in a double-blind, randomized placebo-controlled multicentre study minocycline showed no statistically significant beneficial effect when added to subcutaneous IFN-1a therapy [111].

3.6 Ion Channel Blockade

Following up on the findings that demyelination leads to an altered energy demand and changes in intracellular ion homeostasis in neurons, ion channels have become potential drug targets in the treatment of MS. Several ion channel blockers already in use for other medical conditions are now being investigated in CNS autoimmunity. Evidence from animal studies has shown beneficial effect in rats with chronic EAE for up to 180 days after treatment with phenytoin [112], a Na^+ channel blocker commonly used for epilepsy. Interestingly, when the study was repeated using either phenytoin or carbamazepine, another antiepileptic with Na^+ channel blocker capacities, the animals became acutely worse after the withdrawal of either drug [113], indicating that more work needs to be done to understand the consequences of the long-term effects of Na^+ channel blockers and of their withdrawal in MS. Two other Na^+ blocking agents, the anti-arrhythmic agent flecainide and the antiepileptic lamotrigine, have now been shown to improve axonal survival and decrease in disability in EAE-affected rats [114, 115]. However, in a phase 2 study in patients with secondary progressive multiple sclerosis (SPMS), lamotrigine showed an increase of cerebral volume loss, which was not clinically relevant, but could not be explained [116]. Another clinical study investigated the antiepileptic drug topiramate, which has partial Na^+ channel blocking capabilities, in combination with IFN-β in patients with RRMS. The results of this trial are currently awaited. Although evidence for a direct neuroprotective effect of Na^+ channel blockers is still missing and additional immunomodulatory actions on microglia and macrophages have been suggested [117], studies on axons subjected to anoxia in vitro [48] and immediately after exposure to elevated levels of nitric oxide in vivo [118] demonstrated, at least in part, the involvement of a direct effect on axons.

As the increased intracellular Ca^{2+} levels contribute to axonal damage through activation of different enzymes, in particular proteases, blockade of voltage gated Ca^{2+} channels (VGCC) is a potentially promising target. In a study of EAE-affected rats, the effect of bepridil, a broad-spectrum Ca^{2+} channel blocker was compared with the dihydropyridine, nitrendipine, which is a blocker of L-type VGCCs. Both drugs prevented axonal loss and disability in treated animals [119]. There are however no clinical trials in MS patients existing at the moment.

3.7 Cannabinoid Receptors

Cannabis is used by MS patients for relief from a variety of symptoms [120], despite equivocal results of several clinical trials [121]. Improved knowledge about the major psychoactive ingredient of cannabis, delta-9-tetrahydrocannabinol, and its CB1 and CB2 receptors has resulted in an increase of experimental data from MS animal models. In vitro evidence suggests that cannabinoids have an effect on several potential mechanisms of axonal injury, including glutamate release [122], oxidative free radicals as well as damaging Ca^{2+} influx [123]. Furthermore, exogenous agonists of the cannabinoid CB1 receptor have possible neuroprotective effects in EAE animal models [124], and strategies to increase the endogenous cannabinoid anandamide also appear to attenuate the clinicopathological features of EAE [125, 126]. Despite these promising results, neuroprotective effects in MS by cannabinoids and the modulation of the endocannabinoid system have still to be established.

3.8 Estrogen Receptors

Estrogens represent another potential treatment of MS that is thought to rely on inhibition of NF-κB activation. It is well known that many MS patients have a reduced frequency of relapses during pregnancy [127, 128], due, at least in part, to altered levels of sex hormones such as estriol, estradiol and progesterone. Studies point to regulatory effects of estriol on T cell migration and cytokine profile by inhibition of NF-κB [129]. The effects of estrogens during pregnancy may be mediated by induction of indolamine 2,3-dioxygenase (IDO) in dendritic cells [130]. Given that IDO is considered to be a major regulator of the NF-κB pathway, it is likely that estrogens would have downregulatory effects on immune cells, although their effects on cells of the CNS remain to be elucidated. A pilot trial of oral estriol treatment in RRMS patients decreased CEL numbers and volumes on MRI and increased the cognitive function as measured by the PASAT [131]. There are now several phase 2 trials underway to further determine the efficacy of estrogen treatment in MS.

3.9 α4 Integrin

α4 integrin is a molecule expressed on lymphocytes, monocytes and hematopoietic cells that forms heterodimers with β1 and β7 integrin [132]. The heterodimer α4β1 integrin (very late activating antigen-4 (VLA-4)) is the counter-receptor for vascular adhesion molecule-1 (VCAM-1) and the key player in the transmigration of lymphocytes across the blood–brain barrier, as was demonstrated almost two decades ago in the EAE model [133]. Natalizumab is a long-acting humanized recombinant monoclonal antibody that binds to α4 subunit of α4β1 and α4β7 integrins. On the basis of results from phase 3 trials, this agent was approved for patients with highly active RRMS in 2006 [134, 135]. Natalizumab is very effective in preventing the occurrence of clinical relapses and new brain lesions on MRI. However, the substantial mortality and morbidity associated with progressive multifocal leukoencephalopathy (PML) has restricted the use of natalizumab [136].

Firategrast, an orally bioavailable small molecule α4 integrin antagonist could with a shorter half-life than natalizumab provide an attractive alternative for MS patients. As the increased risk of PML might be associated with α4 integrin antagonism, a faster elimination of the drug could enhance safety. Firategrast has been effective in a number of animal models of inflammation, including the relapsing remitting Biozzi mouse EAE model. On the basis of these preclinical investigations, the efficacy of firategrast in RRMS was evaluated in a phase 2 clinical trial. A 49% reduction in the cumulative number of new CEL was seen for the firategrast-treated group versus the placebo group [137]. Firategrast was generally well tolerated, and no cases of PML or evidence of JC-virus reactivation were recorded. However, whether firategrast possesses substantial advantages over natalizumab with regard to PML remains to be elucidated.

4   Conclusion

In the past, the immunological aspects of MS were extensively studied, elucidating the immune system's involvement in the development and enhancement of the myelin-targeted inflammatory attack. The rediscovery of the neuronal pathology of MS and its importance regarding disability in patients with MS has now shifted attention to the neurobiological consequences of autoimmune demyelination. Deeper molecular insights into the mechanisms of inflammatory neuronal injury in MS helped identifying molecular targets for the development of novel treatment strategies. Future studies will be needed to translate these concepts into more efficient therapy of MS.

References

1. Siffrin V, Vogt J, Radbruch H, Nitsch R, Zipp F (2010) Multiple sclerosis—candidate mechanisms underlying CNS atrophy. Trends Neurosci 33(4):202–210

2. Charcot J (1880) Leçons sur les maladies du système nerveux faites à la Salpetrière, 4th edn. Bourneville, Paris

3. Aktas O, Ullrich O, Infante-Duarte C, Nitsch R, Zipp F (2007) Neuronal damage in brain inflammation. Arch Neurol 64 (2):185–189

4. Inglese M, Ge Y, Filippi M, Falini A, Grossman RI, Gonen O (2004) Indirect evidence for early widespread gray matter involvement in relapsing-remitting multiple sclerosis. NeuroImage 21(4):1825–1829

5. Ferguson B, Matyszak MK, Esiri MM, Perry VH (1997) Axonal damage in acute multiple sclerosis lesions. Brain 120(Pt 3):393–399

6. Kuhlmann T, Lingfeld G, Bitsch A, Schuchardt J, Bruck W (2002) Acute axonal damage in multiple sclerosis is most extensive in early disease stages and decreases over time. Brain 125(Pt 10):2202–2212

7. Trapp BD, Peterson J, Ransohoff RM, Rudick R, Mork S, Bo L (1998) Axonal transection in the lesions of multiple sclerosis. N Engl J Med 338(5):278–285

8. Bjartmar C, Kinkel RP, Kidd G, Rudick RA, Trapp BD (2001) Axonal loss in normal-appearing white matter in a patient with acute MS. Neurology 57(7):1248–1252

9. Bo L, Vedeler CA, Nyland HI, Trapp BD, Mork SJ (2003) Subpial demyelination in the cerebral cortex of multiple sclerosis patients. J Neuropathol Exp Neurol 62 (7):723–732

10. Kutzelnigg A, Lucchinetti CF, Stadelmann C, Bruck W, Rauschka H, Bergmann M et al (2005) Cortical demyelination and diffuse white matter injury in multiple sclerosis. Brain 128(Pt 11):2705–2712

11. Traugott U, Reinherz EL, Raine CS (1983) Multiple sclerosis. Distribution of T cells, T cell subsets and Ia-positive macrophages in lesions of different ages. J Neuroimmunol 4 (3):201–221

12. Bitsch A, Schuchardt J, Bunkowski S, Kuhlmann T, Bruck W (2000) Acute axonal injury in multiple sclerosis. Correlation with demyelination and inflammation. Brain 123 (Pt 6):1174–1183

13. Trapp BD, Stys PK (2009) Virtual hypoxia and chronic necrosis of demyelinated axons in multiple sclerosis. Lancet Neurol 8 (3):280–291

14. Mager R, Meuth SG, Krauchi K, Schmidlin M, Muller-Spahn F, Falkenstein M (2009) Mismatch and conflict: neurophysiological and behavioral evidence for conflict priming. J Cogn Neurosci 21 (11):2185–2194

15. Nimmerjahn A, Kirchhoff F, Helmchen F (2005) Resting microglial cells are highly dynamic surveillants of brain parenchyma in vivo. Science 308(5726):1314–1318

16. Remington LT, Babcock AA, Zehntner SP, Owens T (2007) Microglial recruitment, activation, and proliferation in response to primary demyelination. Am J Pathol 170 (5):1713–1724

17. Trapp BD, Wujek JR, Criste GA, Jalabi W, Yin X, Kidd GJ et al (2007) Evidence for synaptic stripping by cortical microglia. Glia 55(4):360–368

18. Hohlfeld R (1997) Biotechnological agents for the immunotherapy of multiple sclerosis. Principles, problems and perspectives. Brain 120(Pt 5):865–916

19. Ransohoff RM, Perry VH (2009) Microglial physiology: unique stimuli, specialized responses. Annu Rev Immunol 27:119–145

20. Vogt J, Paul F, Aktas O, Muller-Wielsch K, Dorr J, Dorr S et al (2009) Lower motor neuron loss in multiple sclerosis and experimental autoimmune encephalomyelitis. Ann Neurol 66(3):310–322

21. Heppner FL, Greter M, Marino D, Falsig J, Raivich G, Hovelmeyer N et al (2005) Experimental autoimmune encephalomyelitis repressed by microglial paralysis. Nat Med 11(2):146–152

22. Babbe H, Roers A, Waisman A, Lassmann H, Goebels N, Hohlfeld R et al (2000) Clonal expansions of CD8(+) T cells dominate the T cell infiltrate in active multiple sclerosis lesions as shown by micromanipulation and single cell polymerase chain reaction. J Exp Med 192(3):393–404

23. Jacobsen M, Cepok S, Quak E, Happel M, Gaber R, Ziegler A et al (2002) Oligoclonal expansion of memory CD8+ T cells in cerebrospinal fluid from multiple sclerosis patients. Brain 125(Pt 3):538–550

24. Hu D, Ikizawa K, Lu L, Sanchirico ME, Shinohara ML, Cantor H (2004) Analysis of regulatory CD8 T cells in Qa-1-deficient mice. Nat Immunol 5(5):516–523

25. Jiang H, Curran S, Ruiz-Vazquez E, Liang B, Winchester R, Chess L (2003) Regulatory CD8+ T cells fine-tune the myelin basic protein-reactive T cell receptor V beta repertoire during experimental autoimmune encephalomyelitis. Proc Natl Acad Sci U S A 100(14):8378–8383

26. Lu L, Ikizawa K, Hu D, Werneck MB, Wucherpfennig KW, Cantor H (2007) Regulation of activated CD4+ T cells by NK cells via the Qa-1-NKG2A inhibitory pathway. Immunity 26(5):593–604

27. Medana IM, Gallimore A, Oxenius A, Martinic MM, Wekerle H, Neumann H (2000) MHC class I-restricted killing of neurons by virus-specific CD8+ T lymphocytes is effected through the Fas/FasL, but not the perforin pathway. Eur J Immunol 30(12):3623–3633

28. Meuth SG, Herrmann AM, Simon OJ, Siffrin V, Melzer N, Bittner S et al (2009) Cytotoxic CD8+ T cell-neuron interactions: perforin-dependent electrical silencing precedes but is not causally linked to neuronal cell death. J Neurosci 29(49):15397–15409

29. Medana I, Martinic MA, Wekerle H, Neumann H (2001) Transection of major histocompatibility complex class I-induced neurites by cytotoxic T lymphocytes. Am J Pathol 159(3):809–815

30. International Multiple Sclerosis Genetics C, Wellcome Trust Case Control C, Sawcer S, Hellenthal G, Pirinen M, Spencer CC et al (2011) Genetic risk and a primary role for cell-mediated immune mechanisms in multiple sclerosis. Nature 476(7359):214–219

31. Aktas O, Waiczies S, Smorodchenko A, Dorr J, Seeger B, Prozorovski T et al (2003) Treatment of relapsing paralysis in experimental encephalomyelitis by targeting Th1 cells through atorvastatin. J Exp Med 197(6):725–733

32. Aktas O, Smorodchenko A, Brocke S, Infante-Duarte C, Schulze Topphoff U, Vogt J et al (2005) Neuronal damage in autoimmune neuroinflammation mediated by the death ligand TRAIL. Neuron 46(3):421–432

33. Nitsch R, Bechmann I, Deisz RA, Haas D, Lehmann TN, Wendling U et al (2000) Human brain-cell death induced by tumour-necrosis-factor-related apoptosis-inducing ligand (TRAIL). Lancet 356(9232):827–828

34. Siffrin V, Radbruch H, Glumm R, Niesner R, Paterka M, Herz J et al (2010) In vivo imaging of partially reversible th17 cell-induced neuronal dysfunction in the course of encephalomyelitis. Immunity 33(3):424–436

35. Birkner K, Wasser B, Ruck T, Thalman C, Luchtman D, Pape K et al (2020) beta1-integrin- and KV1.3 channel-dependent signaling stimulates glutamate release from Th17 cells. J Clin Invest 130(2):715–732

36. Arnold DL, Riess GT, Matthews PM, Francis GS, Collins DL, Wolfson C et al (1994) Use of proton magnetic resonance spectroscopy for monitoring disease progression in multiple sclerosis. Ann Neurol 36(1):76–82

37. De Stefano N, Matthews PM, Fu L, Narayanan S, Stanley J, Francis GS et al (1998) Axonal damage correlates with disability in patients with relapsing-remitting multiple sclerosis. Results of a longitudinal magnetic resonance spectroscopy study. Brain 121(Pt 8):1469–1477

38. Bostock H, Sears TA (1978) The internodal axon membrane: electrical excitability and continuous conduction in segmental demyelination. J Physiol 280:273–301

39. Felts PA, Baker TA, Smith KJ (1997) Conduction in segmentally demyelinated mammalian central axons. J Neurosci 17(19):7267–7277

40. Dutta R, McDonough J, Yin X, Peterson J, Chang A, Torres T et al (2006) Mitochondrial dysfunction as a cause of axonal degeneration in multiple sclerosis patients. Ann Neurol 59(3):478–489

41. Su KG, Banker G, Bourdette D, Forte M (2009) Axonal degeneration in multiple sclerosis: the mitochondrial hypothesis. Curr Neurol Neurosci Rep 9(5):411–417

42. Mahad D, Ziabreva I, Lassmann H, Turnbull D (2008) Mitochondrial defects in acute multiple sclerosis lesions. Brain 131(Pt 7):1722–1735

43. Mahad DJ, Ziabreva I, Campbell G, Lax N, White K, Hanson PS et al (2009) Mitochondrial changes within axons in multiple sclerosis. Brain 132(Pt 5):1161–1174

44. Lucchinetti C, Bruck W, Parisi J, Scheithauer B, Rodriguez M, Lassmann H (2000) Heterogeneity of multiple sclerosis lesions: implications for the pathogenesis of demyelination. Ann Neurol 47(6):707–717

45. Lindquist S, Bodammer N, Kaufmann J, Konig F, Heinze HJ, Bruck W et al (2007) Histopathology and serial, multimodal magnetic resonance imaging in a multiple sclerosis variant. Mult Scler 13(4):471–482

46. Yang XO, Nurieva R, Martinez GJ, Kang HS, Chung Y, Pappu BP et al (2008) Molecular antagonism and plasticity of regulatory and inflammatory T cell programs. Immunity 29(1):44–56

47. Nikolaeva MA, Mukherjee B, Stys PK (2005) Na+−dependent sources of intra-axonal Ca^{2+} release in rat optic nerve during in vitro chemical ischemia. J Neurosci 25(43):9960–9967

48. Stys PK, Waxman SG, Ransom BR (1992) Ionic mechanisms of anoxic injury in mammalian CNS white matter: role of Na^+ channels and $Na(+)-Ca^{2+}$ exchanger. J Neurosci 12(2):430–439

49. Stys PK, Jiang Q (2002) Calpain-dependent neurofilament breakdown in anoxic and ischemic rat central axons. Neurosci Lett 328(2):150–154

50. Stys PK, Ransom BR, Waxman SG, Davis PK (1990) Role of extracellular calcium in anoxic injury of mammalian central white matter. Proc Natl Acad Sci U S A 87(11):4212–4216

51. Nave KA, Trapp BD (2008) Axon-glial signaling and the glial support of axon function. Annu Rev Neurosci 31:535–561

52. Trapp BD, Nave KA (2008) Multiple sclerosis: an immune or neurodegenerative disorder? Annu Rev Neurosci 31:247–269

53. Wilkins A, Chandran S, Compston A (2001) A role for oligodendrocyte-derived IGF-1 in trophic support of cortical neurons. Glia 36 (1):48–57

54. Wilkins A, Majed H, Layfield R, Compston A, Chandran S (2003) Oligodendrocytes promote neuronal survival and axonal length by distinct intracellular mechanisms: a novel role for oligodendrocyte-derived glial cell line-derived neurotrophic factor. J Neurosci 23 (12):4967–4974

55. Griffiths I, Klugmann M, Anderson T, Yool D, Thomson C, Schwab MH et al (1998) Axonal swellings and degeneration in mice lacking the major proteolipid of myelin. Science 280(5369):1610–1613

56. Barnett MH, Prineas JW (2004) Relapsing and remitting multiple sclerosis: pathology of the newly forming lesion. Ann Neurol 55 (4):458–468

57. Waiczies S, Prozorovski T, Infante-Duarte C, Hahner A, Aktas O, Ullrich O et al (2005) Atorvastatin induces T cell anergy via phosphorylation of ERK1. J Immunol 174 (9):5630–5635

58. Albert MA, Danielson E, Rifai N, Ridker PM, Investigators P (2001) Effect of statin therapy on C-reactive protein levels: the pravastatin inflammation/CRP evaluation (PRINCE): a randomized trial and cohort study. JAMA 286 (1):64–70

59. Kobashigawa JA, Katznelson S, Laks H, Johnson JA, Yeatman L, Wang XM et al (1995) Effect of pravastatin on outcomes after cardiac transplantation. N Engl J Med 333 (10):621–627

60. Kwak B, Mulhaupt F, Myit S, Mach F (2000) Statins as a newly recognized type of immunomodulator. Nat Med 6(12):1399–1402

61. Stanislaus R, Pahan K, Singh AK, Singh I (1999) Amelioration of experimental allergic encephalomyelitis in Lewis rats by lovastatin. Neurosci Lett 269(2):71–74

62. Youssef S, Stuve O, Patarroyo JC, Ruiz PJ, Radosevich JL, Hur EM et al (2002) The HMG-CoA reductase inhibitor, atorvastatin, promotes a Th2 bias and reverses paralysis in central nervous system autoimmune disease. Nature 420(6911):78–84

63. Paul F, Waiczies S, Wuerfel J, Bellmann-Strobl J, Dorr J, Waiczies H et al (2008) Oral high-dose atorvastatin treatment in relapsing-remitting multiple sclerosis. PLoS One 3(4):e1928

64. Togha M, Karvigh SA, Nabavi M, Moghadam NB, Harirchian MH, Sahraian MA et al (2010) Simvastatin treatment in patients with relapsing-remitting multiple sclerosis receiving interferon beta 1a: a double-blind randomized controlled trial. Mult Scler 16 (7):848–854

65. Garcia-Roman N, Alvarez AM, Toro MJ, Montes A, Lorenzo MJ (2001) Lovastatin induces apoptosis of spontaneously immortalized rat brain neuroblasts: involvement of nonsterol isoprenoid biosynthesis inhibition. Mol Cell Neurosci 17(2):329–341

66. Bosel J, Gandor F, Harms C, Synowitz M, Harms U, Djoufack PC et al (2005) Neuroprotective effects of atorvastatin against glutamate-induced excitotoxicity in primary cortical neurones. J Neurochem 92 (6):1386–1398

67. Baeuerle PA, Henkel T (1994) Function and activation of NF-kappa B in the immune system. Annu Rev Immunol 12:141–179

68. Li Q, Verma IM (2002) NF-kappaB regulation in the immune system. Nat Rev Immunol 2(10):725–734

69. Viatour P, Merville MP, Bours V, Chariot A (2005) Phosphorylation of NF-kappaB and IkappaB proteins: implications in cancer and inflammation. Trends Biochem Sci 30 (1):43–52

70. Hilliard B, Samoilova EB, Liu TS, Rostami A, Chen Y (1999) Experimental autoimmune encephalomyelitis in NF-kappa B-deficient mice:roles of NF-kappa B in the activation and differentiation of autoreactive T cells. J Immunol 163(5):2937–2943

71. Pahan K, Schmid M (2000) Activation of nuclear factor-kB in the spinal cord of experimental allergic encephalomyelitis. Neurosci Lett 287(1):17–20

72. van Loo G, De Lorenzi R, Schmidt H, Huth M, Mildner A, Schmidt-Supprian M et al (2006) Inhibition of transcription factor NF-kappaB in the central nervous system ameliorates autoimmune encephalomyelitis in mice. Nat Immunol 7(9):954–961

73. Gveric D, Kaltschmidt C, Cuzner ML, Newcombe J (1998) Transcription factor NF-kappaB and inhibitor I kappaBalpha are localized in macrophages in active multiple sclerosis lesions. J Neuropathol Exp Neurol 57(2):168–178

74. Bonetti B, Stegagno C, Cannella B, Rizzuto N, Moretto G, Raine CS (1999)

Activation of NF-kappaB and c-jun transcription factors in multiple sclerosis lesions. Implications for oligodendrocyte pathology. Am J Pathol 155(5):1433–1438

75. Lock C, Hermans G, Pedotti R, Brendolan A, Schadt E, Garren H et al (2002) Gene-microarray analysis of multiple sclerosis lesions yields new targets validated in autoimmune encephalomyelitis. Nat Med 8 (5):500–508

76. Dasgupta S, Jana M, Zhou Y, Fung YK, Ghosh S, Pahan K (2004) Antineuroinflammatory effect of NF-kappaB essential modifier-binding domain peptides in the adoptive transfer model of experimental allergic encephalomyelitis. J Immunol 173 (2):1344–1354

77. Aktas O, Prozorovski T, Smorodchenko A, Savaskan NE, Lauster R, Kloetzel PM et al (2004) Green tea epigallocatechin-3-gallate mediates T cellular NF-kappa B inhibition and exerts neuroprotection in autoimmune encephalomyelitis. J Immunol 173 (9):5794–5800

78. Sethi G, Ahn KS, Aggarwal BB (2008) Targeting nuclear factor-kappa B activation pathway by thymoquinone: role in suppression of antiapoptotic gene products and enhancement of apoptosis. Mol Cancer Res 6 (6):1059–1070

79. Mohamed A, Afridi DM, Garani O, Tucci M (2005) Thymoquinone inhibits the activation of NF-kappaB in the brain and spinal cord of experimental autoimmune encephalomyelitis. Biomed Sci Instrum 41:388–393

80. Cohen JA, Chun J (2011) Mechanisms of fingolimod's efficacy and adverse effects in multiple sclerosis. Ann Neurol 69 (5):759–777

81. Graeler M, Goetzl EJ (2002) Activation-regulated expression and chemotactic function of sphingosine 1-phosphate receptors in mouse splenic T cells. FASEB J 16 (14):1874–1878

82. Jaillard C, Harrison S, Stankoff B, Aigrot MS, Calver AR, Duddy G et al (2005) Edg8/S1P5: an oligodendroglial receptor with dual function on process retraction and cell survival. J Neurosci 25(6):1459–1469

83. Yamagata K, Tagami M, Torii Y, Takenaga F, Tsumagari S, Itoh S et al (2003) Sphingosine 1-phosphate induces the production of glial cell line-derived neurotrophic factor and cellular proliferation in astrocytes. Glia 41 (2):199–206

84. Herr DR, Chun J (2007) Effects of LPA and S1P on the nervous system and implications for their involvement in disease. Curr Drug Targets 8(1):155–167

85. Foster CA, Howard LM, Schweitzer A, Persohn E, Hiestand PC, Balatoni B et al (2007) Brain penetration of the oral immuno-modulatory drug FTY720 and its phosphorylation in the central nervous system during experimental autoimmune encephalomyelitis: consequences for mode of action in multiple sclerosis. J Pharmacol Exp Ther 323 (2):469–475

86. Oo ML, Thangada S, Wu MT, Liu CH, Macdonald TL, Lynch KR et al (2007) Immuno-suppressive and anti-angiogenic sphingosine 1-phosphate receptor-1 agonists induce ubiquitinylation and proteasomal degradation of the receptor. J Biol Chem 282 (12):9082–9089

87. Webb M, Tham CS, Lin FF, Lariosa-Willingham K, Yu N, Hale J et al (2004) Sphingosine 1-phosphate receptor agonists attenuate relapsing-remitting experimental autoimmune encephalitis in SJL mice. J Neuroimmunol 153(1–2):108–121

88. Cohen JA, Barkhof F, Comi G, Hartung HP, Khatri BO, Montalban X et al (2010) Oral fingolimod or intramuscular interferon for relapsing multiple sclerosis. N Engl J Med 362(5):402–415

89. Kappos L, Radue EW, O'Connor P, Polman C, Hohlfeld R, Calabresi P et al (2010) A placebo-controlled trial of oral fingolimod in relapsing multiple sclerosis. N Engl J Med 362(5):387–401

90. Tavares A, Barret O, Alagile D (2014) Brain distribution of MS565, an imaging analogue of siponimod (BAF312), in non-human primates. Neurology 82(10 Supplement)

91. Kappos L, Bar-Or A, Cree BAC, Fox RJ, Giovannoni G, Gold R et al (2018) Siponimod versus placebo in secondary progressive multiple sclerosis (EXPAND): a double-blind, randomised, phase 3 study. Lancet 391 (10127):1263–1273

92. Lemberger T, Desvergne B, Wahli W (1996) Peroxisome proliferator-activated receptors: a nuclear receptor signaling pathway in lipid physiology. Annu Rev Cell Dev Biol 12:335–363

93. Feinstein DL, Galea E, Gavrilyuk V, Brosnan CF, Whitacre CC, Dumitrescu-Ozimek L et al (2002) Peroxisome proliferator-activated receptor-gamma agonists prevent experimental autoimmune encephalomyelitis. Ann Neurol 51(6):694–702

94. Clark RB, Bishop-Bailey D, Estrada-Hernandez T, Hla T, Puddington L, Padula SJ (2000) The nuclear receptor PPAR gamma and immunoregulation: PPAR gamma mediates inhibition of helper T cell responses. J Immunol 164(3):1364–1371

95. Natarajan C, Muthian G, Barak Y, Evans RM, Bright JJ (2003) Peroxisome proliferator-activated receptor-gamma-deficient heterozygous mice develop an exacerbated neural antigen-induced Th1 response and experimental allergic encephalomyelitis. J Immunol 171(11):5743–5750

96. Niino M, Iwabuchi K, Kikuchi S, Ato M, Morohashi T, Ogata A et al (2001) Amelioration of experimental autoimmune encephalomyelitis in C57BL/6 mice by an agonist of peroxisome proliferator-activated receptor-gamma. J Neuroimmunol 116(1):40–48

97. Diab A, Deng C, Smith JD, Hussain RZ, Phanavanh B, Lovett-Racke AE et al (2002) Peroxisome proliferator-activated receptor-gamma agonist 15-deoxy-Delta(12,14)-prostaglandin J(2) ameliorates experimental autoimmune encephalomyelitis. J Immunol 168(5):2508–2515

98. Natarajan C, Bright JJ (2002) Peroxisome proliferator-activated receptor-gamma agonists inhibit experimental allergic encephalomyelitis by blocking IL-12 production, IL-12 signaling and Th1 differentiation. Genes Immun 3(2):59–70

99. Diab A, Hussain RZ, Lovett-Racke AE, Chavis JA, Drew PD, Racke MK (2004) Ligands for the peroxisome proliferator-activated receptor-gamma and the retinoid X receptor exert additive anti-inflammatory effects on experimental autoimmune encephalomyelitis. J Neuroimmunol 148(1–2):116–126

100. Bernardo A, Bianchi D, Magnaghi V, Minghetti L (2009) Peroxisome proliferator-activated receptor-gamma agonists promote differentiation and antioxidant defenses of oligodendrocyte progenitor cells. J Neuropathol Exp Neurol 68(7):797–808

101. Schmidt S, Moric E, Schmidt M, Sastre M, Feinstein DL, Heneka MT (2004) Anti-inflammatory and antiproliferative actions of PPAR-gamma agonists on T lymphocytes derived from MS patients. J Leukoc Biol 75(3):478–485

102. Shukla DK, Kaiser CC, Stebbins GT, Feinstein DL (2010) Effects of pioglitazone on diffusion tensor imaging indices in multiple sclerosis patients. Neurosci Lett 472(3):153–156

103. Heneka MT, Landreth GE, Hull M (2007) Drug insight: effects mediated by peroxisome proliferator-activated receptor-gamma in CNS disorders. Nat Clin Pract Neurol 3(9):496–504

104. Kaiser CC, Shukla DK, Stebbins GT, Skias DD, Jeffery DR, Stefoski D et al (2009) A pilot test of pioglitazone as an add-on in patients with relapsing remitting multiple

sclerosis. J Neuroimmunol 211(1–2):124–130

105. Pershadsingh HA, Heneka MT, Saini R, Amin NM, Broeske DJ, Feinstein DL (2004) Effect of pioglitazone treatment in a patient with secondary multiple sclerosis. J Neuroinflammation 1(1):3

106. Yong VW (2005) Metalloproteinases: mediators of pathology and regeneration in the CNS. Nat Rev Neurosci 6(12):931–944

107. Correale J (2003) Bassani Molinas Mde L. temporal variations of adhesion molecules and matrix metalloproteinases in the course of MS. J Neuroimmunol 140(1–2):198–209

108. Brundula V, Rewcastle NB, Metz LM, Bernard CC, Yong VW (2002) Targeting leukocyte MMPs and transmigration: minocycline as a potential therapy for multiple sclerosis. Brain 125(Pt 6):1297–1308

109. Metz LM, Zhang Y, Yeung M, Patry DG, Bell RB, Stoian CA et al (2004) Minocycline reduces gadolinium-enhancing magnetic resonance imaging lesions in multiple sclerosis. Ann Neurol 55(5):756

110. Zabad RK, Metz LM, Todoruk TR, Zhang Y, Mitchell JR, Yeung M et al (2007) The clinical response to minocycline in multiple sclerosis is accompanied by beneficial immune changes: a pilot study. Mult Scler 13(4):517–526

111. Sørensen PS, Sellebjerg F, Lycke J, Färkkilä M, Créange A, Lund CG, Schluep M, Frederiksen JL, Stenager E, Pfleger C, Garde E, Kinnunen E, Marhardt K; RECYCLINE Study Investigators. (2016) Minocycline added to subcutaneous interferon β-1a in multiple sclerosis: randomized RECYCLINE study. Eur J Neurol 23(5):861–870. https://doi.org/10.1111/ene.12953

112. Black JA, Liu S, Hains BC, Saab CY, Waxman SG (2006) Long-term protection of central axons with phenytoin in monophasic and chronic-relapsing EAE. Brain 129(Pt 12):3196–3208

113. Black JA, Liu S, Carrithers M, Carrithers LM, Waxman SG (2007) Exacerbation of experimental autoimmune encephalomyelitis after withdrawal of phenytoin and carbamazepine. Ann Neurol 62(1):21–33

114. Bechtold DA, Kapoor R, Smith KJ (2004) Axonal protection using flecainide in experimental autoimmune encephalomyelitis. Ann Neurol 55(5):607–616

115. Bechtold DA, Miller SJ, Dawson AC, Sun Y, Kapoor R, Berry D et al (2006) Axonal protection achieved in a model of multiple sclerosis using lamotrigine. J Neurol 253(12):1542–1551

116. Kapoor R, Furby J, Hayton T, Smith KJ, Altmann DR, Brenner R et al (2010) Lamotrigine for neuroprotection in secondary progressive multiple sclerosis: a randomised, double-blind, placebo-controlled, parallel-group trial. Lancet Neurol 9(7):681–688

117. Craner MJ, Damarjian TG, Liu S, Hains BC, Lo AC, Black JA et al (2005) Sodium channels contribute to microglia/macrophage activation and function in EAE and MS. Glia 49(2):220–229

118. Kapoor R, Davies M, Blaker PA, Hall SM, Smith KJ (2003) Blockers of sodium and calcium entry protect axons from nitric oxide-mediated degeneration. Ann Neurol 53 (2):174–180

119. Brand-Schieber E, Werner P (2004) Calcium channel blockers ameliorate disease in a mouse model of multiple sclerosis. Exp Neurol 189(1):5–9

120. Clark AJ, Ware MA, Yazer E, Murray TJ, Lynch ME (2004) Patterns of cannabis use among patients with multiple sclerosis. Neurology 62(11):2098–2100

121. Pryce G, Baker D (2005) Emerging properties of cannabinoid medicines in management of multiple sclerosis. Trends Neurosci 28 (5):272–276

122. Fujiwara M, Egashira N (2004) New perspectives in the studies on endocannabinoid and cannabis: abnormal behaviors associate with CB1 cannabinoid receptor and development of therapeutic application. J Pharmacol Sci 96 (4):362–366

123. Kreitzer AC, Regehr WG (2001) Cerebellar depolarization-induced suppression of inhibition is mediated by endogenous cannabinoids. J Neurosci 21(20):RC174

124. Pryce G, Ahmed Z, Hankey DJ, Jackson SJ, Croxford JL, Pocock JM et al (2003) Cannabinoids inhibit neurodegeneration in models of multiple sclerosis. Brain 126 (Pt 10):2191–2202

125. Eljaschewitsch E, Witting A, Mawrin C, Lee T, Schmidt PM, Wolf S et al (2006) The endocannabinoid anandamide protects neurons during CNS inflammation by induction of MKP-1 in microglial cells. Neuron 49 (1):67–79

126. Ligresti A, Cascio MG, Pryce G, Kulasegram S, Beletskaya I, De Petrocellis L et al (2006) New potent and selective inhibitors of anandamide reuptake with antispastic activity in a mouse model of multiple sclerosis. Br J Pharmacol 147(1):83–91

127. Confavreux C, Hutchinson M, Hours MM, Cortinovis-Tourniaire P, Moreau T (1998) Rate of pregnancy-related relapse in multiple sclerosis. Pregnancy in multiple sclerosis group. N Engl J Med 339(5):285–291

128. Vukusic S, Confavreux C (2006) Pregnancy and multiple sclerosis: the children of PRIMS. Clin Neurol Neurosurg 108(3):266–270

129. Zang YC, Halder JB, Hong J, Rivera VM, Zhang JZ (2002) Regulatory effects of estriol on T cell migration and cytokine profile: inhibition of transcription factor NF-kappa B. J Neuroimmunol 124(1–2):106–114

130. Zhu WH, Lu CZ, Huang YM, Link H, Xiao BG (2007) A putative mechanism on remission of multiple sclerosis during pregnancy: estrogen-induced indoleamine 2,3-dioxygenase by dendritic cells. Mult Scler 13(1):33–40

131. Sicotte NL, Liva SM, Klutch R, Pfeiffer P, Bouvier S, Odesa S et al (2002) Treatment of multiple sclerosis with the pregnancy hormone estriol. Ann Neurol 52(4):421–428

132. Fontoura P (2010) Monoclonal antibody therapy in multiple sclerosis: paradigm shifts and emerging challenges. MAbs 2 (6):670–681

133. Yednock TA, Cannon C, Fritz LC, Sanchez-Madrid F, Steinman L, Karin N (1992) Prevention of experimental autoimmune encephalomyelitis by antibodies against alpha 4 beta 1 integrin. Nature 356(6364):63–66

134. Polman CH, O'Connor PW, Havrdova E, Hutchinson M, Kappos L, Miller DH et al (2006) A randomized, placebo-controlled trial of natalizumab for relapsing multiple sclerosis. N Engl J Med 354(9):899–910

135. Rudick RA, Stuart WH, Calabresi PA, Confavreux C, Galetta SL, Radue EW et al (2006) Natalizumab plus interferon beta-1a for relapsing multiple sclerosis. N Engl J Med 354(9):911–923

136. Hunt D, Giovannoni G (2012) Natalizumab-associated progressive multifocal leucoencephalopathy: a practical approach to risk profiling and monitoring. Pract Neurol 12 (1):25–35

137. Miller DH, Weber T, Grove R, Wardell C, Horrigan J, Graff O et al (2012) Firategrast for relapsing remitting multiple sclerosis: a phase 2, randomised, double-blind, placebo-controlled trial. Lancet Neurol 11 (2):131–139

Chapter 19

Translational Aspects of Immunotherapeutic Targets in Multiple Sclerosis

Vinzenz Fleischer

Abstract

In this chapter, selected immunopharmacological targets that are currently part of multiple sclerosis (MS) therapy are described, with particular attention paid to translational aspects in the development and real-world use of the drugs. The therapeutics are classified based on the primary way in which they affect the immune system, bearing in mind, however, that most of the drugs exhibit pleiotropic effects. Although there are differences between the pathophysiology of the animal model of MS, called experimental autoimmune encephalomyelitis (EAE), and the human disease MS, EAE has held its ground in neuroimmunological research and is commonly accepted as the most appropriate animal model for the preclinical development of therapies for MS. Here, we review selected approved therapies with regard to their translational role in advancing MS research. In particular, we highlight the mode of action within MS pathophysiology and show the contribution of animal models to a better understanding of drug mechanisms. Finally, we touch on the clinical efficacy and safety profile of the drugs demonstrated in large and pivotal phase III clinical trials.

Key words Immunotherapy, Mode of action, Phase III clinical trials, Efficacy, Safety

---

1 Introduction

With advances in our understanding of both immune and pathophysiological processes in MS, new treatment approaches continue to be developed [1]. Some have been derived from therapeutic approaches to similar autoimmune disease entities, while others have passed through the complete translational process, from animal studies to pivotal phase III clinical trials [2]. All drugs target the immune system, but with various mechanisms of action and different modes of application, as well as distinct clinical efficacies and side effects. The established injectable medications exhibit a rather multifaceted effect on the immune system. Glatiramer acetate (GA) and interferon beta (IFN-β) are characterized by their only moderate effect on disease activity but also by their relatively safe risk profile, based on the long clinical experience [3, 4]. Another

Sergiu Groppa and Sven G. Meuth (eds.), *Translational Methods for Multiple Sclerosis Research*, Neuromethods, vol. 166, https://doi.org/10.1007/978-1-0716-1213-2_19, © Springer Science+Business Media, LLC, part of Springer Nature 2021

more specific way of guiding the misguided immune system is the inhibition of leukocyte trafficking leading to compartmentaliza- tion. Natalizumab elegantly closes the blood–brain barrier, stop- ping the transmigration of immune cells across the vascular endothelium into the CNS and very effectively controlling the disease [5]. However, patients need to take the risk of potentially lethal progressive multifocal leukoencephalopathy (PML) during treatment into account. Fingolimod and siponimod inhibits leuko- cyte trafficking by specifically capturing lymphocytes within the peripheral lymph nodes manifesting in marked lymphopenia which in turn opens the doors for potentially fatal opportunistic infections [6]. Another promising treatment approach uses lym- phocyte depletion, through either continuous or interval depletion. Both teriflunomide and dimethyl fumarate are drugs that are not entirely new for clinicians as they originate from the treatment of rheumatoid arthritis and psoriasis, respectively. Both oral drugs have the advantage that the adverse risk profile is known and can be roughly transferred to MS patients. The reduction of disease activity is comparable between the two drugs and is of average efficacy compared to other MS drugs [7, 8]. Finally, the novel interval depletion treatment approaches with ocrelizumab, ofatu- mumab amd alemtuzumab are only readministered every 6 or 12 months, respectively, meaning fewer hospital appointments for patients [9, 10]. In addition, oral cladribine given as short courses over two annual cycles also depletes lymphocytes. However, the effects of these drugs on the immune system are strong, as shown by a marked reduction of lymphocytes and their subpopulations in the peripheral blood of patients. Nevertheless, ocrelizumab was the first drug approved for the progressive phase of the disease [11].

2   Multifaceted Immunomodulation

2.1   Glatiramer Acetate (GA)

GA is a synthetic copolymer that is composed of four amino acids (glutamate, lysine, alanine, and tyrosine) in a specific ratio. Initially this drug was developed as a synthetic analogue of the myelin basic protein (MBP) to induce experimental autoimmune encephalomy- elitis (EAE). Paradoxically, the administration to animals revealed the exact opposite effect by highly suppressing and even preventing EAE by GA [12]. However, the precise effects modulated by GA are complex and have not been elucidated in detail. On the cellular and humoral levels, GA cross-reacts with MBP-associated peptides as an altered peptide ligand, whereby GA first develops its immu- nomodulatory potential [13]. GA binds to MHC class II molecules on antigen-presenting cells, while competing with MBP for this binding site [14]. As a consequence, the actual T cell response is altered which has been repeatedly shown by the effective suppression of myelin-reactive T cells by GA [15, 16]. Besides its suppression on auto-reactive inflammatory T cells, GA further modulates microglia, macrophages and dendritic cells, and drives them into anti-inflammatory responses [17, 18] suggesting that the effect of GA is multifaceted and may affect multiple cell types. In addition, the impact of GA on B cell mediated pathogenesis within MS has gained attention in recent years. In particular, reduced numbers of B cells have been observed, as well as a shift from a pro-inflammatory to an anti-inflammatory B cell phenotype [19]. Finally, a neuroprotective effect through augmentation of neurotropic factors produced directly by CNS-resident and infiltrating T cells [20] and through remyelination processes [21] have been shown in EAE. The fruitful development of GA from studies in EAE highlights the strength of this model in evaluating potential immunotherapies.

In the pivotal phase III clinical trial, was a 29% reduction of the annualized relapse rate (ARR) in MS patients on GA with daily subcutaneous injections compared to placebo [3]. Despite the promising observations on neuroprotection in the EAE model, GA has no significant effect on clinical disability progression in progressive form of the disease [22]. Due to the good safety profile and the long-term efficacy, GA has been well established in the clinical setting. The side effects include local injection reactions, allergic reactions, and a rare systemic post-injection reaction (flushing) resolving spontaneously and rapidly after injection. One key advantage of GA over other immunomodulatory drugs is its indication for MS patients during pregnancy, since GA has shown no adverse effects on fetal outcomes [23].

2.2 Interferon Beta (IFN-β)

IFN-β is a cytokine that binds to the type-I IFN receptor on human cells and is produced by lymphocytes, macrophages, and endothelial cells [24]. The precise mode of action of IFN-β is still not completely understood, but there are many purported mechanisms of action [25, 26]. IFN-β plays a crucial role in the regulation of the immune system, by the activation of signal-transduction pathways, triggering a wide range of immunomodulatory and antiproliferative effects [27]. These include primarily a downregulation of antigen presenting cells and a reduction of dendritic cells. Furthermore, IFN-β therapy reduces CD4+ and CD8+ T cell reactivity [28] and the differentiation of CD4+ T cells shifts from a Th1 to Th2 phenotype promoting the secretion of more anti-inflammatory cytokines and chemokines [29]. Moreover, the number of IL-17 producing Th17 cells decreases leading to the apoptosis of auto-reactive T cells [2, 24]. In mice, IFN-β could successfully ameliorate neurological symptoms, disease progression and, indeed, the development of EAE [30]. In a series of experiments in mice, it was identified that mice with Th1-induced EAE benefit from IFN-β treatment with a reduction in the degree of disability, whereas mice with Th17-induced EAE do not respond, and their disease worsens [31]. In humans, a high IL-17 concentration in the serum of relapsing-remitting MS (RRMS) patients was associated with non-responsiveness to therapy with IFN-β [31].

IFN-β is available as IFN-β 1a and IFN-β 1b, which differ in their amino-acid composition. Each preparation further differs in terms of the route of administration and frequency of application. In RRMS patients, IFN-β 1b showed a 34% reduction in the ARR [4] and a delay of conversion from clinically isolated syndrome to definite MS [32]. The IFN-β 1a formulations showed similar results with approximately one-third reduction of the relapse rate in comparison to the placebo group [33]. Two studies focused on the progressive phase of the disease and demonstrated conflicting results with respect to the effect of IFN-β on clinical disease progression [34, 35]. Ultimately, IFN-β has also been approved for the secondary progressive phase, but only in MS patients with both EDSS progression and evidence of inflammatory activity. Prominent side effects of IFN-β include allergic reactions, influenza-like symptoms, injection-site reactions, depression and elevated liver enzymes. Notably, due to its immunogenic potential, neutralizing antibodies may occur and were shown to be clinically relevant, as they can reduce the beneficial effect of IFN-β on inflammatory activity [36].

3   Inhibition of Leukocyte Trafficking

3.1   Fingolimod/Siponimod

Fingolimod is a functional antagonist of the sphingosin-1-phosphate receptor (S1PR) and prevents lymphocyte trafficking from secondary lymphoid organs into the systemic blood circulation [37, 38]. Stimulation of S1PR by binding fingolimod results in rapid desensitization, associated in part with depletion of receptor from the cell surface relying on endocytosis of G-protein coupled receptors [39]. This effect is mainly seen in the retention of $CD4^+$ and $CD8^+$ naïve lymphocytes and central memory T cells in lymphoid organs. Studies have also investigated in more detail the lymphocyte subtype profile and have detected that fingolimod increases the number and enhances the functional activity of regulatory T cells in vitro. These fingolimod-treated cells possessed the activity to downregulate the pro-inflammatory reactivity of lymphocytes [40]. Moreover, fingolimod was shown to be effective in preventing the development of MBP-induced EAE in rats [41]. S1PRs are expressed in different densities on endothelial cells, neurons, and glia cells, suggesting an effect on promoting myelin repair [42, 43]. However, the most likely explanation for the effectiveness of this drug is the inhibition of T cell infiltration into the CNS by sequestration of T cells in secondary lymphoid tissue [44].

Fingolimod was the first oral drug that was approved for the treatment of RRMS. Research evidence was generated from two phase III clinical trials, one which compared fingolimod to placebo [6] and the other to IFN-β 1a [45]. The comparison to placebo showed a decrease in the ARR of 50%, whereas the comparison to weekly intramuscular IFN-β 1a showed a relative risk reduction of 52%. Even though total MRI-derived brain atrophy was slower in the fingolimod group over the observation period in these trials, a subsequent trial with fingolimod in primary progressive MS stated no significant effect on clinical disability progression [46].

The first oral application may lead to bradycardia in most patients as a common adverse event of fingolimod. Hence, at the start of treatment, patients are monitored closely within the first 6 h after the first dose in a clinical setting to check for cardiac-rhythm abnormalities. This side effect can be ascribed to the presence of S1PRs on atrial myocytes [47]. Further adverse events include macular edema, disseminated varicella infection, cutaneous malignancies, and opportunistic infections [48]. Owing to the inhibition of lymphocyte trafficking, the range from mild to severe lymphopenia characterizes the blood count under fingolimod treatment, although lymphocyte count is not associated with the level of clinical response [49]. The development of a more specific ligand targeting the S1PR exclusively on lymphoid organs has led to siponimod being selective for the S1PR subset 1 and 5. In a phase III clinical trial in secondary progressive MS patients, siponimod reduced the risk of disability progression with a safety profile similar to that of other S1PR modulators [50].

3.2 Natalizumab

In both, MS and its animal model, circulating leukocytes penetrate the blood–brain barrier and damage myelin, resulting in impaired nerve conduction and paralysis. In the early 1990s, the adhesion receptors that mediate the attachment of circulating leukocytes to the inflamed brain endothelium in EAE were identified [51]. Binding was inhibited by antibodies against the integrin molecule α4 subunit, but not by antibodies against other adhesion receptors. When tested in vivo, anti-α4 integrin effectively prevented the accumulation of leukocytes in the human brain and the development of EAE. Encouraged by these pivotal findings, therapies designed to interfere with α4 integrin were developed. Natalizumab was the first humanized recombinant monoclonal antibody that targets the α4 subunit of the α4 integrins (α4β1 and α4β7). These adhesion molecules are expressed on the surface of circulating peripheral and tissue-resident lymphocytes [52], can be blocked by natalizumab and thus prevent T cells from entering the CNS via the blood–brain barrier [53]. In detail, α4 integrins are glycoproteins that promote cell adhesion and mediate leukocyte rolling on the vascular endothelium prior to extravasation [54, 55]. In summary, the application of the anti-$\alpha 4\beta 1$ antibody natalizumab implicates that auto-aggressive T cells are unable to penetrate the blood–brain barrier and thus unable to attack myelin protein.

Natalizumab has been shown to be highly effective in RRMS, which was depicted in two phase III clinical trials. In comparison to placebo, natalizumab reduced the ARR by 68% [5], whereas the second clinical trial revealed a reduction of 55% with intramuscular IFN-β 1a as an active comparator [56]. Both studies also demonstrated a significant reduction in MRI parameters within the group of MS patients receiving natalizumab, including new or enlarging T2-weighted hyperintense lesions and contrast-enhancing lesions. Despite its efficacy, long-term administration of natalizumab is limited due to a potentially lethal reactivation of the John Cunningham (JC) virus leading to progressive multifocal leukoencephalopathy (PML) resulting from diminished immune surveillance [57]. This severe side effect, which is well known in immune deficient or depressed patients, has led to a temporary withdrawal from the market and a soon come back with a rigorous safety plan to inform patients and doctors about the risk of PML. Testing for JCV serology is recommended every 6 months for patients with negative results due to a yearly seroconversion rate of up to 10% [58]. Finally, the presence of antibodies against natalizumab that can be formed in response to treatment, can lead to a decreased efficacy and also to increased infusion-related adverse events [59]. Hence, natalizumab-treated MS patients with successive disease activity despite treatment or pronounced and ongoing infusion reactions should be tested for the presence of natalizumab antibodies.

4   Continuous Lymphocyte Depletion

4.1   Teriflunomide

Teriflunomide is the active metabolite of the antirheumatic drug leflunomide. Teriflunomide selectively and reversibly inhibits the enzyme dihydroorotate dehydrogenase (DHODH), which is a key mitochondrial enzyme, and mediates de novo pyrimidine synthesis and DNA replication in actively proliferating T and B lymphocytes. As teriflunomide spares resting or hematopoietic cell lines, it can be considered as a cytostatic drug to leukocytes, rather than cytotoxic [60]. A shift to a selective T cell subset composition and receptor repertoire diversity under teriflunomide treatment was proven in patients with MS [61]. In this context, it was further discovered that the inhibition of DHODH corrects metabolic disturbances in T cells, which primarily affects active high-affinity T cell clones suggesting a recovery of an altered T cell receptor repertoire in autoimmunity [62]. In addition, teriflunomide further decreases microglia proliferation and induces IL-10 production, whereas it inhibits the production of the pro-inflammatory IL-17

[63, 64]. Administration of teriflunomide to EAE rats mimicking the inflammatory features of RRMS, improved disease outcomes, delayed disease onset and reduced cumulative disease scores [65]. Moreover, in a virus-induced animal model of MS, teriflunomide showed a reduction of excitotoxicity in the thalamus and basal ganglia by lowering glutamate levels [66].

Teriflunomide was approved in 2013 in Europe for RRMS and is administered orally once daily. One phase III clinical trial had IFN-β 1a as the active comparator [67], whereas the other three phase III trials were placebo-controlled [7, 68, 69]. In comparison to placebo, teriflunomide reduced the ARR by 31–36%. Furthermore, the treatment in patients with a first demyelinating event suggestive of MS showed a risk reduction of 43% in terms of the conversion to clinically definite MS [69].

Taken together, the efficacy of teriflunomide is comparable to that of IFN-β or dimethyl fumarate [67, 70]. Main adverse events include liver enzyme increase, in particular alanine aminotransferase, reversible hair thinning and gastrointestinal side effects; the latter were the most common reason for discontinuation [71]. Due to the enterohepatic recirculation it remains traceable within the blood up to 8 months after cessation. The mechanisms of action and the long turnaround time should be taken into consideration for young women as it has shown to be embryo toxic in animals [72].

4.2 Dimethyl Fumarate (DMF)

Oral dimethyl fumarate (DMF) was approved for the disease-modifying treatment of RRMS by the FDA in 2013 [8], though oral fumaric acid esters have been well known for more than 50 years for the treatment of psoriasis [73]. Anti-inflammatory as well as neuroprotective effects of DMF have been postulated [8, 74]. These include reduction of lymphocyte counts based on an apoptosis-related mechanism and alterations in cytokine profiles [75, 76], impaired migratory activity of immune cells at the blood–brain barrier [77] and activation of the nuclear erythroid 2-related factor 2 (Nrf2) transcriptional pathway leading to an antioxidant response [78, 79]. In DMF-treated MS patients, T and B cell subpopulations are altered by shifting the balance towards a stronger immunocompetence [80, 81]. A pro-tolerogenic cell subset shift was shown to be associated with no evidence of disease activity in MS patients [82]. Moreover, it was observed that DMF responders were not only characterized by a broad and stronger reduction of all lymphocyte subsets, but also by a preferential loss of CD8$^+$ T cells, indicating that they play a key role in the DMF therapeutic mechanism of action [83]. However, a complete explanation of how DMF acts to reduce overall disease activity has not been fully elucidated. Although the use of DMF did not grow out originally from its efficacy in EAE, the formulation was recently found to be effective in attenuating EAE disease severity as well as macrophage infiltration into the CNS [74].

In two randomized placebo-controlled phase III clinical trials, the clinical efficacy of DMF in treating RRMS was demonstrated by reducing the ARR by 53% and 44%, respectively [8, 84]. DMF also reduced the number of new or enlarging T2-weighted hyperintense lesions on cranial MRI. Lymphopenia was one of the most common adverse events. The absolute white blood cell counts decreased over the first year of DMF treatment and then plateaued. Thereby, the absolute lymphocyte counts typically dropped by about 30%, whereas 6% of participants developed lymphopenia grade 3 ($<$500 cells/μl) under DMF treatment [8, 84]. Recently, a study reported that half of the patients treated with DMF over 1 year developed lymphocyte counts below the lower limit of normal, with $CD8^+$ T cells preferentially lost [85]. The reasons underlying a more pronounced lymphopenia remain unclear but should be emphasized because lymphopenia under fumaric ester treatment has been associated with the occurrence of rare cases of PML in both MS and psoriasis patients [86–89]. Further common side effects of DMF include flushing after administration, vomiting, gastrointestinal complaints and increased liver enzymes.

5   Interval Lymphocyte Depletion

5.1   Alemtuzumab

Alemtuzumab is a humanized monoclonal IgG antibody targeting CD52, a surface protein with widely unknown function, which is expressed mainly on circulating B and T cells [2, 90, 91]. Administration of alemtuzumab results in a quick and long-lasting depletion of $CD52^+$ cells by cell-mediated and complement-dependent cytolysis, followed by a slow repopulation arising from unaffected hematopoietic precursor cells [2, 92]. The lowest observed values occur within days after starting treatment. Quantitative and qualitative alterations of the various immune cells are observed, contributing to a rebalancing of autoimmune processes [91]. The repopulation of the immune cell subsets is preceded by fully reemerging B-cells after 6 months, followed by $CD8^+$ T cells after more than 2 years and $CD4^+$ T cells after approximately 5 years [93]. In addition, CD52 is known to be involved in the migration and activation of T lymphocytes [94, 95]. Investigations of alemtuzumab in animal models were hampered by the lack of cross-reactivity between human and mouse CD52. In humanized CD52 transgenic mice, lymphocyte depletion after alemtuzumab was not observable in spleen, lymph nodes, bone marrow, and thymus compared to blood levels [96]. This might be a plausible explanation that alemtuzumab-treated MS patients were still able to exert protective B and T cell responses following vaccination and have revealed a relatively low incidence of severe infections.

The efficacy of alemtuzumab for treating RRMS was shown in two pivotal phase III clinical trials, in which intramuscular IFN-β 1a served as an active comparator [9, 97]. Both trials demonstrated superiority of alemtuzumab with regard to the ARR (55% and 49% reduction in relapses in the respective trials). Also, the secondary MRI outcomes (new or enlarging T2-weighted hyperintense lesions or contrast-enhancing lesions) improved in both studies in comparison to IFN-β 1a. Despite this strong reduction in relapse rate and MRI disease activity, alemtuzumab was not able to prevent clinical disease progression in secondary progressive MS patients [93]. The safety profile of alemtuzumab has been investigated over several years and is mainly characterized by secondary autoimmune disorders typically arising during reconstitution of the lymphocyte repertoire. Thus, besides common infusion-related reactions, upcoming autoimmune-mediated adverse events raised concerns in treating physicians and patients. The most common adverse events affect the thyroid (more than 30%), serious idiopathic thrombocytopenia (about 1%) and rare cases of anti-glomerular basement membrane nephropathy [97].

5.2 Ocrelizumab

For a long time, T cells were thought to be the driving immune cells in the immunopathogenesis of MS [98]. However, the presence of oligoclonal bands in the cerebrospinal fluid from MS patients as well as the detection of B cells in demyelinating plaques have indicated that B cells have a more prominent role in MS pathogenesis beyond T cell-mediated inflammation [99]. The beneficial effects of B cell depletion in MS patients produced by anti-CD20 therapies (rituximab and ocrelizumab) substantiated this observation [10, 11, 100, 101]. B cells contribute to the pathogenesis through antibody production [102], antigen presentation [103] and secretion of co-stimulatory cytokines, presumably promoting the pro-inflammatory differentiation of responding T cells [104]. In particular, B cells were also considered as a source of cytokines which diffuse from the meninges into the cortex forming ectopic lymphoid follicles [105]. The hypothesis that B cell antigen presentation is important for disease pathology was strengthened in mice, as the capacity to induce disease in the EAE mouse model was shown to require B cell expression of MHC class II [106]. CD20 is expressed on maturing B cells just before the cells become terminally differentiated plasma cells. Hence, monoclonal antibodies against CD20 deplete immature and mature B cells but spare plasma cells and hematopoietic stem cells [107].

Having seen a beneficial effect of the successor rituximab [100], the more humanized CD20 antibody ocrelizumab was tested against intramuscular IFN-β 1a in RRMS in a large phase III clinical trial [10]. Ocrelizumab has demonstrated a reduction in the frequency of clinical relapses up to 46% in comparison to the active comparator. Moreover, a placebo-controlled phase III clinical trial among patients with primary progressive MS revealed that ocrelizumab was associated with lower rates of clinical and MRI progression than placebo [11]. The safety profile includes mainly infusion-associated reactions, in particular after the first infusion. Ocrelizumab is administered intravenously every 6 months, during which time the majority of patients are continuously depleted of blood B cells during treatment. However, a translational approach in the animal model has unveiled that distinct subpopulations of B cells differ in their sensitivity to anti-CD20 treatment, suggesting that differentiated B cells persisting in secondary lymphoid organs contribute to the recovering B cells [108].

6 Conclusion

In summary, many immunomodulatory drugs in MS have emerged from the EAE model before being translated into humans. Conversely, pathophysiological questions and observations from clinical trials have been translated back into animal models to gain a deeper understanding of mechanisms of action and to figure out causal chains for unsuspected adverse events. There have been surprises along the way, starting with the unexpected therapeutic effect of GA in the animal model of MS at a time when MS drug development was still in its infancy, and ending up with the successful anti-CD20 therapy. Even though the latter has even been effective in patients with progressive MS, a key unanswered question is how to treat disease progression that occurs as a result of relentless neurodegeneration. Over the last two decades, MS has moved from a disease with limited therapeutic options to one with many treatment possibilities, but also still many unexplored roads ahead. Translational research approaches may facilitate the application of basic scientific discoveries in clinical settings (and back) to prevent and treat human diseases like MS.

References

1. Baecher-Allan C, Kaskow BJ, Weiner HL (2018) Multiple sclerosis: mechanisms and immunotherapy. Neuron 97(4):742–768

2. Rommer PS, Milo R, Han MH, Satyanarayan S, Sellner J, Hauer L et al (2019) Immunological aspects of approved MS therapeutics. Front Immunol 10:1564

3. Johnson KP, Brooks BR, Cohen JA, Ford CC, Goldstein J, Lisak RP et al (1995) Copolymer 1 reduces relapse rate and improves disability in relapsing-remitting multiple sclerosis: results of a phase III multicenter, double-blind placebo-controlled trial. The copolymer 1 multiple sclerosis study group. Neurology 45(7):1268–1276

4. (1995) Interferon beta-1b in the treatment of multiple sclerosis: final outcome of the randomized controlled trial. The IFNB Multiple Sclerosis Study Group and The University of British Columbia MS/MRI Analysis Group. Neurology 45(7):1277–1285

5. Polman CH, O'Connor PW, Havrdova E, Hutchinson M, Kappos L, Miller DH et al (2006) A randomized, placebo-controlled trial of natalizumab for relapsing multiple sclerosis. N Engl J Med 354(9):899–910

6. Calabresi PA, Radue EW, Goodin D, Jeffery D, Rammohan KW, Reder AT et al (2014) Safety and efficacy of fingolimod in patients with relapsing-remitting multiple sclerosis (FREEDOMS II): a double-blind, randomised, placebo-controlled, phase 3 trial. Lancet Neurol 13(6):545–556

7. O'Connor P, Wolinsky JS, Confavreux C, Comi G, Kappos L, Olsson TP et al (2011) Randomized trial of oral teriflunomide for relapsing multiple sclerosis. N Engl J Med 365(14):1293–1303

8. Gold R, Kappos L, Arnold DL, Bar-Or A, Giovannoni G, Selmaj K et al (2012) Placebo-controlled phase 3 study of oral BG-12 for relapsing multiple sclerosis. N Engl J Med 367(12):1098–1107

9. Cohen JA, Coles AJ, Arnold DL, Confavreux C, Fox EJ, Hartung HP et al (2012) Alemtuzumab versus interferon beta 1a as first-line treatment for patients with relapsing-remitting multiple sclerosis: a randomised controlled phase 3 trial. Lancet 380 (9856):1819–1828

10. Hauser SL, Bar-Or A, Comi G, Giovannoni G, Hartung HP, Hemmer B et al (2017) Ocrelizumab versus interferon Beta-1a in relapsing multiple sclerosis. N Engl J Med 376(3):221–234

11. Montalban X, Hauser SL, Kappos L, Arnold DL, Bar-Or A, Comi G et al (2017) Ocrelizumab versus placebo in primary progressive multiple sclerosis. N Engl J Med 376 (3):209–220

12. Teitelbaum D, Arnon R, Sela M (1997) Copolymer 1: from basic research to clinical application. Cell Mol Life Sci 53(1):24–28

13. Dhib-Jalbut S (2003) Glatiramer acetate (Copaxone) therapy for multiple sclerosis. Pharmacol Ther 98(2):245–255

14. Fridkis-Hareli M, Teitelbaum D, Gurevich E, Pecht I, Brautbar C, Kwon OJ et al (1994) Direct binding of myelin basic protein and synthetic copolymer 1 to class II major histocompatibility complex molecules on living antigen-presenting cells--specificity and promiscuity. Proc Natl Acad Sci U S A 91 (11):4872–4876

15. Teitelbaum D, Milo R, Arnon R, Sela M (1992) Synthetic copolymer 1 inhibits human T-cell lines specific for myelin basic protein. Proc Natl Acad Sci U S A 89 (1):137–141

16. Teitelbaum D, Fridkis-Hareli M, Arnon R, Sela M (1996) Copolymer 1 inhibits chronic relapsing experimental allergic encephalomyelitis induced by proteolipid protein (PLP) peptides in mice and interferes with PLP-specific T cell responses. J Neuroimmunol 64(2):209–217

17. Vieira PL, Heystek HC, Wormmeester J, Wierenga EA, Kapsenberg ML (2003) Glatiramer acetate (copolymer-1, copaxone) promotes Th2 cell development and increased IL-10 production through modulation of dendritic cells. J Immunol 170 (9):4483–4488

18. Weber MS, Starck M, Wagenpfeil S, Meinl E, Hohlfeld R, Farina C (2004) Multiple sclerosis: glatiramer acetate inhibits monocyte reactivity in vitro and in vivo. Brain 127 (Pt 6):1370–1378

19. Kuerten S, Jackson LJ, Kaye J, Vollmer TL (2018) Impact of glatiramer acetate on B cell-mediated pathogenesis of multiple sclerosis. CNS Drugs 32(11):1039–1051

20. Aharoni R, Eilam R, Domev H, Labunskay G, Sela M, Arnon R (2005) The immunomodulator glatiramer acetate augments the expression of neurotrophic factors in brains of experimental autoimmune encephalomyelitis mice. Proc Natl Acad Sci U S A 102 (52):19045–19050

21. Aharoni R (2014) Immunomodulation neuroprotection and remyelination—the fundamental therapeutic effects of glatiramer acetate: a critical review. J Autoimmun 54:81–92

22. Wolinsky JS, Narayana PA, O'Connor P, Coyle PK, Ford C, Johnson K et al (2007) Glatiramer acetate in primary progressive multiple sclerosis: results of a multinational, multicenter, double-blind, placebo-controlled trial. Ann Neurol 61(1):14–24

23. Sandberg-Wollheim M, Neudorfer O, Grinspan A, Weinstock-Guttman B, Haas J, Izquierdo G et al (2018) Pregnancy outcomes from the branded glatiramer acetate pregnancy database. Int J MS Care 20(1):9–14

24. Haji Abdolvahab M, Mofrad MR, Schellekens H (2016) Interferon Beta: from molecular level to therapeutic effects. Int Rev Cell Mol Biol 326:343–372

25. Benveniste EN, Qin H (2007) Type I interferons as anti-inflammatory mediators. Sci STKE 2007(416):pe70

26. Guo B, Chang EY, Cheng G (2008) The type I IFN induction pathway constrains Th17-mediated autoimmune inflammation in mice. J Clin Invest 118(5):1680–1690

27. Markowitz CE (2007) Interferon-beta: mechanism of action and dosing issues. Neurology 68(24 Suppl 4):S8–S11

28. Zafranskaya M, Oschmann P, Engel R, Weishaupt A, van Noort JM, Jomaa H et al (2007) Interferon-beta therapy reduces CD4

+ and CD8+ T-cell reactivity in multiple sclerosis. Immunology 121(1):29–39

29. Wandinger KP, Sturzebecher CS, Bielekova B, Detore G, Rosenwald A, Staudt LM et al (2001) Complex immunomodulatory effects of interferon-beta in multiple sclerosis include the upregulation of T helper 1-associated marker genes. Ann Neurol 50(3):349–357

30. Yu M, Nishiyama A, Trapp BD, Tuohy VK (1996) Interferon-beta inhibits progression of relapsing-remitting experimental autoimmune encephalomyelitis. J Neuroimmunol 64(1):91–100

31. Axtell RC, de Jong BA, Boniface K, van der Voort LF, Bhat R, De Sarno P et al (2010) T helper type 1 and 17 cells determine efficacy of interferon-beta in multiple sclerosis and experimental encephalomyelitis. Nat Med 16(4):406–412

32. Jacobs LD, Beck RW, Simon JH, Kinkel RP, Brownscheidle CM, Murray TJ et al (2000) Intramuscular interferon beta-1a therapy initiated during a first demyelinating event in multiple sclerosis. CHAMPS study group. N Engl J Med 343(13):898–904

33. Kappos L, Freedman MS, Polman CH, Edan G, Hartung HP, Miller DH et al (2007) Effect of early versus delayed interferon beta-1b treatment on disability after a first clinical event suggestive of multiple sclerosis: a 3-year follow-up analysis of the BENEFIT study. Lancet 370(9585):389–397

34. (1998) Placebo-controlled multicentre randomised trial of interferon beta-1b in treatment of secondary progressive multiple sclerosis. European Study Group on interferon beta-1b in secondary progressive MS. Lancet 352(9139):1491–1497

35. Panitch H, Miller A, Paty D, Weinshenker B (2004) North American study group on interferon beta-1b in secondary progressive MS. interferon beta-1b in secondary progressive MS: results from a 3-year controlled study. Neurology 63(10):1788–1795

36. Hemmer B, Stuve O, Kieseier B, Schellekens H, Hartung HP (2005) Immune response to immunotherapy: the role of neutralising antibodies to interferon beta in the treatment of multiple sclerosis. Lancet Neurol 4(7):403–412

37. Matloubian M, Lo CG, Cinamon G, Lesneski MJ, Xu Y, Brinkmann V et al (2004) Lymphocyte egress from thymus and peripheral lymphoid organs is dependent on S1P receptor 1. Nature 427(6972):355–360

38. Cyster JG, Schwab SR (2012) Sphingosine-1-phosphate and lymphocyte egress from lymphoid organs. Annu Rev Immunol 30:69–94

39. Reeves PM, Kang YL, Kirchhausen T (2016) Endocytosis of ligand-activated sphingosine 1-phosphate receptor 1 mediated by the Clathrin-pathway. Traffic 17(1):40–52

40. Zhou PJ, Wang H, Shi GH, Wang XH, Shen ZJ, Xu D (2009) Immunomodulatory drug FTY720 induces regulatory CD4(+)CD25(+) T cells in vitro. Clin Exp Immunol 157(1):40–47

41. Fujino M, Funeshima N, Kitazawa Y, Kimura H, Amemiya H, Suzuki S et al (2003) Amelioration of experimental autoimmune encephalomyelitis in Lewis rats by FTY720 treatment. J Pharmacol Exp Ther 305(1):70–77

42. Balatoni B, Storch MK, Swoboda EM, Schonborn V, Koziel A, Lambrou GN et al (2007) FTY720 sustains and restores neuronal function in the DA rat model of MOG-induced experimental autoimmune encephalomyelitis. Brain Res Bull 74(5):307–316

43. Foster CA, Howard LM, Schweitzer A, Persohn E, Hiestand PC, Balatoni B et al (2007) Brain penetration of the oral immunomodulatory drug FTY720 and its phosphorylation in the central nervous system during experimental autoimmune encephalomyelitis: consequences for mode of action in multiple sclerosis. J Pharmacol Exp Ther 323(2):469–475

44. Kataoka H, Sugahara K, Shimano K, Teshima K, Koyama M, Fukunari A et al (2005) FTY720, sphingosine 1-phosphate receptor modulator, ameliorates experimental autoimmune encephalomyelitis by inhibition of T cell infiltration. Cell Mol Immunol 2(6):439–448

45. Cohen JA, Barkhof F, Comi G, Hartung HP, Khatri BO, Montalban X et al (2010) Oral fingolimod or intramuscular interferon for relapsing multiple sclerosis. N Engl J Med 362(5):402–415

46. Lublin F, Miller DH, Freedman MS, Cree BAC, Wolinsky JS, Weiner H et al (2016) Oral fingolimod in primary progressive multiple sclerosis (INFORMS): a phase 3, randomised, double-blind, placebo-controlled trial. Lancet 387(10023):1075–1084

47. Camm J, Hla T, Bakshi R, Brinkmann V (2014) Cardiac and vascular effects of fingolimod: mechanistic basis and clinical implications. Am Heart J 168(5):632–644

48. Huwiler A, Zangemeister-Wittke U (2018) The sphingosine 1-phosphate receptor modulator fingolimod as a therapeutic agent: recent findings and new perspectives. Pharmacol Ther 185:34–49

49. Fragoso YD, Spelman T, Boz C, Alroughani R, Lugaresi A, Vucic S et al (2018) Lymphocyte count in peripheral blood is not associated with the level of clinical response to treatment with fingolimod. Mult Scler Relat Disord 19:105–108

50. Kappos L, Bar-Or A, Cree BAC, Fox RJ, Giovannoni G, Gold R et al (2018) Siponimod versus placebo in secondary progressive multiple sclerosis (EXPAND): a double-blind, randomised, phase 3 study. Lancet 391 (10127):1263–1273

51. Yednock TA, Cannon C, Fritz LC, Sanchez-Madrid F, Steinman L, Karin N (1992) Prevention of experimental autoimmune encephalomyelitis by antibodies against alpha 4 beta 1 integrin. Nature 356(6364):63–66

52. Hemler ME (1990) VLA proteins in the integrin family: structures, functions, and their role on leukocytes. Annu Rev Immunol 8:365–400

53. Kivisakk P, Healy BC, Viglietta V, Quintana FJ, Hootstein MA, Weiner HL et al (2009) Natalizumab treatment is associated with peripheral sequestration of proinflammatory T cells. Neurology 72(22):1922–1930

54. Stuve O, Bennett JL (2007) Pharmacological properties, toxicology and scientific rationale for the use of natalizumab (Tysabri) in inflammatory diseases. CNS Drug Rev 13(1):79–95

55. Sheremata WA, Minagar A, Alexander JS, Vollmer T (2005) The role of alpha-4 integrin in the aetiology of multiple sclerosis: current knowledge and therapeutic implications. CNS Drugs 19(11):909–922

56. Rudick RA, Stuart WH, Calabresi PA, Confavreux C, Galetta SL, Radue EW et al (2006) Natalizumab plus interferon beta-1a for relapsing multiple sclerosis. N Engl J Med 354(9):911–923

57. Koralnik IJ (2004) New insights into progressive multifocal leukoencephalopathy. Curr Opin Neurol 17(3):365–370

58. Schwab N, Schneider-Hohendorf T, Pignolet B, Breuer J, Gross CC, Gobel K et al (2016) Therapy with natalizumab is associated with high JCV seroconversion and rising JCV index values. Neurol Neuroimmunol Neuroinflamm 3(1):e195

59. Calabresi PA, Giovannoni G, Confavreux C, Galetta SL, Havrdova E, Hutchinson M et al (2007) The incidence and significance of anti-natalizumab antibodies: results from AFFIRM and SENTINEL. Neurology 69 (14):1391–1403

60. Goodin DS (2014) Multiple sclerosis and related disorders, vol xvii. Elsevier, Edinburgh ; New York, p 715

61. Li L, Liu J, Delohery T, Zhang D, Arendt C, Jones C (2013) The effects of teriflunomide on lymphocyte subpopulations in human peripheral blood mononuclear cells in vitro. J Neuroimmunol 265(1-2):82–90

62. Klotz L, Eschborn M, Lindner M, Liebmann M, Herold M, Janoschka C et al (2019) Teriflunomide treatment for multiple sclerosis modulates T cell mitochondrial respiration with affinity-dependent effects. Sci Transl Med 11(490):eaao5563

63. Wostradowski T, Prajeeth CK, Gudi V, Kronenberg J, Witte S, Brieskorn M et al (2016) In vitro evaluation of physiologically relevant concentrations of teriflunomide on activation and proliferation of primary rodent microglia. J Neuroinflammation 13(1):250

64. Gonzalez-Alvaro I, Ortiz AM, Dominguez-Jimenez C, Aragon-Bodi A, Diaz Sanchez B, Sanchez-Madrid F (2009) Inhibition of tumour necrosis factor and IL-17 production by leflunomide involves the JAK/STAT pathway. Ann Rheum Dis 68(10):1644–1650

65. Merrill JE, Hanak S, Pu SF, Liang J, Dang C, Iglesias-Bregna D et al (2009) Teriflunomide reduces behavioral, electrophysiological, and histopathological deficits in the dark agouti rat model of experimental autoimmune encephalomyelitis. J Neurol 256(1):89–103

66. Modica CM, Schweser F, Sudyn ML, Bertolino N, Preda M, Polak P et al (2017) Effect of teriflunomide on cortex-basal ganglia-thalamus (CxBGTh) circuit glutamatergic dysregulation in the Theiler's murine encephalomyelitis virus mouse model of multiple sclerosis. PLoS One 12(8):e0182729

67. Vermersch P, Czlonkowska A, Grimaldi LM, Confavreux C, Comi G, Kappos L et al (2014) Teriflunomide versus subcutaneous interferon beta-1a in patients with relapsing multiple sclerosis: a randomised, controlled phase 3 trial. Mult Scler 20(6):705–716

68. Confavreux C, O'Connor P, Comi G, Freedman MS, Miller AE, Olsson TP et al (2014) Oral teriflunomide for patients with relapsing multiple sclerosis (TOWER): a randomised, double-blind, placebo-controlled, phase 3 trial. Lancet Neurol 13(3):247–256

69. Miller AE, Wolinsky JS, Kappos L, Comi G, Freedman MS, Olsson TP et al (2014) Oral teriflunomide for patients with a first clinical

episode suggestive of multiple sclerosis (TOPIC): a randomised, double-blind, placebo-controlled, phase 3 trial. Lancet Neurol 13(10):977–986

70. Kalincik T, Kubala Havrdova E, Horakova D, Izquierdo G, Prat A, Girard M et al (2019) Comparison of fingolimod, dimethyl fumarate and teriflunomide for multiple sclerosis. J Neurol Neurosurg Psychiatry 90 (4):458–468

71. Comi G, Freedman MS, Kappos L, Olsson TP, Miller AE, Wolinsky JS et al (2016) Pooled safety and tolerability data from four placebo-controlled teriflunomide studies and extensions. Mult Scler Relat Disord 5:97–104

72. Miller AE (2017) Oral teriflunomide in the treatment of relapsing forms of multiple sclerosis: clinical evidence and long-term experience. Ther Adv Neurol Disord 10 (12):381–396

73. Balak DM, Fallah Arani S, Hajdarbegovic E, Hagemans CA, Bramer WM, Thio HB et al (2016) Efficacy, effectiveness and safety of fumaric acid esters in the treatment of psoriasis: a systematic review of randomized and observational studies. Br J Dermatol 175 (2):250–262

74. Schilling S, Goelz S, Linker R, Luehder F, Gold R (2006) Fumaric acid esters are effective in chronic experimental autoimmune encephalomyelitis and suppress macrophage infiltration. Clin Exp Immunol 145 (1):101–107

75. Treumer F, Zhu K, Glaser R, Mrowietz U (2003) Dimethylfumarate is a potent inducer of apoptosis in human T cells. J Invest Dermatol 121(6):1383–1388

76. Schimrigk S, Brune N, Hellwig K, Lukas C, Bellenberg B, Rieks M et al (2006) Oral fumaric acid esters for the treatment of active multiple sclerosis: an open-label, baseline-controlled pilot study. Eur J Neurol 13 (6):604–610

77. Vandermeeren M, Janssens S, Borgers M, Geysen J (1997) Dimethylfumarate is an inhibitor of cytokine-induced E-selectin, VCAM-1, and ICAM-1 expression in human endothelial cells. Biochem Biophys Res Commun 234(1):19–23

78. Metz I, Traffehn S, Strassburger-Krogias K, Keyvani K, Bergmann M, Nolte K et al (2015) Glial cells express nuclear nrf2 after fumarate treatment for multiple sclerosis and psoriasis. Neurol Neuroimmunol Neuroinflamm 2(3): e99

79. Scannevin RH, Chollate S, Jung MY, Shackett M, Patel H, Bista P et al (2012) Fumarates promote cytoprotection of central nervous system cells against oxidative stress via the nuclear factor (erythroid-derived 2)-like 2 pathway. J Pharmacol Exp Ther 341 (1):274–284

80. Li R, Rezk A, Ghadiri M, Luessi F, Zipp F, Li H et al (2017) Dimethyl fumarate treatment mediates an anti-inflammatory shift in B cell subsets of patients with multiple sclerosis. J Immunol 198(2):691–698

81. Ghadiri M, Rezk A, Li R, Evans A, Luessi F, Zipp F et al (2017) Dimethyl fumarate-induced lymphopenia in MS due to differential T-cell subset apoptosis. Neurol Neuroimmunol Neuroinflamm 4(3):e340

82. Medina S, Villarrubia N, Sainz de la Maza S, Lifante J, Costa-Frossard L, Roldan E et al (2018) Optimal response to dimethyl fumarate associates in MS with a shift from an inflammatory to a tolerogenic blood cell profile. Mult Scler 24(10):1317–1327

83. Fleischer V, Friedrich M, Rezk A, Buhler U, Witsch E, Uphaus T et al (2018) Treatment response to dimethyl fumarate is characterized by disproportionate CD8+ T cell reduction in MS. Mult Scler 24(5):632–641

84. Fox RJ, Miller DH, Phillips JT, Hutchinson M, Havrdova E, Kita M et al (2012) Placebo-controlled phase 3 study of oral BG-12 or glatiramer in multiple sclerosis. N Engl J Med 367(12):1087–1097

85. Spencer CM, Crabtree-Hartman EC, Lehmann-Horn K, Cree BA, Zamvil SS (2015) Reduction of CD8(+) T lymphocytes in multiple sclerosis patients treated with dimethyl fumarate. Neurol Neuroimmunol Neuroinflamm 2(3):e76

86. Longbrake EE, Cross AH (2015) Dimethyl fumarate associated lymphopenia in clinical practice. Mult Scler 21(6):796–797

87. Ermis U, Weis J, Schulz JB (2013) PML in a patient treated with fumaric acid. N Engl J Med 368(17):1657–1658

88. Havrdova E, Hutchinson M, Kurukulasuriya NC, Raghupathi K, Sweetser MT, Dawson KT et al (2012) Oral BG-12 (dimethyl fumarate) for relapsing-remitting multiple sclerosis: a review of DEFINE and CONFIRM. Evaluation of: Gold R, Kappos L, Arnold D, et al. Placebo-controlled phase 3 study of oral BG-12 for relapsing multiple sclerosis. N Engl J Med 367:1098–1107; and Fox RJ, Miller DH, Phillips JT, et al. Placebo-controlled phase 3 study of oral BG-12 or glatiramer in multiple sclerosis. N Engl J Med 2012;367:1087–97. Expert opinion on pharmacotherapy. 2013;14(15):2145–56

89. Rosenkranz T, Novas M, Terborg C (2015) PML in a patient with lymphocytopenia treated with dimethyl fumarate. N Engl J Med 372(15):1476–1478

90. Havrdova E, Horakova D, Kovarova I (2015) Alemtuzumab in the treatment of multiple sclerosis: key clinical trial results and considerations for use. Ther Adv Neurol Disord 8 (1):31–45

91. Ruck T, Bittner S, Wiendl H, Meuth SG (2015) Alemtuzumab in multiple sclerosis: mechanism of action and beyond. Int J Mol Sci 16(7):16414–16439

92. Li Z, Richards S, Surks HK, Jacobs A, Panzara MA (2018) Clinical pharmacology of alemtuzumab, an anti-CD52 immunomodulator, in multiple sclerosis. Clin Exp Immunol 194 (3):295–314

93. Coles AJ, Cox A, Le Page E, Jones J, Trip SA, Deans J et al (2006) The window of therapeutic opportunity in multiple sclerosis: evidence from monoclonal antibody therapy. J Neurol 253(1):98–108

94. Rowan WC, Hale G, Tite JP, Brett SJ (1995) Cross-linking of the CAMPATH-1 antigen (CD52) triggers activation of normal human T lymphocytes. Int Immunol 7(1):69–77

95. Masuyama J, Yoshio T, Suzuki K, Kitagawa S, Iwamoto M, Kamimura T et al (1999) Characterization of the 4C8 antigen involved in transendothelial migration of CD26(hi) T cells after tight adhesion to human umbilical vein endothelial cell monolayers. J Exp Med 189(6):979–990

96. Hu Y, Turner MJ, Shields J, Gale MS, Hutto E, Roberts BL et al (2009) Investigation of the mechanism of action of alemtuzumab in a human CD52 transgenic mouse model. Immunology 128(2):260–270

97. Coles AJ, Twyman CL, Arnold DL, Cohen JA, Confavreux C, Fox EJ et al (2012) Alemtuzumab for patients with relapsing multiple sclerosis after disease-modifying therapy: a randomised controlled phase 3 trial. Lancet 380(9856):1829–1839

98. Weber MS, Hemmer B (2010) Cooperation of B cells and T cells in the pathogenesis of multiple sclerosis. Results Probl Cell Differ 51:115–126

99. Disanto G, Morahan JM, Barnett MH, Giovannoni G, Ramagopalan SV (2012) The evidence for a role of B cells in multiple sclerosis. Neurology 78(11):823–832

100. Hauser SL, Waubant E, Arnold DL, Vollmer T, Antel J, Fox RJ et al (2008) B-cell depletion with rituximab in relapsing-remitting multiple sclerosis. N Engl J Med 358(7):676–688

101. Hauser SL, Belachew S, Kappos L (2017) Ocrelizumab in primary progressive and relapsing multiple sclerosis. N Engl J Med 376(17):1694

102. Lyons JA, Ramsbottom MJ, Cross AH (2002) Critical role of antigen-specific antibody in experimental autoimmune encephalomyelitis induced by recombinant myelin oligodendrocyte glycoprotein. Eur J Immunol 32(7):1905–1913

103. Mathias A, Perriard G, Canales M, Soneson C, Delorenzi M, Schluep M et al (2017) Increased ex vivo antigen presentation profile of B cells in multiple sclerosis. Mult Scler 23(6):802–809

104. Harp CT, Ireland S, Davis LS, Remington G, Cassidy B, Cravens PD et al (2010) Memory B cells from a subset of treatment-naive relapsing-remitting multiple sclerosis patients elicit CD4(+) T-cell proliferation and IFN-gamma production in response to myelin basic protein and myelin oligodendrocyte glycoprotein. Eur J Immunol 40 (10):2942–2956

105. Wekerle H (2017) B cells in multiple sclerosis. Autoimmunity 50(1):57–60

106. Molnarfi N, Schulze-Topphoff U, Weber MS, Patarroyo JC, Prod'homme T, Varrin-Doyer M et al (2013) MHC class II-dependent B cell APC function is required for induction of CNS autoimmunity independent of myelin-specific antibodies. J Exp Med 210 (13):2921–2937

107. Leandro MJ (2013) B-cell subpopulations in humans and their differential susceptibility to depletion with anti-CD20 monoclonal antibodies. Arthritis Res Ther 15(Suppl 1):S3

108. Hausler D, Hausser-Kinzel S, Feldmann L, Torke S, Lepennetier G, Bernard CCA et al (2018) Functional characterization of reappearing B cells after anti-CD20 treatment of CNS autoimmune disease. Proc Natl Acad Sci U S A 115(39):9773–9778

Chapter 20

Therapeutic Strategies to Potentially Cure Multiple Sclerosis: Insights into the Mechanisms of Autologous Hematopoietic Stem Cell Transplantation

Leoni Rolfes and Marc Pawlitzki

Abstract

Immune reconstitution therapies (IRTs) offer the potential to induce long-lasting remission or even permanent absence of disease activity in patients with multiple sclerosis (MS). Due to the extensive depletion of immune cell subsets, the immune system can subsequently renew itself. Autologous hematopoietic stem cell transplantation (AHSCT) is frequently categorized as IRT. Originally developed for the treatment of hematological malignancies, AHSCT has been adapted to treat severe immune-mediated disorders, such as highly active MS. Indeed, studies have demonstrated that AHSCT can persistently suppress MS disease activity, outperforming clinical results achieved with other disease-modifying therapies for MS. One of the main causes for the reserved implementation of AHSCT in clinical routine is the risk of adverse events and transplant-related mortality. However, the mortality rate has decreased to 0.3% in studies conducted since 2005—an improvement compared to earlier studies (3.6%).

Immunological studies support the fact that AHSCT induces qualitative resetting of the immune system, especially of the adaptive immune cell repertoires, while the underlying mechanisms are far from being completely understood. In-depth knowledge of immunological changes after AHSCT and their correlation with clinical outcomes is essential to guide the optimal use of AHSCT as a durable therapeutic strategy.

In this chapter, we describe the procedure of AHSCT, including evidence of optimal treatment methodology and patient selection. In addition, we review the efficacy, risk of adverse events, and potential mechanisms of AHSCT in the context of immune system renewal and durable disease remission in MS.

Key words Autologous hematopoietic stem cell transplantation, Multiple sclerosis, Treatment

1 Introduction

In the last decade, the aim of multiple sclerosis (MS) treatment has changed from simply reducing relapses to achieving complete and long-lasting disease control [1]. However, despite considerable advances in disease-modifying treatments (DMTs) over the past 20 years, no curative therapy is yet available for MS. The concept of drug-free durable remission has evolved with the development of so-called immune reconstitution therapies (IRTs). Indeed, it has

Sergiu Groppa and Sven G. Meuth (eds.), *Translational Methods for Multiple Sclerosis Research*, Neuromethods, vol. 166,
https://doi.org/10.1007/978-1-0716-1213-2_20, © Springer Science+Business Media, LLC, part of Springer Nature 2021

been postulated that intense short-term immunosuppression enables the rebuilding of the immune system, leading to an eradication of the self-reactive immune repertoire. Autologous hematopoietic stem cell transplantation (AHSCT) is suggested as a method to achieve this goal [2]. The subsequent immune reconstitution induces immune resetting, responsible for the anti-inflammatory effects of this treatment approach [3, 4]. AHSCT has been developed for hematological indications and, after the original report of its feasibility, was repurposed for treating severe autoimmune diseases [5]. Consequently, until July 2019, approximately 1500 patients with MS were subjected to AHSCT [2]. Growing evidence suggests AHSCT is highly efficacious in reducing MS disease activity (clinical relapses and radiological activity), with sustained remission for >10 years in some cohorts [6, 7]. Several studies gave proof that the procedure leads to neurological improvement also in patients with relapsing-remitting MS (RRMS) [8]. Moreover, long-term observations indicate that AHSCT attenuates disability progression in progressive disease forms such as primary progressive (PPMS) and secondary progressive MS (SPMS) [9, 10]. Comparing the efficacy of AHSCT to other standard DMTs in patients with RRMS, a recently published randomized controlled trial demonstrated a significantly lower incidence of confirmed disability in AHSCT-treated patients [11].

Although AHSCT is highly effective, the related mortality rate was initially high, leading to considerable caution in recommending this treatment, but has since significantly declined over the last decade [2, 11–15].

This chapter provides a clear and informative description of the treatment procedures behind AHSCT. We provide an overview of current knowledge on the mechanism of action, clinical efficacy, and common adverse events as well as an evolving paradigm for patient selection.

2 Methodology

AHSCT entails four key stages: (1) mobilization of hematopoietic stem cells (HSCs), (2) HSCs collection, (3) HSCs conditioning, and (4) reinfusion of HSCs and blood/immune recovery (Fig. 1). In principle, HSCs can be collected from either a healthy individual (allogeneic transplantation) or the patient himself (autologous transplantation). Allogeneic transplantation offers the advantage of complete eradication of autoreactive immune cells combined with the regeneration of a healthy immune system with reestablished cellular immune tolerance networks against autoantigens. However, it carries the risk of graft-versus-host disease, reactivation of viral infections, and infections with opportunistic pathogens.

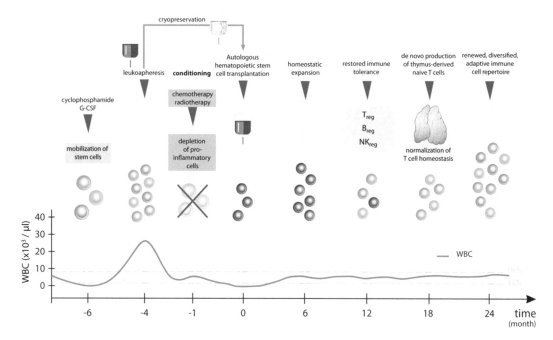

Fig. 1 The procedure of autologous hematopoietic stem cell transplantation (AHSCT). AHSCT includes the following key steps: hematopoietic cell mobilization, harvesting, selection, and cryopreservation; preparative conditioning with chemotherapy with or without irradiation; infusion of stem cell graft; and supportive care after transplantation. Immune reconstitution is characterized by a restored immune tolerance and an extensive renewal of the T cell repertoire. The white blood cell count (WBC) is in the normal range just a few months after AHSCT. G-CSF = granulocyte colony-stimulating factor. (Modified from: Muraro PA et al., 2017, Nature Reviews Neurology)

Therefore, allogeneic stem cell transplantation is only rarely considered for chronic inflammatory diseases [16].

Autologous transplantation in a first step requires the mobilization of CD34$^+$ HSCs from bone marrow, which is usually obtained through the daily administration of granulocyte-colony stimulating factor (G-CSF; 5–10 µg/kg), either alone or with cytotoxic chemotherapy such as cyclophosphamide (2–4 g/m^2) [17]. Cyclophosphamide depletes (autoreactive) lymphocytes in both the peripheral blood and cerebrospinal fluid (CSF). It reduces B and T cells, though preferentially affecting CD4$^+$ T cells. Generally, association with a priming chemotherapy is preferred to increase treatment efficacy and reduce the risk of treatment adverse events, including the reduced potential of autoreactive cells in the graft [12].

After an average of 10 days, the mobilized HSCs are harvested from peripheral blood by leukapheresis, with an optimal target of 5 × 10^6/kg of body weight and a minimum safety threshold of 2 × 10^6/kg in case of graft failure [2]. The HSCs are cryopreserved and frozen, to be stored until the patient is ready for transplantation. The graft can be manipulated by ex vivo selection of CD34$^+$

Table 1
Conditioning regimens used for AHSCT in multiple sclerosis

Intensity	Definition	References
High	TBI, cyclophosphamide, and ATG	[5]
	Busulfan, cyclophosphamide, and ATG	[21]
Intermediate		
Myeloablative	BEAM and ATG	[38, 55]
Nonmyeloablative	• Cyclophosphamide and ATG	[11]
Low	• Cyclophosphamide	[56]
	• Fludarabine-based	[57]
	Melphalan	[56]

ATG anti-thymocyte globulin, *BEAM* carmustine, etoposide, cytosine arabinoside, and melphalan, *TBI* total body irradiation

cells. Although CD34$^+$ selection might influence immune reconstitution, convincing evidence of a relevant clinical benefit compared with unmanipulated cells is missing. Controversially, the procedure is potentially linked to an increased rate of secondary autoimmune diseases [18, 19].

Before transplantation, ablation of the hemato-lymphopoietic system is achieved by high-dose chemotherapy; this stage is known as the preparative or conditioning regime. The most appropriate conditioning in AHSCT remains an open question owing to a lack of comparative data. Conditioning regimens can be classified according to the European Society for Blood and Marrow Transplantation (EBMT) guidelines as high-intensity, intermediate-intensity, and low-intensity (Table 1) [20].

High-intensity chemotherapy protocols usually include the use of cyclophosphamide combined with either the alkylating antineoplastic agent busulfan (Bu-Cy) [21] or total body irradiation (which is no longer standard use for MS) [5], without or combined with anti-thymocyte globulin (ATG). High-intensity regimens were more frequently used in earlier trials [19, 22–24], partly due to sporadic reports of autoimmune disease reactivation following AHSCT when using less intense protocols [25]. However, the available data do not reveal any advantage for high-intensity regimens; and reports of an unfavorable toxicity profile with these regimens have been described [12, 21].

The intermediate-intensity regime is the one most frequently used to treat patients with MS. BEAM is a myeloablative protocol, representing a combination of the chemotherapeutics bis-chloroethylnitrosourea (carmustine), etoposide, cytosine arabinoside, and melphalan, either alone or combined with ATG

[12]. Furthermore, cyclophosphamide-based regimens, which are also followed by ATG, provide so-called nonmyeloablative conditioning [11]. In a nonmyeloablative regime, recovery of hematopoiesis could occur spontaneously, albeit delayed, and HSCs are routinely reinfused to shorten the aplastic phase and attendant risks. Although nonmyeloablative conditioning is better tolerated, it is potentially less effective at reducing MS activity than myeloablative regimes [12].

Lastly, low-intensity protocols imply chemotherapy only (e.g., cyclophosphamide or melphalan), not combined with ATG serotherapy.

In general, ATG is administered towards the end of the AHSCT protocol to decimate T cells that survive chemotherapy, but due to its long half-life, it depletes and also prevents the engraftment of T cells present in the autologous graft, reducing the risk of graft-versus-host disease [26]. Also, beyond lymphocyte depletion, some evidence indicates that ATG alone has immunomodulatory effects, such as enhancing the number and function of regulatory T cells [27]. However, available ATG formulations differ in their ability to deplete lymphocyte subsets and enhance the expansion of regulatory immune cells [27, 28]. Rabbit ATG has been used extensively, while some data is available on the use of horse ATG [29]. Administration of the latter was associated with a significantly higher rate of allergic reactions compared to rabbit ATG (23.8% vs. 0%), in a study including MS patients treated in the years 2001–2006 [29]. However, the influence of ATG use on immunological and clinical outcomes in patients with MS is poorly understood and requires further investigation [1, 30].

Immediately after completing the conditioning regimen, patients develop pancytopenia and a transient bone marrow aplasia, and the stored HSCs are finally administered via intravenous infusion. This enables immune system reconstitution, repopulation of bone marrow cells, and recovery of hematopoiesis.

Besides the four key steps, patients require close ongoing monitoring to ensure a safe recovery, including supportive care to reduce serious complications. In detail, prophylactic antibiotic (e.g., cephalosporin, vancomycin, and trimethoprim–sulfamethoxazole [11] started at HSC reinfusion or added for a febrile episode), antiviral (e.g., acyclovir started on admission and continued for 1 year [11]), and antifungal regimes (e.g., fluconazole for 3 months) are used to minimize infections. In addition, extensive immune suppression, particularly with the inclusion of ATG in the conditioning regimens [31], bears the risk of viral reactivations. Especially, regular screening for viral loads of cytomegalovirus (CMV) and Epstein–Barr virus (EBV) are essential so that preemptive treatment can be instituted promptly [31].

3   Immunological Mechanism of Self-Tolerance Following AHSCT

In contrast to other autoimmune diseases, the autoantigen(s) in MS are still poorly defined, precluding a target-specific therapeutic approach and implying that extended immunosuppression might be favorable. However, widespread and prolonged immunosuppression does not automatically lead to immune renewal and is associated with severe side effects [32]. In contrast, the more selective new depleting immune therapies are gaining importance in MS due to their presumed immune reconstitution [1, 33]. This refers to the partly persistent dynamic changes of the immune cell populations towards a less pronounced pro-inflammatory immune signature [33]. After AHSCT, the B cell repertoire reached normal levels, whereas the sustained modification of T cell populations was characterized by expansion of immune-regulatory and depletion of pro-inflammatory T cells as well as a decrease of proinflammatory cytokine levels, similar to the aforementioned immune cell depleting strategies.

However, in contrast to the latter, both the intensity of on-time immune depletion, eliminating auto-aggressive immune cells irrespective of antigenic specificity, and the reinfusion of the autologous hematopoietic cell product might reestablish immune tolerance suggesting a profound impact that goes beyond the relative changes within immune cell subsets. Therefore, several studies investigated pre- and post-treatment immunophenotyping and quantification of the T cell receptor repertoire in MS patients to gain deeper insights on immune reconstitution.

A normalization of T cell homeostasis after AHSCT has been reported and was predominantly driven by phenotypic renewal of $CD4^+$ T cells, which is characterized by an increased frequency of de novo generated naive $CD4^+$ T cells, originating from the still active thymus, and an overall broader clonal diversity [3, 34]. In addition, the concurrent reduction of memory T cells appears to be beneficial due to their associated pathogenic role in MS [35]. Regarding the clonal specificity, further in-depth analysis revealed a complete renewal of the $CD4^+$ T cell compartment following AHSCT, whereas the effect within the $CD8^+$ subset was less pronounced. The lack of renewal in $CD8^+$ T cell subsets seems, however, not related to treatment response, suggesting lower or no auto-reactive potential.

Interestingly, reduced diversity of the T cell repertoire early after AHSCT was related to treatment failure, which underlines their importance in restoring immune tolerance [36]. Moreover, recent work showed similar broadened effects of T cell repertoire renewal in the intrathecal compartment, with a widespread removal of T cell clones that were exclusively detected in the CSF [37]. Intrathecal changes were also underlined by a significant reduction of CSF immunoglobulin-G levels and the number of oligoclonal bands (OCB) 2 years after AHSCT [38]. Of note, there are contrasting results on persistent OCBs in patients with MS despite AHSCT treatment. However, these studies included small numbers of patients, all of whom had advanced forms of MS, and most CSF samples were collected within 1 year after treatment [39].

The persistent immunological renewal is reflected in changes in gene expression profiles of $CD4^+$ and $CD8^+$ T cells from patients with MS tending towards normalization [40]. Furthermore, expression of the microRNAs miR-16, miR-155, and miR-142-3P, which modulate T cell activity and are highly expressed in MS patients [41], normalized after AHSCT to the same levels as in healthy individuals [12]. Interestingly, the microRNA expression of $CD34^+$ hematopoietic progenitor cells did not differ between MS patients and healthy controls, implying that these cells do not have a preprogrammed pro-inflammatory state [42].

In summary, AHSCT-related immune reconstitution is underpinned by several mechanisms, including the enhancement of thymopoiesis and naive T cell repopulation, clonal renewal and diversification of the T cell repertoire, reduction of some pro-inflammatory cell subsets and cytokine levels, and significant normalization of gene expression.

4   Clinical Studies of AHCT in MS

The potential for curing MS with AHSCT was first described in patients with coincident autoimmune disease and hematologic malignancy, who remained in long-term remission of both diseases after allogeneic transplantation [43].

Initial clinical trials of AHSCT in MS focused almost exclusively on patients with high disability and/or progressive disease course [5, 29, 44, 45]. Contrastingly, studies since 2015 have predominantly included patients with RRMS [11, 13, 14], indicating a pivotal reduction of inflammatory activity [21]. In a multicenter phase II trial, no clinical relapses occurred in any of the 24 RRMS patients during a 13-year follow-up, and none of the 327 post-transplantation magnetic resonance imaging (MRI) scans showed gadolinium-enhancing lesions (GEL) [21].

The ASTIMS phase II randomized clinical trial compared treatment effects of BEAM/ATG AHSCT with mitoxantrone in patients with highly active RRMS or SPMS [6]. The primary study endpoint was achieved, namely, decreased numbers of new cerebral T2 hyperintense lesions over 4 years compared to mitoxantrone treatment. During the entire follow-up, none of the patients in the AHSCT group developed GEL, versus 56% of patients treated with mitoxantrone. Although AHSCT induced a complete suppression of active inflammatory lesions and a significant reduction of relapses, there was no significant effect on disability progression. However, most of the participants (67%) had a SPMS disease course at baseline.

Several trials since 2010 used no evidence of disease activity (NEDA) as primary study outcome [11, 13, 14, 21, 38], which enables an indirect comparison between trials with other DMTs. Particularly, a multi-center phase II clinical trial of AHSCT in patients with aggressive, treatment-resistant RRMS by Nash and colleagues demonstrated NEDA in approximately 70% of patients at 5 years after transplantation [38]. Accordingly, Muraro and colleagues provided a cross-sectional analysis of patients achieving NEDA 2 years post-DMT initiation or transplantation and showed that the proportion was considerably higher in patients who underwent AHSCT (70–92%) compared to placebo (7–16%), interferon-β-1a (13–27%), and patients who received other drugs (22–48%) [12]. This effect is even more impressive since patients who underwent AHSCT had a more aggressive disease (based on relapse rate and MRI activity) at baseline than participants of the other clinical trials [12].

Accordingly, the recently published multicenter MIST trial is the first study to demonstrate superior efficacy of AHSCT versus approved DMTs in a RRMS population in a randomized design [11]. The 3-year interim analysis revealed that disease progression occurred in 3 patients of the AHSCT group and 34 patients of the DMT group. During the first year, disability improved in patients who underwent transplantation (mean EDSS improvement from 3.38 to 2.36), whereas it deteriorated in the DMT group (mean EDSS worsening from 3.31 to 3.98). However, one of the most important limitations of the study by Burt et al. is that the DMT cohort did not have access to the most effective DMTs, such as ocrelizumab, cladribine, and alemtuzumab (either because they were not yet available or due to safety concerns).

5   Adverse Events and Mortality Rates

In contrast to classic DMTs, AHSCT is a one-off treatment that decreases the risk of cumulative adverse effects associated with chronic immunosuppression. However, the risk and severity of adverse events are influenced by the intensity of the procedure.

Since AHSCT leads to prompt and severe immunosuppression, most of the nonneurological adverse effects reported are secondary to this phenomenon and include urinary infections, neutropenic fever, sepsis, pneumonia, fungal infections, and viral reactivations (mainly EBV, CMV, varicella-zoster virus) [46, 47]. Moreover, transient alopecia and amenorrhea are common adverse effects [48].

Furthermore, since AHSCT interferes with the balance between tolerance and autoimmunity, late effects might include secondary autoimmune disease [12]. In two separately published EBMT Registry analyses, the incidence of secondary autoimmune disease was 3.6% and 6.4%, respectively [47, 49]. The most frequent secondary autoimmune phenomenon is thyroiditis (5% of patients over a median follow-up of 6.6 years) but at substantially lower frequencies than seen with other IRTs modalities such as alemtuzumab [9]. Moreover, the occurrence of malignancies after AHSCT is reported in a few studies [9, 11]. In the retrospective analysis by Muraro et al., 9 of 281 treated patients (3.2%) were diagnosed with malignancies, with no organ-specific prevalence detectable [9]. A further potential long-term adverse event is impaired fertility. However, several pregnancies resulting in healthy live births were reported after AHSCT, and no congenital diseases were reported in the children [50].

Importantly, transplant-related mortality is the main concern that has limited the development and use of AHSCT [12]. Early treatment trials from 1995–2000 showed mortality rates of 3.6%, whereas a marked decrease has been reported since, with a significant mortality rate decline to 0.2% in the past 5 years. A meta-analysis of 15 studies [51] and a retrospective long-term analysis of MS patients who underwent AHSCT between 1996 and 2005 [47] correlated clinical and demographic features of patients with transplant-related mortality. Results indicated reduced treatment-related mortality in patients with RRMS, lower age, lower disease disability at baseline, and transplantation after 2005.

6 Optimization and Recommendations

A large long-term study identified key demographic, disease-related, and treatment-related factors associated with progression-free survival of MS patients following AHSCT [12]. In this study, younger age, diagnosis of RRMS, and fewer prior DMTs were associated with better neurological outcomes [12]. Especially patients younger than 40 years who were treated within 5 years of disease onset were described to exhibit better clinical outcomes than older patients with a longstanding disease [47]. As an underlying mechanism, it can be speculated that younger age and shorter disease duration are associated with a higher proportion of active inflammatory disease course. This assumption was strengthened by further studies evidencing that patients with high relapse rates and radiological activity at baseline are likely to benefit from therapy [11, 13, 14, 38]. Since several studies have further indicated beneficial outcomes of AHSCT in progressive forms of MS [6, 52], it can be claimed that the inflammatory activity (relapses or GEL) might be effectively targeted with AHSCT in those patients, similar to findings regarding the B-cell depleting compound ocrelizumab [53] or the sphingosine-1-phosphate receptor modulator siponimod [54]. However, sustained evidence of a beneficial role of AHSCT in patients with progressive MS is still lacking [10, 12].

Increasing clinical experience further enables refinement of the methodology. In this context, the intensity of the procedure provides a predominant role (especially in terms of adverse events) and can be influenced by several main elements: the administration of cyclophosphamide for HSCs mobilization, the use of hematopoietic graft manipulation to enrich HSCs (ex vivo $CD34^+$ selection), the intensity of the conditioning regime, and the administration of in vivo lymphodepleting serotherapy with ATG. Since life-threatening infections were reported in studies using high-intensity conditioning, intermediate- and mild-intensity regimes are predominantly recommended to reduce toxicity [8, 15]. Due to a lack of comprehensive studies, however, no evidence is available on whether the implementation of the aforementioned aspects increases clinical effectiveness. Currently, a trial aiming at refining the treatment procedure by comparing two different conditioning regimens (cyclophosphamide/ATG vs. cyclophosphamide/ATG followed by administration of intravenous immunoglobulin post-transplantation) is ongoing (NCT03342638).

7   Conclusion

The term "immune reconstitution" was originally introduced to describe immunological outcomes of allogeneic and autologous HSCT, including the renewal of adaptive immune cell repertoires [1]. Growing evidence suggests that one-off treatment with AHSCT results in a high percentage of NEDA or even improved disability in a high proportion of properly selected MS patients. However, this approach is associated with risks that are generally greater than those associated with other DMTs. Over the years, safety and toxicity of AHSCT have improved, along with ameliorated efficacy in selected patient populations and refinement of the methodology. Concerning the underlying immunological mechanism, AHSCT is characterized by de novo production of thymus-derived naive T cells and a renewed, diversified, adaptive immune cell repertoire. Studies applying in-depth immunophenotyping approaches to both peripheral blood and CSF are required to improve our understanding of the long-term drug-free remission concept and to completely understand specific immune changes leading to the establishment of CNS immune tolerance during treatment. Such insights will help us to define reliable biomarkers and guide the optimal use of AHCT methodology as a durable therapeutic strategy in MS.

Author Disclosures

Leoni Rolfes: received travel reimbursements from Merck Serono and Sanofi Genzyme, Roche.

Marc Pawlitzki: received travel/accommodation/meeting expenses from Novartis.

References

1. Lunemann JD, Ruck T, Muraro PA, Bar-Or A, Wiendl H (2020) Immune reconstitution therapies: concepts for durable remission in multiple sclerosis. Nat Rev Neurol 16 (1):56–62

2. Sharrack B, Saccardi R, Alexander T, Badoglio M, Burman J, Farge D et al (2020) Autologous haematopoietic stem cell transplantation and other cellular therapy in multiple sclerosis and immune-mediated neurological diseases: updated guidelines and recommendations from the EBMT autoimmune diseases working party (ADWP) and the joint accreditation committee of EBMT and ISCT (JACIE). Bone Marrow Transplant 55 (2):283–306

3. Muraro PA, Douek DC, Packer A, Chung K, Guenaga FJ, Cassiani-Ingoni R et al (2005) Thymic output generates a new and diverse TCR repertoire after autologous stem cell transplantation in multiple sclerosis patients. J Exp Med 201(5):805–816

4. Abrahamsson SV, Angelini DF, Dubinsky AN, Morel E, Oh U, Jones JL et al (2013) Non-myeloablative autologous haematopoietic stem cell transplantation expands regulatory cells and depletes IL-17 producing mucosal-associated invariant T cells in multiple sclerosis. Brain 136(Pt 9):2888–2903

5. Fassas A, Anagnostopoulos A, Kazis A, Kapinas K, Sakellari I, Kimiskidis V et al (1997) Peripheral blood stem cell transplantation in the treatment of progressive multiple sclerosis: first results of a pilot study. Bone Marrow Transplant 20(8):631–638

6. Mancardi GL, Saccardi R, Filippi M, Gualandi F, Murialdo A, Inglese M et al (2001) Autologous hematopoietic stem cell transplantation suppresses Gd-enhanced MRI activity in MS. Neurology 57(1):62–68

7. Saiz A, Blanco Y, Carreras E, Berenguer J, Rovira M, Pujol T et al (2004) Clinical and MRI outcome after autologous hematopoietic stem cell transplantation in MS. Neurology 62 (2):282–284

8. Burt RK, Loh Y, Cohen B, Stefoski D, Balabanov R, Katsamakis G et al (2009) Autologous non-myeloablative haemopoietic stem cell transplantation in relapsing-remitting multiple sclerosis: a phase I/II study. Lancet Neurol 8(3):244–253

9. Muraro PA, Pasquini M, Atkins HL, Bowen JD, Farge D, Fassas A et al (2017) Long-term outcomes after autologous hematopoietic stem cell transplantation for multiple sclerosis. JAMA Neurol 74(4):459–469

10. Mariottini A, Filippini S, Innocenti C, Forci B, Mechi C, Barilaro A et al (2020) Impact of autologous haematopoietic stem cell transplantation on disability and brain atrophy in secondary progressive multiple sclerosis. Mult Scler:1352458520902392

11. Burt RK, Balabanov R, Burman J, Sharrack B, Snowden JA, Oliveira MC et al (2019) Effect of Nonmyeloablative hematopoietic stem cell transplantation vs continued disease-modifying therapy on disease progression in patients with relapsing-remitting multiple sclerosis: a randomized clinical trial. JAMA 321(2):165–174

12. Muraro PA, Martin R, Mancardi GL, Nicholas R, Sormani MP, Saccardi R (2017) Autologous haematopoietic stem cell transplantation for treatment of multiple sclerosis. Nat Rev Neurol 13(7):391–405

13. Nash RA, Hutton GJ, Racke MK, Popat U, Devine SM, Griffith LM et al (2015) High-dose immunosuppressive therapy and autologous hematopoietic cell transplantation for relapsing-remitting multiple sclerosis (HALT-MS): a 3-year interim report. JAMA Neurol 72 (2):159–169

14. Burman J, Iacobaeus E, Svenningsson A, Lycke J, Gunnarsson M, Nilsson P et al (2014) Autologous haematopoietic stem cell transplantation for aggressive multiple sclerosis: the Swedish experience. J Neurol Neurosurg Psychiatry 85(10):1116–1121

15. Burt RK, Balabanov R, Han X, Sharrack B, Morgan A, Quigley K et al (2015) Association of nonmyeloablative hematopoietic stem cell transplantation with neurological disability in patients with relapsing-remitting multiple sclerosis. JAMA 313(3):275–284

16. Dong A, Gao M, Wang Y, Gao L, Zuo C (2016) FDG PET/CT in acute Tumefactive multiple sclerosis occurring in a case of chronic graft-versus-host disease after allogeneic hematopoietic stem cell transplantation. Clin Nucl Med 41(9):e414–e416

17. Jaime-Perez JC, Gomez-Galaviz AC, Turrubiates-Hernandez GA, Picon-Galindo E, Salazar-Riojas R, Mendez-Ramirez N et al (2020) Mobilization kinetics of CD34+ hematopoietic stem cells stimulated by G-CSF and cyclophosphamide in patients with multiple sclerosis who receive an autotransplant. Cytotherapy 22(3):144–148

18. Moore J, Brooks P, Milliken S, Biggs J, Ma D, Handel M et al (2002) A pilot randomized trial comparing CD34-selected versus unmanipulated hemopoietic stem cell transplantation for severe, refractory rheumatoid arthritis. Arthritis Rheum 46(9):2301–2309

19. Fassas A, Anagnostopoulos A, Kazis A, Kapinas K, Sakellari I, Kimiskidis V et al (2000) Autologous stem cell transplantation in progressive multiple sclerosis--an interim analysis of efficacy. J Clin Immunol 20 (1):24–30

20. Snowden JA, Badoglio M, Labopin M, Giebel S, McGrath E, Marjanovic Z et al (2017) Evolution, trends, outcomes, and economics of hematopoietic stem cell transplantation in severe autoimmune diseases. Blood Adv 1(27):2742–2755

21. Atkins HL, Bowman M, Allan D, Anstee G, Arnold DL, Bar-Or A et al (2016) Immunoablation and autologous haemopoietic stem-cell transplantation for aggressive multiple sclerosis: a multicentre single-group phase 2 trial. Lancet 388(10044):576–585

22. Burt RK, Cohen BA, Russell E, Spero K, Joshi A, Oyama Y et al (2003) Hematopoietic stem cell transplantation for progressive multiple sclerosis: failure of a total body irradiation-based conditioning regimen to prevent disease progression in patients with high disability scores. Blood 102(7):2373–2378

23. Nash RA, Bowen JD, McSweeney PA, Pavletic SZ, Maravilla KR, Park MS et al (2003) High-dose immunosuppressive therapy and autologous peripheral blood stem cell transplantation for severe multiple sclerosis. Blood 102 (7):2364–2372

24. Samijn JP, te Boekhorst PA, Mondria T, van Doorn PA, Flach HZ, van der Meche FG et al (2006) Intense T cell depletion followed by autologous bone marrow transplantation for severe multiple sclerosis. J Neurol Neurosurg Psychiatry 77(1):46–50

25. Euler HH, Marmont AM, Bacigalupo A, Fastenrath S, Dreger P, Hoffknecht M et al (1996) Early recurrence or persistence of autoimmune diseases after unmanipulated autologous stem cell transplantation. Blood 88 (9):3621–3625

26. Mancardi G, Saccardi R (2008) Autologous haematopoietic stem-cell transplantation in multiple sclerosis. Lancet Neurol 7 (7):626–636

27. Kekre N, Antin JH (2017) ATG in allogeneic stem cell transplantation: standard of care in 2017? Counterpoint. Blood Adv 1 (9):573–576

28. Palchaudhuri R, Saez B, Hoggatt J, Schajnovitz A, Sykes DB, Tate TA et al (2016) Non-genotoxic conditioning for hematopoietic stem cell transplantation using a hematopoietic-cell-specific internalizing immunotoxin. Nat Biotechnol 34(7):738–745

29. Hamerschlak N, Rodrigues M, Moraes DA, Oliveira MC, Stracieri AB, Pieroni F et al (2010) Brazilian experience with two conditioning regimens in patients with multiple sclerosis: BEAM/horse ATG and CY/rabbit ATG. Bone Marrow Transplant 45(2):239–248

30. Scheinberg P, Nunez O, Weinstein B, Scheinberg P, Biancotto A, Wu CO et al (2011) Horse versus rabbit antithymocyte globulin in acquired aplastic anemia. N Engl J Med 365(5):430–438

31. Ismail A, Sharrack B, Saccardi R, Moore JJ, Snowden JA (2019) Autologous haematopoietic stem cell therapy for multiple sclerosis: a review for supportive care clinicians on behalf of the autoimmune diseases working Party of the European Society for blood and marrow transplantation. Curr Opin Support Palliat Care 13(4):394–401

32. Weiner HL, Cohen JA (2002) Treatment of multiple sclerosis with cyclophosphamide: critical review of clinical and immunologic effects. Mult Scler 8(2):142–154

33. Sellner J, Rommer PS (2020) Immunological consequences of "immune reconstitution therapy" in multiple sclerosis: a systematic review. Autoimmun Rev:102492

34. Douek DC, McFarland RD, Keiser PH, Gage EA, Massey JM, Haynes BF et al (1998) Changes in thymic function with age and during the treatment of HIV infection. Nature 396(6712):690–695

35. Berzins SP, Uldrich AP, Sutherland JS, Gill J, Miller JF, Godfrey DI et al (2002) Thymic regeneration: teaching an old immune system new tricks. Trends Mol Med 8(10):469–476

36. Muraro PA, Robins H, Malhotra S, Howell M, Phippard D, Desmarais C et al (2014) T cell repertoire following autologous stem cell transplantation for multiple sclerosis. J Clin Invest 124(3):1168–1172

37. Harris KM, Lim N, Lindau P, Robins H, Griffith LM, Nash RA et al (2020) Extensive intrathecal T cell renewal following hematopoietic transplantation for multiple sclerosis. JCI Insight 5(2):e127655

38. Nash RA, Hutton GJ, Racke MK, Popat U, Devine SM, Steinmiller KC et al (2017) High-dose immunosuppressive therapy and autologous HCT for relapsing-remitting MS. Neurology 88(9):842–852

39. Mondria T, Lamers CH, te Boekhorst PA, Gratama JW, Hintzen RQ (2008) Bone-marrow transplantation fails to halt intrathecal lymphocyte activation in multiple sclerosis. J Neurol Neurosurg Psychiatry 79(9):1013–1015

40. de Paula ASA, Malmegrim KC, Panepucci RA, Brum DS, Barreira AA, Carlos Dos Santos A et al (2015) Autologous haematopoietic stem cell transplantation reduces abnormalities in the expression of immune genes in multiple sclerosis. Clin Sci (Lond) 128(2):111–120

41. Waschbisch A, Atiya M, Linker RA, Potapov S, Schwab S, Derfuss T (2011) Glatiramer acetate treatment normalizes deregulated microRNA expression in relapsing remitting multiple sclerosis. PLoS One 6(9):e24604

42. Lutterotti A, Jelcic I, Schulze C, Schippling S, Breiden P, Mazzanti B et al (2012) No proinflammatory signature in CD34+ hematopoietic progenitor cells in multiple sclerosis patients. Mult Scler 18(8):1188–1192

43. Nelson JL, Torrez R, Louie FM, Choe OS, Storb R, Sullivan KM (1997) Pre-existing autoimmune disease in patients with long-term survival after allogeneic bone marrow transplantation. J Rheumatol Suppl 48:23–29

44. Fassas A, Kimiskidis VK, Sakellari I, Kapinas K, Anagnostopoulos A, Tsimourtou V et al (2011) Long-term results of stem cell transplantation for MS: a single-center experience. Neurology 76(12):1066–1070

45. Bowen JD, Kraft GH, Wundes A, Guan Q, Maravilla KR, Gooley TA et al (2012) Autologous hematopoietic cell transplantation following high-dose immunosuppressive therapy for advanced multiple sclerosis: long-term results. Bone Marrow Transplant 47(7):946–951

46. Rush CA, Atkins HL, Freedman MS (2019) Autologous hematopoietic stem cell transplantation in the treatment of multiple sclerosis. Cold Spring Harb Perspect Med 9(3):a029082

47. Saccardi R, Kozak T, Bocelli-Tyndall C, Fassas A, Kazis A, Havrdova E et al (2006) Autologous stem cell transplantation for progressive multiple sclerosis: update of the European Group for Blood and Marrow Transplantation autoimmune diseases working party database. Mult Scler 12(6):814–823

48. Maciejewska M, Snarski E, Wiktor-Jedrzejczak WA (2016) Preliminary online study on menstruation recovery in women after autologous hematopoietic stem cell transplant for autoimmune diseases. Exp Clin Transplant 14(6):665–669

49. Daikeler T, Labopin M, Di Gioia M, Abinun M, Alexander T, Miniati I et al (2011) Secondary autoimmune diseases occurring after HSCT for an autoimmune disease: a retrospective study of the EBMT autoimmune disease working party. Blood 118(6):1693–1698

50. Snarski E, Snowden JA, Oliveira MC, Simoes B, Badoglio M, Carlson K et al (2015) Onset and outcome of pregnancy after autologous haematopoietic SCT (AHSCT) for autoimmune diseases: a retrospective study of the EBMT autoimmune diseases working party (ADWP). Bone Marrow Transplant 50(2):216–220

51. Sormani MP, Muraro PA, Schiavetti I, Signori A, Laroni A, Saccardi R et al (2017) Autologous hematopoietic stem cell transplantation in multiple sclerosis: a meta-analysis. Neurology 88(22):2115–2122

52. Chen B, Zhou M, Ouyang J, Zhou R, Xu J, Zhang Q et al (2012) Long-term efficacy of autologous haematopoietic stem cell transplantation in multiple sclerosis at a single institution in China. Neurol Sci 33(4):881–886

53. Montalban X, Hauser SL, Kappos L, Arnold DL, Bar-Or A, Comi G et al (2017) Ocrelizumab versus placebo in primary progressive multiple sclerosis. N Engl J Med 376(3):209–220

54. Kappos L, Bar-Or A, Cree BAC, Fox RJ, Giovannoni G, Gold R et al (2018) Siponimod versus placebo in secondary progressive multiple sclerosis (EXPAND): a double-blind, randomised, phase 3 study. Lancet 391(10127):1263–1273

55. Mancardi GL, Sormani MP, Gualandi F, Saiz A, Carreras E, Merelli E et al (2015) Autologous hematopoietic stem cell transplantation in multiple sclerosis: a phase II trial. Neurology 84(10):981–988

56. Farge D, Labopin M, Tyndall A, Fassas A, Mancardi GL, Van Laar J et al (2010) Autologous hematopoietic stem cell transplantation

for autoimmune diseases: an observational study on 12 years' experience from the European Group for Blood and Marrow Transplantation Working Party on autoimmune diseases. Haematologica 95(2):284–292

57. Rabusin M, Andolina M, Maximova N, Lepore L, Parco S, Tuveri G et al (2000) Immunoablation followed by autologous hematopoietic stem cell infusion for the treatment of severe autoimmune disease. Haematologica 85(11 Suppl):81–85

Chapter 21

Symptomatic MS Therapy

Julia Krämer and Sven G. Meuth

Abstract

Multiple sclerosis (MS) can cause a range of disabling symptoms which may differ greatly from patient to patient and over the course of the disease. While numerous effective immunotherapies are available for MS patients, symptomatic treatment options are scarce. Randomized controlled trials examining symptomatic therapies are lacking and the current data basis is insufficient and inconsistent. This chapter will give an overview of potential pharmacological and nonpharmacological treatment strategies for the most common MS symptoms and present useful multimodal approaches for managing different symptoms in MS patients.

Key words Multiple sclerosis, Symptomatic therapy, Multidisciplinary strategy, Symptoms, Fatigue, Cognitive impairment, Neuropsychiatric syndromes, Bladder dysfunction, Bowel dysfunction, Sexual dysfunction, Movement disorder, Spasticity, Gait disorders, Eye movement disorder

Abbreviations

BF	Biofeedback
BPAD	Bipolar affective disorder
CBT	Cognitive behavioral therapy
CI	Cognitive impairment
EAE	Experimental autoimmune encephalomyelitis
INO	Internuclear ophthalmoplegia
ISC	Intermittent self-catheterization
MDD	Major depressive disorder
NBD	Neurogenic bowel dysfunction
NIBS	Noninvasive brain stimulation
MS	Multiple sclerosis
QoL	Quality of life
RCT	Randomized controlled trials
rTMS	Repetitive transcranial magnetic stimulation
SD	Sexual dysfunction
SE	Side effects
SNRI	Selective serotonin and noradrenergic reuptake inhibitors
SSRI	Selective serotonin reuptake inhibitors
TCA	Tricyclic antidepressants

Sergiu Groppa and Sven G. Meuth (eds.), *Translational Methods for Multiple Sclerosis Research*, Neuromethods, vol. 166, https://doi.org/10.1007/978-1-0716-1213-2_21, © Springer Science+Business Media, LLC, part of Springer Nature 2021

1   Introduction

Despite the availability of numerous effective and well-tolerable immunotherapies, many patients with multiple sclerosis (MS) suffer from severe and disabling symptoms that restrict their professional, social, and private lives and hence affect their quality of life (QoL). These symptoms are however often underreported or overlooked. Therefore, clinicians should actively screen for, address, and educate patients, their families, and caregivers about the multitude of MS-related symptoms. Evidence-based symptomatic treatment options are scarce and studies are urgently needed to improve treatment options. The purpose of this chapter is to give an overview of potential pharmacological and nonpharmacological treatment strategies to reduce MS-related symptoms, prevent secondary complications, and improve QoL.

2   Fatigue

With a prevalence of up to 75%, fatigue is one of the most frequent and disabling symptoms of MS [1, 2]. The treatment is multimodal and includes a combination of pharmacological and nonpharmacological strategies. Before diagnosing primary fatigue, causes of secondary fatigue (e.g. sleep disorders, depression, internal comorbidities such as anemia or thyroid dysfunction) need to be excluded or otherwise treated accordingly; therapies that cause/exacerbate fatigue (e.g. antispasmodics, pain medication) should be interrupted or switched. Nonpharmacological treatment strategies for fatigue comprise avoiding activities at high temperatures, cooling therapies, physical exercise—especially aerobic endurance training, but also yoga, tai chi, aquatherapy, and hippotherapy, psychological interventions such as energy management and conservation programs, cognitive behavioral therapy (CBT), mindfulness-based stress reduction, neuropsychological training, multimodal rehabilitation, and dietetic measures [1–5]. No clear proof of concept exists for complementary and alternative therapies [6]. For pharmacological fatigue treatment, especially modafinil and amantadine are used. However, both therapies are not approved for MS-related fatigue. According to a recent evaluation of amantadine by an expert group of the BfArM (Bundesinstitut für Arzneimittel und Medizinprodukte, the german Federal Institute for Drugs and Medical Devices), the current data basis is insufficient and study results are inconsistent, with only marginal effects and no clear relevance for affected patients with MS (https://www.bfarm.de/SharedDocs/Downloads/DE/Arzneimittel/Zulassung/BereitsZugelAM/offlabel/Bewertungen/Amantadin_Addendum1.pdf?__blob=publicationFile&v=3). Regarding the use of modafinil for treating MS-related fatigue, several meta-analyses and Cochrane reviews concluded that, overall, the supporting evidence is weak [2, 3]. Fatigue symptoms often overlap with symptoms of depression [3]; hence, patients with MS suffering from fatigue and depression should be treated with activating anti-depressants, primarily selective serotonin reuptake inhibitors (SSRI) followed by monoamine oxidase-A inhibitors [1, 2]. Positive effects have been described for pemoline, 3,4-diaminopyridine, 4-aminopyridin, aspirin, acetyl-L-carnitine, ginseng, and vitamin D [2, 3, 7]. However, the current fatigue-related data basis on these drugs is insufficient and they are therefore not recommended as standard treatment. Pemoline was withdrawn from the market due to liver toxicity. No significant effects on fatigue were demonstrated for amphetamine [3]. A recent review and meta-analysis showed that noninvasive brain stimulation (NIBS) is a safe and effective treatment for MS-related fatigue, though to date results have not been replicated in a sufficient number of large-scale randomized controlled trials (RCT) [8].

3   Neuropsychiatric Syndromes

Neuropsychiatric syndromes, particularly depression and anxiety, occur in up to 60% of patients with MS and are among the main contributors to MS-associated morbidity and mortality [5, 9, 10]. There is a lack of well-designed RCT examining specific management strategies for neuropsychiatric syndromes in MS. Major depressive disorder (MDD) is the most common psychiatric disorder associated with MS, with a prevalence of up to 50% during the disease course [5, 11]. Also, the risk of suicide in MS is 7.5 times higher than in the general population [5]. Depression, pain, cognitive impairment (CI), and fatigue are highly correlated in patients with MS [3, 10, 12]. Due to this overlap, the treatment of other symptoms should be considered initially. Current evidence for effective depression treatment in MS is limited, but available data support the effectiveness of standard treatment approaches including both CBT and antidepressants. Based on their favorable side-effect profile, SSRI are an appropriate first-line pharmacological strategy for MDD [3, 11] while mirtazapine, low-dose tricyclic antidepressants (TCA), and selective serotonin and noradrenergic reuptake inhibitors (SNRI) should be considered as second-line treatments due to the increased risk for potential side effects (SE) [3, 11]. TCA and SNRI may be useful for patients with comorbid neuropathic pain, while bupropion has the advantage of a reduced risk of sexual dysfunction [11]. Mindfulness-based therapy and psychological therapy such as CBT were shown to have positive effects on neuropsychiatric syndromes [10], although they cannot be used for patients with significant CI. In cases of moderate to severe depression, a combination of psychological and pharmacological therapy is recommended. The literature on treating MS-related depression with physical exercise is modest and no data are available for treatment with NIBS [10]. The prevalence of bipolar affective disorder (BPAD) in MS is approximately twice as high compared to the general population, with one-third of the cases solely attributable to the SE of steroid or antidepressant treatment [11]. Other agents that have also been implicated in causing episodes of (hypo)mania and depression include baclofen, dantrolene, tizanidine, and psychoactive substances. The treatment of BPAD and psychosis is similar for individuals with MS and those without comorbid MS. However, caution is required due to an increased potential for adverse effects. SSRI should be considered as first-line therapy for treating anxiety disorders. Nonpharmacological strategies include stress management and CBT. Patients with MS have been reported to have a 13.6% lifetime prevalence of alcohol abuse. Psychoactive substance misuse or addiction has been noted to be present at increased rates in patients with MS, and is found among younger patients, those still employed, or those who have less severe MS symptoms [11].

4   Cognitive Impairment

CI is a key symptom of MS, affecting 43–70% of patients [13–15]. The cognitive domains most frequently impaired in MS are complex attention, information processing speed, memory, and executive function. The development of appropriate treatment strategies that address CI in MS is still in its infancy [14]. Therefore, the therapeutic management of CI in MS is challenging [16]. Pharmacological interventions (DMTs and symptomatic treatment) and nonpharmacological interventions (cognitive rehabilitation and exercise training) can be distinguished [16]. Several DMTs are known to have beneficial effects on cognitive performance relative to placebo or delayed therapy onset [17]. However, methodological limitations render it difficult to draw any firm conclusions [14, 18]. Moreover, the absence of head-to-head clinical trials does not support for the use of one immunomodulatory agent over another. Based on reviews of several methodologically weak studies [18, 19], there is no evidence to date to support symptomatic pharmacological treatment of CI in MS [14]. Substances that were most heavily tested include acetylcholinesterase inhibitors (donepezil, rivastigmine), stimulants (pemoline, amantadine, methylphenidate, modafinil, L-amphetamine sulfate, and lisdexamfetamine dimesylate), a NMDA receptor antagonist (memantine), and *ginkgo biloba* [5, 16, 18–20]. Recently, treatment with 4-aminopyridine was demonstrated to have positive effects on different cognitive domains of MS patients [7, 21].

In contrast, there is evidence that multidisciplinary cognitive rehabilitation interventions may offer modest benefits for improving cognition in MS [19, 20, 22, 23], although training must be deficit-specific [13, 14]. Cognitive rehabilitation includes a wide variety of approaches and techniques, for example computer-assisted programs, textbook exercises, story memory technique, learning compensatory strategies, and psychotherapeutic support with integration of relatives [13, 16]. A recent Cochrane review found low-level evidence for positive effects of neuropsychological/cognitive rehabilitation in cognition-impaired patients with MS, pointing out heterogeneity in interventions and outcome measures [23]. There is also emerging evidence suggesting that physical exercise, mainly aerobic exercise, has pro-cognitive effects [17, 19, 20, 24–26]. Only one study reported improved working memory performance after a single session of high-frequency repetitive transcranial magnetic stimulation (rTMS) over the right dorsolateral prefrontal cortex [27].

5   Bladder, Bowel, and Sexual Dysfunction

Bladder, bowel, and sexual dysfunction are common symptoms in patients with MS [28]. They are present early in the disease course and sometimes reported among the first MS-related symptoms [29, 30]. While a wide range of bladder symptoms is present in over 80% of patients with MS, detrusor overactivity is the most frequently reported urodynamic abnormality [29]. Clinical evaluation of bladder related symptoms should include a bladder diary, urinalysis, uroflowmetry followed by measurement of post-void residual urine volume, ultrasonography, assessment of renal function, QoL assessments, and in some cases urodynamic investigations and/or cystoscopy (Fig. 1) (http://uroweb.org/wp-content/uploads/21-Neuro-Urology_LR2.pdf). Managing affected patients requires a multidisciplinary approach. General advice for storage symptoms includes intake of 1.5–2 l fluids per day and reducing caffeine intake to below 100 mg per day. Also, physical therapy (pelvic floor exercises) with and without neuromuscular electrical stimulation and EMG biofeedback (BF) can be effective for treating bladder related symptoms in patients with MS with mild disability [29]. Antimuscarinics are the first-line treatment for patients with storage symptoms, despite limited evidence of efficacy or effectiveness in patients with MS and their potential cognitive SE [31]. Other substances for improving bladder emptying are α-blockers that relax the bladder constrictor, antispasmodics, SSRI, and cannabinoids [32]. If antimuscarinics are ineffective or poorly tolerated, a range of other approaches such as intradetrusor botulinum toxin A injections, tibial and pudendal nerve stimulation, and sacral neuromodulation are available, with varying levels of evidence of their efficacy in patients

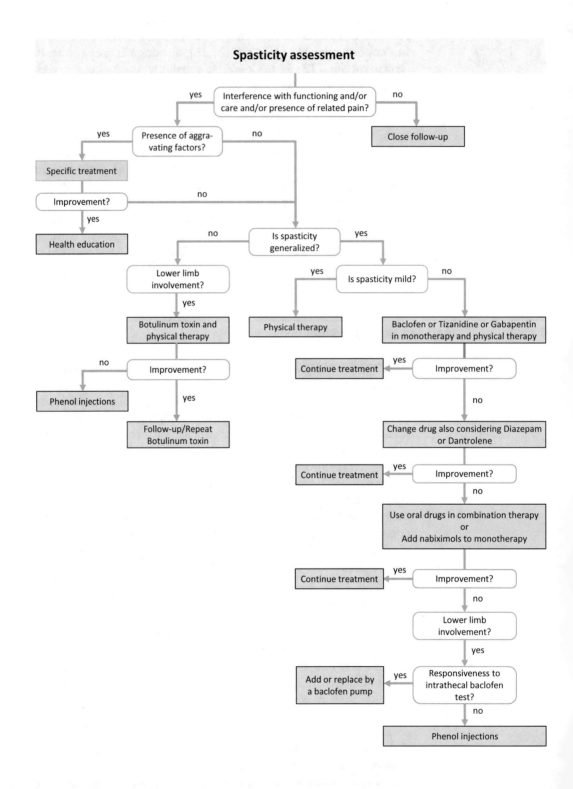

Fig. 1 Algorithm for managing bladder symptoms in MS patients (modified from Phé et al. [29])

with MS. Surgical options include augmentation cystoplasty, cutaneous continent diversion, and ileal conduit surgery, and should be performed only when conservative therapies have failed; or when intermittent self-catheterization (ISC) through the urethra is not possible; and in patients with serious complications such as sepsis, urethral or perineal fistulae, renal failure, or severe urinary incontinence. Stress urinary incontinence owing to sphincter deficiency remains a therapeutic challenge and is only managed surgically if conservative measures (behavioral modifications, alteration of fluid and voiding habits, devices) have failed. ISC is the preferred option for managing incomplete bladder emptying and urinary retention in patients with voiding symptoms but also in those with storage symptoms. For patients with a substantially increased post-void residual volume who are unwilling or unable to conduct ISC, or who have incontinence that is refractory to treatment, a long-term indwelling transurethral or suprapubic catheter may be used with preference of a suprapubic catheter. Detrusor sphincter dyssynergia has been successfully treated with intraurethral botulinum toxin A injections and transurethral sphincterotomy [29, 33].

Depending on the population studied, between 39% and 73% of patients with MS experience neurogenic bowel dysfunction (NBD) (i.e., constipation and/or fecal incontinence) [34]. NBD can be present regardless of disease duration, disease type, or level of disability. At present, the options for managing bowel dysfunction in patients with MS remain unsatisfactory. Engaging the patient and the primary caregiver, changing lifestyle factors, establishing a bowel regime, and initiating constipation and laxative therapy should be the initial steps. If these attempts fail or do not lead to the desired results, other measures such as BF and transanal irrigation should be included. A stoma can improve QoL and is not necessarily the last option. Antegrade colonic enemas can also provide effective relief, while sacral neuromodulation has not yet proven its role [34].

Sexual dysfunction (SD) is common in MS (affecting approximately 33–75% of females and 47–75% of males) and can occur throughout the disease course [35]. Primary SD is a result of neurologic damage to the brain or spinal cord. Secondary SD is a result of other MS-related problems such as neurogenic bladder disorders or spasticity. Finally, tertiary SD occurs as a reaction to larger psychosocial effects. Managing SD includes discontinuation of medication that causes or strengthens erectile dysfunction, treating symptoms that complicate sexual intercourse or disrupt intimacy, diagnosing and treating partnership conflicts, and initiating CBT. Sexual satisfaction can be improved with pelvic floor exercises with neuromuscular electrical stimulation or transcutaneous stimulation of N. tibial or pudendal nerve stimulation. PDE-5-inhibitors are applied to treat erectile dysfunction. If they are not effective or not tolerated, prostaglandin can be tried instead. Invasive procedures include "intracavernous injection therapy" with prostaglandin or its transurethral application. Hormones (e.g., tibolone) can be tried in case of libido loss or dyspareunia [35, 36].

6 Movement Disorders

Movement disorders are common, even in early stages of the disease. They include restless legs syndrome, tremor, ataxia, tonic and hemifacial spasms, myoclonus, focal dystonia, spontaneous clonus, fasciculations, pseudoathetosis, hyperekplexia, and paroxysmal choreoathetosis [37, 38]. Up to 80% of patients with MS experience tremor or ataxia at some point during their disease [39, 40]. Nonpharmacological approaches for handling MS ataxia and tremor include occupational therapy and physiotherapy regimens, appropriate medical aids, autogenic training, and progressive muscle relaxation according to Jacobson [40, 41]. Pharmacological approaches for improving ataxic symptoms are generally disappointing [41]. Short-term (1–15 min) local ice treatment has proven to reduce temporary intention tremor [42]. Small open-label studies and case reports on tremor patients with MS have suggested treatment benefits from a range of drugs. However, drugs are rarely effective and often associated with SE [40, 41]. For clonazepam, ondansetron, dolasetron, physostigmine, isoniazid, levetiracetam, and nabiximols, no positive data have been reported [40]. Only 4-aminopyridine could improve upper limb tremor in a patient with progressive MS [43]. Several RCT with cannabis extracts have concluded that cannabinoids appear to have no beneficial effect on MS tremor [41]. Surgery has variable outcomes depending on the severity of the underlying ataxia [44]. Surgical options include gamma knife surgery, thalamotomy, or deep brain stimulation depending on the individual circumstances. In one neurosurgical study of thalamotomy versus thalamic stimulation, the existing tremor was immediately eliminated by both thalamotomy and thalamic stimulation. However, the tremor returned in almost all patients 6 months post-surgery (albeit less severe than preoperatively) and general disability scores were unchanged [45]. Transcranial focused ultrasound is a newer, less invasive method that precisely places focal thermal lesions in the brain [38].

7 Spasticity

Spasticity affects more than 80% of patients with MS at some point of the disease and is associated with impaired ambulation, pain, and the development of contractures [46–50]. Optimal management of spasticity requires a multidisciplinary team, regular follow-up, and a multimodal approach that combines nonpharmacological and pharmacological interventions [51]. Nonpharmacological spasticity treatment generally includes avoidance of trigger factors, regular physiotherapy (e.g., muscle elongation mainly of the lower extremities, Bobath and Vojta therapy, proprioceptive neuromuscular facilitation, motor-driven movement exercises, BF), orthopedic surgery (static or dynamic splints, air splints), occupational therapy, and other physical therapies [50–52]. For patients with moderate to severe generalized spasticity, baclofen, tizanidine, or gabapentin are the first-line options (Fig. 2). Due to the potential risk of dose-related SE, treatment should be initiated at low doses and gradually titrated upwards. Due to the frequency of SE, diazepam or dantrolene should only be considered if no clinical improvement is seen with first-line treatment (Fig. 2). Nabiximols (a 1:1 extract of cannabidiol (CB1/CB2 full receptor agonist) and Δ9-THC (partial CB1>CB2 receptor agonist)), has a positive effect when used as add-on therapy in patients with MS who poorly respond to and/or tolerate first-line oral treatment (Fig. 2). It is necessary to evaluate the therapeutic response to nabiximols after 4 weeks since less than 50% of patients are responders. A recent review of high-quality studies support the use of cannabinoids for reducing the severity of self-reported spasticity due to MS [53]. The benefit of cannabinoids was also supported by data on progressive experimental autoimmune encephalomyelitis (EAE) mice. Here, THC, the synthetic CB1/CB2 receptor agonist W55,212-2, methanandamide, and JWH-133 were able to ameliorate not only disease progression and neurological disability but also spasticity and tremor [54]. Especially the CB1-system seems to be related to the control of spasticity since the administration of CB1 receptor antagonists or genetic depletion of CB1 leads to a significant worsening of spasticity and tremor in the aforementioned animal model [54, 55]. Despite limited evidence, invasive treatment options for lower limb spasticity are botulinum toxin A or phenol injections and intrathecal baclofen or phenol (Fig. 2) [51]. Peripheral surgery is reserved for a few special cases [50].

8   Eye Movement Abnormalities

One-third of patients with MS develop eye-movement disorders in the course of the disease, more than half of the patients develop nystagmus. The most common abnormality of ocular motility is internuclear ophthalmoplegia (INO). Others are cranial nerve palsies (VI > III > IV), nuclear syndromes such as horizontal gaze palsy and one-and-a-half syndrome, dorsal midbrain syndromes, skew deviation, all forms of nystagmus with acquired pendular nystagmus as the most common type, and saccadic intrusions [4, 56]. The most effective therapy for pendular nystagmus

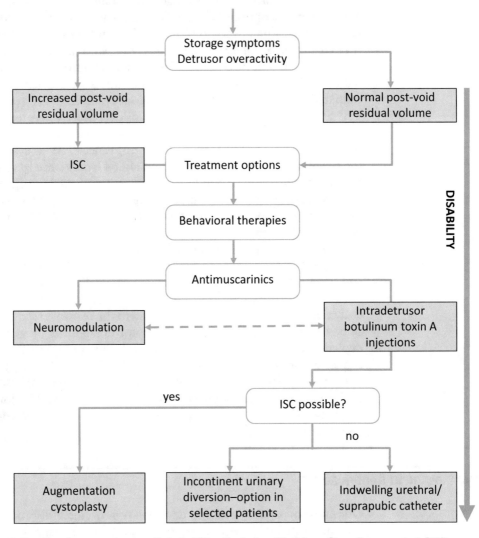

Assessment

- History
- Assessment of symptoms and quality of life
- Bladder diary
- Physical examination
- Urinalysis/culture

- Urea, creatinine, and creatinine clearance if necessary
- Urinary tract imaging
- Measuring post-void residual volume
- Uroflowmetry
- Urodynamics and/or cystoscopy

Fig. 2 Algorithm for managing spasticity in MS patients (modified from Otero-Romero et al. [51])

has been the use of memantine, gabapentin, baclofen, blockers of alpha-2-delta calcium channels and glutamate receptors [56, 57]. Downbeat nystagmus has been shown to respond to clonazepam, baclofen, and gabapentin, upbeat nystagmus responds to fampridine and baclofen, central positional nystagmus responds to fampridine, and periodic alternating nystagmus responds to baclofen [36, 56]. 4-Aminopyridin was shown to improve horizontal saccadic conjugacy in MS patients suffering from chronic INO [58]. Nonpharmacological interventions such as the use of base-down prisms for downbeat nystagmus or prisms to compensate for skew deviation should always be considered [56].

9 Gait Impairment

50–90% of patients with MS develop balance and gait dysfunctions that are most typically secondary to weakness, abnormal tone, visual disturbances, coordination disorders, cerebellar and sensory ataxia, spasticity, motor fatigue, or impaired epicritic sensitivity [4, 59–61]. Interventions include both rehabilitation and pharmacological therapies. The cornerstone of rehabilitation interventions is exercise [60]. Several studies have evaluated the efficacy of a variety of exercises for improving balance, walking, and falls in patients with MS [59, 60], whereby those related to sensory facilitation and task-specific dual-task training are most effective [59]. However, the small size of most studies and the lack of consistency in interventions prevent the optimal selection of specific strategies and clear conclusions on their effectiveness [60]. Compensatory rehabilitation strategies can also improve balance and gait and may prevent falls in people with MS. These include the use of assistive devices (such as canes, walkers, and crutches), lower-extremity braces, rigid and nonrigid ankle–foot orthoses, functional transcutaneous electrical stimulation devices, textured insoles, taping, and wheelchairs [59]. 4-aminopyridin, a voltage-sensitive potassium channel blocker (K_V channel), is approved for patients with MS with impaired mobility. It was shown to increase the walking speed in the short-term and even up to 1 year in approximately 40% of patients and to have strong effects on the ability to walk short and middle distances and on perceived walking capacity [62, 63]. However, given the small number of studies and the disparity between them, it is not possible to draw any conclusions on treatment efficacy beyond the first year [63]. Treatment with 4-aminopyridine was also shown to improve walking ability and motor coordination in MOG_{35-55} immunized C57Bl/6 mice but not the disease course as previously thought [64].

Results of studies evaluating oral or intrathecal baclofen for treating MS spasticity-related gait impairment are equivocal. In contrast, nabiximols was shown to consistently improve spasticity-related gait impairment in patients with MS with resistant spasticity [61]. New, longer-term pharmacological agents continue to be developed, including VSN16R which was shown to alleviate neuronal hyperexcitability and spasticity in EAE mice [65].

References

1. Patejdl R, Penner IK, Noack TK, Zettl UK (2015) Fatigue in patients with multiple sclerosis--pathogenesis, clinical picture, diagnosis and treatment. Fortschr Neurol Psychiatr 83:211–220. https://doi.org/10.1055/s-0034-1399353

2. Penner IK, Paul F (2017) Fatigue as a symptom or comorbidity of neurological diseases. Nat Rev Neurol 13:662–675. https://doi.org/10.1038/nrneurol.2017.117

3. Brenner P, Piehl F (2016) Fatigue and depression in multiple sclerosis: pharmacological and non-pharmacological interventions. Acta Neurol Scand 134(Suppl 200):47–54. https://doi.org/10.1111/ane.12648

4. Frohman TC, Castro W, Shah A et al (2011) Symptomatic therapy in multiple sclerosis. Ther Adv Neurol Disord 4:83–98. https://doi.org/10.1177/1756285611400658

5. Miller E, Morel A, Redlicka J, Miller I, Saluk J (2018) Pharmacological and non-pharmacological therapies of cognitive impairment in multiple sclerosis. Curr Neuropharmacol 16:475–483. https://doi.org/10.2174/1570159X15666171109132650

6. Yadav V, Narayanaswami P (2014) Complementary and alternative medical therapies in multiple sclerosis--the American Academy of Neurology guidelines: a commentary. Clin Ther 36:1972–1978. https://doi.org/10.1016/j.clinthera.2014.10.011

7. Broicher SD, Filli L, Geisseler O et al (2018) Positive effects of fampridine on cognition, fatigue and depression in patients with multiple sclerosis over 2 years. J Neurol 265:1016–1025. https://doi.org/10.1007/s00415-018-8796-9

8. Liu M, Fan S, Xu Y, Cui L (2019) Non-invasive brain stimulation for fatigue in multiple sclerosis patients: a systematic review and meta-analysis. Mult Scler Relat Disord 36:101375. https://doi.org/10.1016/j.msard.2019.08.017

9. Marrie RA, Reingold S, Cohen J et al (2015) The incidence and prevalence of psychiatric disorders in multiple sclerosis: a systematic review. Mult Scler 21:305–317. https://doi.org/10.1177/1352458514564487

10. Solaro C, Gamberini G, Masuccio FG (2018) Depression in multiple sclerosis: epidemiology, aetiology, diagnosis and treatment. CNS Drugs 32:117–133. https://doi.org/10.1007/s40263-018-0489-5

11. Murphy R, O'Donoghue S, Counihan T et al (2017) Neuropsychiatric syndromes of multiple sclerosis. J Neurol Neurosurg Psychiatry 88:697–708. https://doi.org/10.1136/jnnp-2016-315367

12. Penner IK (2016) Evaluation of cognition and fatigue in multiple sclerosis: daily practice and future directions. Acta Neurol Scand 134 (Suppl 200):19–23. https://doi.org/10.1111/ane.12651

13. Henze T, Feneberg W, Flachenecker P et al (2018) New aspects of symptomatic MS treatment: Part 6 - cognitive dysfunction and rehabilitation. Nervenarzt 89:453–459. https://doi.org/10.1007/s00115-017-0443-7

14. Penner IK (2017) Cognition in multiple sclerosis. Neurodegener Dis Manag 7:19–21. https://doi.org/10.2217/nmt-2017-0036

15. Samkoff LM, Goodman AD (2011) Symptomatic management in multiple sclerosis. Neurol Clin 29:449–463. https://doi.org/10.1016/j.ncl.2011.01.008

16. Grzegorski T, Losy J (2017) Cognitive impairment in multiple sclerosis—a review of current knowledge and recent research. Rev Neurosci 28:845–860. https://doi.org/10.1515/revneuro-2017-0011

17. Pflugshaupt T, Geisseler O, Nyffeler T, Linnebank M (2016) Cognitive impairment in multiple sclerosis: clinical manifestation, neuroimaging correlates, and treatment. Semin Neurol 36:203–211. https://doi.org/10.1055/s-0036-1579696

18. Amato MP, Langdon D, Montalban X et al (2013) Treatment of cognitive impairment in multiple sclerosis: position paper. J Neurol 260:1452–1468. https://doi.org/10.1007/s00415-012-6678-0

19. Cotter J, Muhlert N, Talwar A, Granger K (2018) Examining the effectiveness of acetylcholinesterase inhibitors and stimulant-based medications for cognitive dysfunction in multiple sclerosis: a systematic review and meta-analysis. Neurosci Biobehav Rev 86:99–107. https://doi.org/10.1016/j.neubiorev.2018.01.006

20. Lovera J, Kovner B (2012) Cognitive impairment in multiple sclerosis. Curr Neurol Neurosci Rep 12:618–627. https://doi.org/10.1007/s11910-012-0294-3

21. Arreola-Mora C, Silva-Pereyra J, Fernandez T, Paredes-Cruz M, Bertado-Cortes B, Grijalva I (2019) Effects of 4-aminopyridine on attention and executive functions of patients with multiple sclerosis: randomized, double-blind, placebo-controlled clinical trial. Preliminary report. Mult Scler Relat Disord 28:117–124. https://doi.org/10.1016/j.msard.2018.12.026

22. das Nair R, Martin KJ, Lincoln NB (2016) Memory rehabilitation for people with multiple sclerosis. Cochrane Database Syst Rev 3: CD008754. https://doi.org/10.1002/14651858.CD008754.pub3

23. Rosti-Otajarvi EM, Hamalainen PI (2014) Neuropsychological rehabilitation for multiple sclerosis. Cochrane Database Syst Rev 2: CD009131. https://doi.org/10.1002/14651858.CD009131.pub3

24. Sandroff BM, Motl RW, Scudder MR, DeLuca J (2016) Systematic, evidence-based review of exercise, physical activity, and physical fitness effects on cognition in persons with multiple sclerosis. Neuropsychol Rev 26:271–294. https://doi.org/10.1007/s11065-016-9324-2

25. Motl RW, Sandroff BM, Kwakkel G et al (2017) Exercise in patients with multiple sclerosis. The lancet neurology 16:848–856. https://doi.org/10.1016/S1474-4422(17)30281-8

26. Dalgas U (2017) Exercise therapy in multiple sclerosis and its effects on function and the brain. Neurodegener Dis Manag 7:35–40. https://doi.org/10.2217/nmt-2017-0040

27. Hulst HE, Goldschmidt T, Nitsche MA et al (2017) rTMS affects working memory performance, brain activation and functional connectivity in patients with multiple sclerosis. J Neurol Neurosurg Psychiatry 88:386–394. https://doi.org/10.1136/jnnp-2016-314224

28. Wang G, Marrie RA, Fox RJ et al (2018) Treatment satisfaction and bothersome bladder, bowel, sexual symptoms in multiple sclerosis. Mult Scler Relat Disord 20:16–21. https://doi.org/10.1016/j.msard.2017.12.006

29. Phe V, Chartier-Kastler E, Panicker JN (2016) Management of neurogenic bladder in patients with multiple sclerosis. Nat Rev Urol 13:275–288. https://doi.org/10.1038/nrurol.2016.53

30. Almeida MN, Silvernale C, Kuo B, Staller K (2019) Bowel symptoms predate the diagnosis among many patients with multiple sclerosis: a 14-year cohort study. Neurogastroenterol Motil 31:e13592. https://doi.org/10.1111/nmo.13592

31. Morrow SA, Rosehart H, Sener A, Welk B (2018) Anti-cholinergic medications for bladder dysfunction worsen cognition in persons with multiple sclerosis. J Neurol Sci 385:39–44. https://doi.org/10.1016/j.jns.2017.11.028

32. Henze T, Feneberg W, Flachenecker P et al (2018) What is new in symptomatic MS treatment: Part 3-bladder dysfunction. Der Nervenarzt 89:184–192. https://doi.org/10.1007/s00115-017-0440-x

33. Tornic J, Panicker JN (2018) The Management of Lower Urinary Tract Dysfunction in multiple sclerosis. Curr Neurol Neurosci Rep 18:54. https://doi.org/10.1007/s11910-018-0857-z

34. Preziosi G, Gordon-Dixon A, Emmanuel A (2018) Neurogenic bowel dysfunction in patients with multiple sclerosis: prevalence, impact, and management strategies. Degener Neurol Neuromuscul Dis 8:79–90. https://doi.org/10.2147/DNND.S138835

35. Delaney KE, Donovan J (2017) Multiple sclerosis and sexual dysfunction: a need for further education and interdisciplinary care. NeuroRehabilitation 41:317–329. https://doi.org/10.3233/NRE-172200

36. Henze T, Feneberg W, Flachenecker P et al (2018) New aspects of symptomatic MS treatment: Part 4-sexual dysfunction and eye movement disorders. Der Nervenarzt 89:193–197. https://doi.org/10.1007/s00115-017-0441-9

37. Abboud H, Yu XX, Knusel K, Fernandez HH, Cohen JA (2019) Movement disorders in early MS and related diseases: a prospective observational study. Neurol Clin Pract 9:24–31. https://doi.org/10.1212/CPJ.0000000000000560

38. Deuschl G (2016) Movement disorders in multiple sclerosis and their treatment. Neurodegener Dis Manag 6:31–35. https://doi.org/10.2217/nmt-2016-0053

39. Mills RJ, Yap L, Young CA (2007) Treatment for ataxia in multiple sclerosis. Cochrane Database Syst Rev 1:CD005029. https://doi.org/10.1002/14651858.CD005029.pub2

40. Henze T, Feneberg W, Flachenecker P et al (2017) What is new in symptomatic MS treatment: Part 1-introduction and methodical approach, ataxia and tremor. Der Nervenarzt 88:1421–1427. https://doi.org/10.1007/s00115-017-0438-4

41. Wilkins A (2017) Cerebellar dysfunction in multiple sclerosis. Front Neurol 8:312. https://doi.org/10.3389/fneur.2017.00312

42. Feys P, Helsen W, Liu X et al (2005) Effects of peripheral cooling on intention tremor in multiple sclerosis. Journal of neurology, neurosurgery, and psychiatry 76:373–379. https://doi.org/10.1136/jnnp.2004.044305

43. Schniepp R, Jakl V, Wuehr M et al (2013) Treatment with 4-aminopyridine improves upper limb tremor of a patient with multiple sclerosis: a video case report. Mult Scler 19:506–508. https://doi.org/10.1177/1352458512461394

44. Soh D, Fasano A (2017) Multiple sclerosis tremor: are technical advances enough? Lancet

Neurol 16:678–679. https://doi.org/10. 1016/S1474-4422(17)30198-9

45. Hassan A, Ahlskog JE, Rodriguez M, Matsumoto JY (2012) Surgical therapy for multiple sclerosis tremor: a 12-year follow-up study. Eur J Neurol 19:764–768. https://doi.org/10. 1111/j.1468-1331.2011.03626.x

46. Patejdl R, Zettl UK (2017) Spasticity in multiple sclerosis: contribution of inflammation, autoimmune mediated neuronal damage and therapeutic interventions. Autoimmun Rev 16:925–936. https://doi.org/10.1016/j. autrev.2017.07.004

47. Izquierdo G (2017) Multiple sclerosis symptoms and spasticity management: new data. Neurodegener Dis Manag 7:7–11. https:// doi.org/10.2217/nmt-2017-0034

48. Amatya B, Khan F, La Mantia L, Demetrios M, Wade DT (2013) Non pharmacological interventions for spasticity in multiple sclerosis. Cochrane Database Syst Rev 2:CD009974. https://doi.org/10.1002/14651858. CD009974.pub2

49. Urits I, Adamian L, Fiocchi J et al (2019) Advances in the understanding and Management of Chronic Pain in multiple sclerosis: a comprehensive review. Curr Pain Headache Rep 23:59. https://doi.org/10.1007/ s11916-019-0800-2

50. Dressler D, Bhidayasiri R, Bohlega S et al (2017) Botulinum toxin therapy for treatment of spasticity in multiple sclerosis: review and recommendations of the IAB-interdisciplinary working Group for Movement Disorders task force. J Neurol 264:112–120. https://doi. org/10.1007/s00415-016-8304-z

51. Otero-Romero S, Sastre-Garriga J, Comi G et al (2016) Pharmacological management of spasticity in multiple sclerosis: Systematic review and consensus paper. Mult Scler 22:1386–1396. https://doi.org/10.1177/ 1352458516643600

52. Henze T, Feneberg W, Flachenecker P et al (2017) What is new in symptomatic MS treatment: Part 2-gait disorder and spasticity. Der Nervenarzt 88:1428–1434. https://doi.org/ 10.1007/s00115-017-0439-3

53. Rice J, Cameron M (2018) Cannabinoids for treatment of MS symptoms: state of the evidence. Curr Neurol Neurosci Rep 18:50. https://doi. org/10.1007/s11910-018-0859-x

54. Baker D, Pryce G, Croxford JL et al (2000) Cannabinoids control spasticity and tremor in a multiple sclerosis model. Nature 404:84–87. https://doi.org/10.1038/35003583

55. Pryce G, Baker D (2007) Control of spasticity in a multiple sclerosis model is mediated by CB1, not CB2, cannabinoid receptors. Br J Pharmacol 150:519–525. https://doi.org/ 10.1038/sj.bjp.0707003

56. Serra A, Chisari CG, Matta M (2018) Eye movement abnormalities in multiple sclerosis: pathogenesis, modeling, and treatment. Front Neurol 9:31. https://doi.org/10.3389/fneur. 2018.00031

57. Strupp M, Brandt T (2006) Pharmacological advances in the treatment of neuro-otological and eye movement disorders. Curr Opin Neurol 19:33–40

58. Serra A, Skelly MM, Jacobs JB, Walker MF, Cohen JA (2014) Improvement of internuclear ophthalmoparesis in multiple sclerosis with dalfampridine. Neurology 83:192–194. https:// doi.org/10.1212/WNL.0000000000000567

59. Stevens V, Goodman K, Rough K, Kraft GH (2013) Gait impairment and optimizing mobility in multiple sclerosis. Phys Med Rehabil Clin N Am 24:573–592. https://doi.org/ 10.1016/j.pmr.2013.07.002

60. Cameron MH, Nilsagard Y (2018) Balance, gait, and falls in multiple sclerosis. Handb Clin Neurol 159:237–250. https://doi.org/ 10.1016/B978-0-444-63916-5.00015-X

61. Arroyo Gonzalez R (2018) A review of the effects of baclofen and of THC:CBD oromucosal spray on spasticity-related walking impairment in multiple sclerosis. Expert Rev Neurother 18:785–791. https://doi.org/10. 1080/14737175.2018.1510772

62. Valet M, Quoilin M, Lejeune T et al (2019) Effects of Fampridine in people with multiple sclerosis: a systematic review and meta-analysis. CNS Drugs 33(11):1087–1099. https://doi. org/10.1007/s40263-019-00671-x

63. Lecat M, Decavel P, Magnin E, Lucas B, Gremeaux V, Sagawa Y (2017) Multiple sclerosis and clinical gait analysis before and after Fampridine: a systematic review. Eur Neurol 78:272–286. https://doi.org/10.1159/ 000480729

64. Gobel K, Wedell JH, Herrmann AM et al (2013) 4-Aminopyridine ameliorates mobility but not disease course in an animal model of multiple sclerosis. Experimental neurology 248:62–71. https://doi.org/10.1016/j. expneurol.2013.05.016

65. Baker D, Pryce G, Visintin C et al (2017) Big conductance calcium-activated potassium channel openers control spasticity without sedation. Br J Pharmacol 174:2662–2681. https://doi.org/10.1111/bph.13889

INDEX

Sergiu Groppa and Sven G. Meuth (eds.), *Translational Methods for Multiple Sclerosis Research*, Neuromethods, vol. 166, https://doi.org/10.1007/978-1-0716-1213-2, © Springer Science+Business Media, LLC, part of Springer Nature 2021

Printed in the United States
by Baker & Taylor Publisher Services